Oriental Foods and Herbs

ACS SYMPOSIUM SERIES **859**

Oriental Foods and Herbs

Chemistry and Health Effects

Chi-Tang Ho, EDITOR
Rutgers The State University of New Jersey

Jen-Kun Lin, EDITOR
National Taiwan University

Qun Yi Zheng, EDITOR
Pure World Botanicals, Inc.

Sponsored by the
ACS Division of Agricultural and Food Chemistry

American Chemical Society, Washington, DC

Library of Congress Cataloging-in-Publication Data

Oriental foods and herbs : chemistry and health effects / Chi-Tang Ho, editor, Jen-Kun Lin, editor, Qun Yi Zheng, editor ; sponsored by the ACS Division of Agricultural and Food Chemistry.

    p. cm.—(ACS symposium series ; 859)

    "Sponsored by the ACS Divisions of Physical Chemistry and Colloid and Surface Chemistry."

    Includes bibliographical references and index.

    ISBN 0-8412-3841-3

    1. Functional foods—China—Congresses. 2. Cookery, Chinese—Health aspects—Congresses. 3. Materca medica, Vegetables—Congresses. 4. Phytochemicals—Congresses.

    I. Ho, Chi-Tang, 1944- II. Lin, Jen-Kun, Paschalis. III. Zheng, Qun Yi, 1957- IV. American Chemical Society. Division of Agricultural and Food Chemistry. V. Series.

QP144.F85O755   2003
613.2—dc21   2003052008

The paper used in this publication meets the minimum requirements of American National Standard for Information Sciences—Permanence of Paper for Printed Library Materials, ANSI Z39.48-1984.

Copyright © 2003 American Chemical Society

Distributed by Oxford University Press

All Rights Reserved. Reprographic copying beyond that permitted by Sections 107 or 108 of the U.S. Copyright Act is allowed for internal use only, provided that a per-chapter fee of $24.75 plus $0.75 per page is paid to the Copyright Clearance Center, Inc., 222 Rosewood Drive, Danvers, MA 01923, USA. Republication or reproduction for sale of pages in this book is permitted only under license from ACS. Direct these and other permission requests to ACS Copyright Office, Publications Division, 1155 16th St., N.W., Washington, DC 20036.

The citation of trade names and/or names of manufacturers in this publication is not to be construed as an endorsement or as approval by ACS of the commercial products or services referenced herein; nor should the mere reference herein to any drawing, specification, chemical process, or other data be regarded as a license or as a conveyance of any right or permission to the holder, reader, or any other person or corporation, to manufacture, reproduce, use, or sell any patented invention or copyrighted work that may in any way be related thereto. Registered names, trademarks, etc., used in this publication, even without specific indication thereof, are not to be considered unprotected by law.

PRINTED IN THE UNITED STATES OF AMERICA

Foreword

The ACS Symposium Series was first published in 1974 to provide a mechanism for publishing symposia quickly in book form. The purpose of the series is to publish timely, comprehensive books developed from ACS sponsored symposia based on current scientific research. Occasionally, books are developed from symposia sponsored by other organizations when the topic is of keen interest to the chemistry audience.

Before agreeing to publish a book, the proposed table of contents is reviewed for appropriate and comprehensive coverage and for interest to the audience. Some papers may be excluded to better focus the book; others may be added to provide comprehensiveness. When appropriate, overview or introductory chapters are added. Drafts of chapters are peer-reviewed prior to final acceptance or rejection, and manuscripts are prepared in camera-ready format.

As a rule, only original research papers and original review papers are included in the volumes. Verbatim reproductions of previously published papers are not accepted.

ACS Books Department

Contents

Preface xi

Overview

1. Oriental Herbal Products: The Basis for Development
   of Dietary Supplements and New Medicines
   in the 21st Century 2
   Kuo-Hsiung Lee, Hideji Itokawa, and Mutsuo Kozuka

2. Modernization of Traditional Chinese Herbal Medicine 32
   Yuh-Chiang Shen, Wen-Fei Chiou, Guei-Jane Wang,
   Cheng-Jen Chou, and Chieh-Fu Chen

3. Targeting Inflammation Using Nutraceuticals 48
   Mohamed M. Rafi, Prem N. Yadav, and Il-Kyung Maeng

Biological Activity

4. Carnosol from Rosemary Suppresses Inducible Nitric
   Oxide Synthase through Down-Regulating NF κB
   in Murine Macrophages 66
   Jen-Kun Lin, Ai-Hsiang Lo, Chi-Tang Ho,
   and Shoei-Yn Lin-Shiau

5. Hypolipidemic Effect and Antiatherogenic Potential
   of Pu-Erh Tea 87
   Lucy Sun Hwang, Lan-Chi Lin, Nien-Tsu Chen,
   Huei-Chiuan Liuchang, and Ming-Shi Shiao

6. Protective Effects of Baicalein and Wogonin
   against Mutagen-Induced Genotoxicities 104
   Yune-Fang Ueng, Chi-Chuo Shyu, Tsung-Yun Liu,
   Yoshimitsu Oda, Sang Shin Park, Yun-Lian Lin, and Chieh-Fu Chen

7. Biological Activities of Flavonoids Isolated from Chinese Herb Huang Qui: Inhibition of NO and PGE$_2$ Production by Flavonoids 113
   Yen-Chou Chen, Shing-Chuan Shen, and Foun-Lin Hsu

8. Induction of Apoptosis by Rosemary Polyphenols in HL-60 Cells 121
   Shoei-Yn Lin-Shiau, Ai-Hsiang Lo, Chi-Tang Ho, and Jen-Kun Lin

9. Anticaries Effect of Wasabi Components 142
   Hideki Masuda, Toshio Inoue, and Yoko Kobayashi

10. Bioactivity of Lycopene-Rich Carotenoid Concentrate Extracted from Tomatoes 154
    John Shi

Antioxidants

11. Antioxidants in Herbs of Okinawa Islands 166
    Nobuji Nakatani

12. Antioxidants from Some Tropical Spices 176
    Hiroe Kikuzaki

13. Antioxidant Capacity of Berry Crops and Herbs 190
    Shiow Y. Wang

14. Antioxidant Properties of Hsian-tsao (*Mesona procumbens* Hemsl.) 202
    Gow-Chin Yen, Chien-Ya Hung, and Yen-Ju Chen

15. Effect of Different Heating Processes on Cytotoxic and Free Radical Scavenging Properties of Onion Powder 215
    Hui-Yin Fu and Tzou-Chi Huang

16. Identification of Antioxidants from Du-Zhong (*Eucommia ulmoides* Oliver) Directed by DPPH Free Radical-Scavenging Activity 224
    Yong Chen, Nanqun Zhu, and Chi-Tang Ho

Phytochemistry

17. *Schisandra chinensis:* Chemistry and Analysis 234
    Mingfu Wang, Qing-Li Wu, Yaakov Tadmor, James E. Simon,
    Shengmin Sang, and Chi-Tang Ho

18. Phytochemical and Biological Studies on *Evodia lepta* 247
    Guolin Li, Dayuan Zhu, and Ravindra K. Pandey

19. Unique Chemistry of Aged Garlic Extract 258
    Kenjiro Ryu and Robert T. Rosen

20. Three New Sesquiterpene Lactones from *Inula britannica* 271
    Naisheng Bai, Bing-Nan Zhou, Li Zhang, Shengmin Sang,
    Kan He, and Qun Yi Zheng

21. Chemistry and Bioactivity of the Seeds of *Vaccaria segetalis* 279
    Shengmin Sang, Aina Lao, Zhongliang Chen, Jun Uzawa,
    and Yasuo Fujimoto

22. Studies on the Chemical Constituents of Loquat Leaves
    (*Eriobotrya japonica*) 292
    Qing-Li Wu, Mingfu Wang, James E. Simon, Shi-Chun Yu,
    Pei-Gen Xiao, and Chi-Tang Ho

23. Changes in Some Components of Tea Fungus
    Fermented Black Tea 307
    Hui-Yin Fu and Den-En Shieh

24. Chemistry and Bioactivity of *Allium tubersom* Seeds 317
    Shengmin Sang, Aina Lao, and Zhongliang Chen

Indexes

Author Index 333

Subject Index 334

Preface

In recent years, a growing interest in Oriental foods and herbs, as well as the health aspects of Oriental diets, has occurred. Certainly, the link between certain commonly consumed Oriental foods and beverages and potential health benefits generates a great consumer interest. Some notable examples are soy foods, green tea, and garlic. Recent studies show that soy foods appear to have an important role in the prevention of cancer, cardiovascular disease, and osteoporosis. In studies with animals, green tea, the most consumed type of tea in China and Japan, has been demonstrated repeatedly to have anticarcinogenic activity. Various beneficial health properties of garlic have also been well documented.

In addition, recently increasing demands of consumers for alternative and preventive health management have stimulated a rapid and exponential growth of the supplement and nutraceutical market. All these supplements and nutraceuticals are derived from plants, particularly herbal materials. The long history of using herbs for medical and culinary purposes in Oriental culture has been a valuable source for the development of these supplement and nutraceutical products.

The symposium upon which this book is based was developed to bring together the leading scientists from Taiwan, Japan, the United States, and Canada to discuss and share information about their current research on chemistry and biological activity of Oriental foods and herbs. It is based on the two-day symposium entitled *Oriental Foods and Herbs: Chemistry and Health Effects* that was part of the 223rd National American Chemical Society (ACS) meeting in Orlando, Florida, April 2002.

This book has been arranged into four sections. In the overview section, Lee gives an in-depth introduction to Oriental herbal products, followed by a chapter on the standardization of functional foods using medicinal herbs, and, finally a chapter discusses the use of modern biochemical tools to study the antiinflammatory activity of certain phytochemicals derived from Oriental herbs. The following two sections cover the biological and antioxidant activities of selected Oriental foods and herbs. The last section reviews and discusses the phytochemical analysis of several Oriental herbs and teas.

We acknowledge with great appreciation financial assistance from the ACS Division of Agricultural and Food Chemistry. Finally, we thank all the authors for their contributions, efforts, and cooperation in the preparation of this book.

Chi-Tang Ho
Department of Food Science
Rutgers, The State University of New Jersey
New Brunswick, NJ 08901

Jen-Kun Lin
Institute of Biochemistry
College of Medicine
National Taiwan University
Number 1, Section 1, Jen-Ai Road
Taipei, Taiwan

Qun Yi Zheng
Pure World Botanicals, Inc.
375 Huyler Street
South Hackensack, NJ 07606

Overview

Chapter 1

Oriental Herbal Products: The Basis for Development of Dietary Supplements and New Medicines in the 21st Century

Kuo-Hsiung Lee, Hideji Itokawa, and Mutsuo Kozuka

Natural Products Laboratory, School of Pharmacy, University of North Carolina, Chapel Hill, NC 27599-7360

Oriental herbal products, including herbal and traditional medicines, are a treasured legacy of Asian peoples. Multicomponent-processed natural products, primarily medicinal herbs, are used both as daily functional foods and as drugs to maintain health and treat disease states. Oriental herbal products have a prominent role in today's nutraceutical market as dietary supplements for many health issues. These products have also already played an important role as the basis of numerous new leads for modern "single-component" drug discovery and development and will undoubtedly continue to do so in the pharmaceutical market of the 21st century. However, not only single bioactive lead compounds, but also active fractions and effective and safe herbal prescriptions, must be investigated through modern scientific technology. Efficacy and safety must be assured by developing qualitative and quantitative analyses for quality control in order to develop new, high quality dietary supplements, therapies, and world-class drugs.

In China and throughout the Orient, including Japan, Korea, Taiwan, and Southeast Asia, herbs and certain foods have been widely used for centuries to promote good health and treat disease. Indeed, herbs are being increasingly used by the general population of both Eastern and Western countries. International health organizations and governmental agencies such as WHO and the US FDA, respectively, have recognized the consumer popularity of herbal products. In response, scientists must recognize the need to assure the efficacy and safety of traditional herbal remedies through modern research. In addition, components of Oriental herbal products and traditional Chinese medicine (TCM) have been isolated or modified to produce modern drugs. Thus, Oriental herbal products should be continually researched as the basis for development of both high quality dietary supplements and new medicines in the 21st century.

Oriental herbal products are commonly used as dietary supplements, including both daily foods (cereals, vegetables, fruits) and "functional" foods or "Yao Shan", which are TCM-based dietary supplement dishes. The latter Chinese eating culture combines TCM and food for replenishment and medical purposes. Yao Shan includes herbal foods, teas, wines, congees, and pills (or powder) and is used for immunopotentiation, improving systemic circulation, disease prevention, and aging control.

TCM is based on natural products, primarily (>80%) plants. In China, many medicinal plants are used as folk drugs (Min Chien Yao); however, TCM also holds a predominant and respected position in medicine and single or, more commonly, multiple formulated herbs are prescribed by TCM doctors as Chinese Materia Medica or traditional drugs (Chung Yao) *(1)*. Chinese medical literature has recorded over 100,000 (I Fang Chi Chieh) different formulas, usually containing 4 to 12 individual herbs with different pharmacological actions.

Consequently, Chinese and Western medicine vary in fundamental theory, practice, and drugs used. While Western medicine uses pure natural or synthetic compounds aimed at a single target, Chinese medicine uses processed crude multicomponent natural products, in various combinations and formulations aimed at multiple targets, to treat the entirety of different symptoms.

The basic theory of TCM is to establish and maintain a holistic balance in the body, which results in good health. By following the principle that "Therapy by Food is Better than Therapy by Medicine", proper daily diet and functional ("Yao Shan") foods help to maintain health and the balance of Yin and Yang. However, as disease is caused by an imbalance of these two forces. TCM doctors prescribe drugs to restore the body's equilibrium. Thus, a Yin-cool drug is advised for a patient with a Yang-fever, but a Yang-warm drug is prescribed for a Yin-cool problem.

By carefully monitoring the body's reactions to herbal products, effective and safe TCM formulations have been developed and recorded in Food Recipe and Chinese Materia Medica books, including three literary classics: Shen Nung

Pen Tsao Ching (The Book of Herbs by Shen Nung), Huang Ti Nei Ching (The Yellow Emperor's Classic on Internal Medicine), and Shang Han Tsa Ping Lun (Treatise on Febrile and Miscellaneous Diseases). In 1590, Pen Tsao Kan Mu (A General Catalog of Herbs) was published and translated into many languages and Chung Hua Pen Tsao (Chinese Materia Medica), which is the most comprehensive and detailed publication on TCM to date, was published in 1999.

Individual herbs/drugs have different properties and effects, which have different results on the body. [In addition to the 4 essences (cold, cool drugs for Yang diseases; warm, hot drugs for Yin problems), other herbal properties include the 5 flavors (pungent, sour, sweet, bitter, salty), 4 directions of action (ascending, floating, descending, sinking), and 7 effects (single, additive, synergic, antagonistic, inhibitive, destructive, opposite).] Herbs in TCM formulations fall into four categories *(2)*.

- Imperial Herb -- the chief herb (main ingredient) in a formula
- Ministerial Herb -- ancillary to the imperial herb, it augments and promotes the action of the chief herb
- Assistant Herb -- reduces the side effects of the imperial herb
- Servant Herb -- harmonizes or coordinates the actions of other herbs

The supporting herbs aid the effectiveness of the principle herb and any change in composition can lead to different pharmacological actions *(3)*. For example, Mahuang Combination and Mahuang Apricot Seed Combination both contain Mahuang, Apricot Seed, and Licorice (in different proportions); however, the former formulation contains Cassia (Ministerial herb), which assists Mahuang (Imperial herb) in liberating heat, while the latter formulation contains Gypsum (Assistant/Imperial herb), which is antagonistic with Mahuang (Imperial/Assistant). Overall, both combination suppress coughing, but Mahuang Combination <u>induces</u> sweating, while Mahuang Apricot Seed Combination <u>suppresses</u> sweating. Thus, the two combinations are used for different therapeutic indications.

Shen Nung also grouped 365 herbs into three classes, Upper, Middle, and Lower, based on herbal toxicities. The nontoxic and rejuvenating Upper class herbs can be taken continuously for a long period and form the main components of the dietary supplement dishes ("Yao Shan"). The Middle class herbs promote mental stability and can have either nontoxic or toxic effects. Thus, more caution must be taken with their use. The Lower Class herbs have toxic properties, cannot be taken for extended periods, and must be properly processed to reduce their toxicity before being used in TCM formulations. Selected Oriental herbal products of all three classes will be discussed in more detail in the next section, including folkloric use, chemical composition, and currently identified biological activities.

Oriental Herbal Products

Upper Class Herbs

Medicinal Mushrooms

Mushrooms, including "Ling Chih", which was recorded as an Upper Class herb in Shen Nung Pen Tsao Ching, are used worldwide for both their food value and their medicinal properties coupled with few deleterious side effects. Five major medicinal mushrooms are discussed below.

Ling Chih [Ganoderma lucidum (Leyss.: Fr.) P.Karst, Polyporaceae].

Ling Chih, Reishi mushroom or "Spirit Plant", is the common Chinese fungus *G. lucidum* and is the leading cultivated medicinal mushroom. It is used as a tonic and sedative and to treat hyperlipidemia, angina pectoris, chronic bronchitis, hepatitis, leukopenia, and autoimmune disease *(4-6)*.

Chemical Composition & Pharmacological Properties

Like many other mushrooms, Ling Chih spores contain immunostimulating polysaccharides, including $(1\rightarrow3)$-β-, $(1\rightarrow4)$-β-, $(1\rightarrow6)$-β-D-glycans and linear $(1\rightarrow3)$-α-D-glucans *(7)*. Ling Chih also has antimicrobial, antiviral, hypoglycemic, antitumor, free radical scavenging and antioxidative (i.e. antiaging) activities *(7)*. The spores also contain bioactive triterpenoids, both ganoderic acids and ganolucidic acid A. These compounds have antitumor and anti-HIV-protease activities, as well as analgesic effects on the CNS *(8)*. Oxygenated triterpenes also exhibit hypolipidemic activity by blocking cholesterol absorption and inhibiting HMG-CoA reductase *(6)*.

Yun-Chih (Coriolus versicolor Quél, Polyporaceae)

The mushroom Yun-Chih is used as an immunostimulating, antitumor, and hypoglycemic agent *(9)*. It contains a protein-bound polysaccharide (PSK, trade name: Krestin). PSK enhances the immune system and consequently has a tumor-retarding effect *(10)*.

Ganoderic Acid A: $R_1 = O=$, $R_2 = \beta\text{-OH}$,
$R_3 = R_5 = H$, $R_4 = \alpha\text{-OH}$
Ganoderic Acid B: $R_1 = R_2 = \beta\text{-OH}$,
$R_3 = R_5 = H$, $R_4 = O=$
Ganoderic Acid G: $R_1 = R_2 = R_3 = \beta\text{-OH}$,
$R_4 = O=$, $R_5 = CH_3$
Ganoderic Acid H: $R_1 = \beta\text{-OH}$, $R_2 = R_4 = O=$,
$R_3 = \beta\text{-OAc}$, $R_5 = CH_3$

Oxygenated Triterpenes
$R_1 = R_2 = \alpha\text{-OAc}$
$R_1 = \alpha\text{-OAc}$, $R_2 = \alpha\text{-OH}$
$R_1 = \beta\text{-OH}$, $R_2 = \alpha\text{-OH}$

Hsian Ku (China) or Shiitake (Japan) [Lentinus edodes (Berk.) Sing,

This Asian fungus is the second most commonly cultivated edible Tricholomataceae] mushroom also used for medicinal purposes *(11)*. It contains immunostimulating polysaccharides, including linear $(1\rightarrow3)\text{-}\beta\text{-D-glucans}$. Lentinan, a polysaccharide (β-D-glucan) fraction from *L. edodes* has been studied more extensively and displays greater antitumor activity than other similar substances *(12)*. Lentinan appears active in certain animals against various tumors *(13)*.

Shen Ku (Agaricus blazei Murill, Agaricaceae)

This rare mushroom is now cultivated in Brazil (as Pa Hsi Mo Ku), Japan (as Hime Matsutake), China, and the US, and is widely popular in Japan, Korea, and China as a dietary supplement. This mushroom displays antiviral, tumor chemopreventive *(14)*, immuno-stimulating, blood sugar lowering, and cholesterol reducing activites *(13)*. In the 1980s, anticancer activity was widely studied. It contains immunostimulating polysaccharides, including $(1\rightarrow6)\text{-}\beta\text{-}$ and $(1\rightarrow3)\text{-}\beta\text{-D-glycans}$ *(15)*.

Ginsengs

Asian Ginseng (Panax ginseng, Araliaceae)

Ginseng is the root of *Panax ginseng* found in China or Korea. In TCM, it is known for being tonic, regenerating, and rejuvenating. Cultivated ginseng has mostly replaced the scarce wild ginseng. Other species are also used including:

American ginseng (*P. quinquefolium*); cultivated in North America.
Japanese ginseng (*P. japonicus*); widely distributed in Japan.
San-chi ginseng (*P. notoginseng*); tonic and hemostatic in China.

Traditionally, this ginseng is used to restore normal pulse, remedy collapse, benefit the spleen and liver, promote production of body fluid, calm nerves, and treat diabetes and cancer.

Chemical Composition (16-18) & Pharmacological Properties

Many compounds have been isolated from the root: polysaccharides, glycopeptides (panaxanes), vitamins, sterols, amino acids and peptides, essential oil, and polyalkynes (panaxynol, panaxytriol). Approximately 30 saponins have been isolated, including oligoglycosides of tetracyclic dammarane aglycones, both a 3β,12β,20(S) trihydroxylated type (protopanaxadiol) and a 3β,6α,12β,20(S) tetrahydroxylated type (protopanaxatriol). The saponins (ginsenosides Ra-h) differ in the mono-, di-, or tri-saccharide nature of the two sugars attached at the C-20 and C-3 or C-6 hydroxy groups. Various anticancer effects have been studied. Ginsenosides Rg3 and Rg5 significantly reduced lung tumor incidence, Rg3, Rg5 and Rh2 (red ginseng) showed anticarcinogenic activity *(19)*, Rg3 inhibited cancer cell invasion and metastasis, Rb$_2$ inhibited tumor angiogenesis, and Rh2 inhibited human ovarian cancer growth in nude mice *(17)*. Ginsenosides Rb$_1$ and Rb$_2$ are metabolized by intestinal bacteria to compound K, also known as M1, which induces apoptosis of tumor cells *(20)*. A case-control study on cancer and ginseng intake was reported *(21)*.

American Ginseng (Panax quinquefolium, Araliaceae)

According to TCM's theory, American ginseng is somewhat "cool" and used to reduce internal heat and promote secretion of body fluids, while Asian ginseng is "warm" and known to replenish vital energy. Although the two species contain many identical components *(22)*, ginsenoside Rf is found in Asian ginseng, but not American ginseng; conversely, 24(R)-pseudoginsenoside F$_{11}$ is abundant in American ginseng, but found only in trace amounts in Asian ginseng *(23-26)*.

	Ginsenoside	R₁	R₂
Protopanaxadiol, R₁ = R₂ = H	Rb1	Glc(2-1)Glc	Glc(6-1)Glc
	Rb2	Glc(2-1)Glc	Glc(6-1)Ara(p)
	Rc	Glc(2-1)Glc	Glc6Ara(f)
	Rd	Glc(2-1)Glc	Glc
	Rg3	Glc(2-1)Glc	H
	Rh2	Glc	H
Protopanaxatriol R₁ = R₂ = H	Re	Glc(2-1)Rha	Glc
	Rf	Glc(2-1)Glc	H
	Rg1	Glc	Glc
	Rg2	Glc(2-1)Rha	H
	Rh1	Glc	H
	Rg5	Glc(2-1)Glc	—

Ginsenoside Rf
(in Oriental Ginseng)

24(R)-Pseudoginsenoside F_{11}
(in American Ginseng)

Quinqueginsin, a 53 kDa homodimeric protein isolated from American ginseng, showed various biological activities, including inhibition of HIV-1 reverse transcriptase *(27)*.

Sanchi Ginseng (Panax notoginseng, Araliaceae)

This ginseng exerts a major effect on the cardiovascular system, resulting in increased coronary flow and decreased blood pressure, and is used to arrest bleeding, remove blood stasis, and relieve pain. In recent studies, a preparation (Sanqi Gaunxin Ning) improved symptoms of angina pectoris by 95.5% and ECG pattern by 83%. Three new ginsenoside type saponins were recently isolated, together with 11 known saponins *(28)* and the small molecule dencichine (which arrests bleeding) *(29, 30)*.

Denchicine   HOOC−C(NH$_2$)(H)−CH$_2$NHCOCOOH

Tan Shen or Sage (*Salvia miltiorrhiza, Labiatae*)

This herb exhibits hypotensive and positive inotropic effects, causes coronary artery vasodilation, and inhibits platelet aggregation. The rhizome and roots of *S. miltiorrhiza* are used in TCM to treat various cardiovascular diseases.

Chemical Composition & Pharmacological Properties

Sodium tanshinone II-A sulfonate (a water soluble sulfonate of the parent diterpenoid) is used to treat angina pectoris and myocardial infarction. It exhibits a strong membrane-stabilizing effect on red blood corpuscles and may act similarly to the drug verapamil. Intravenously, a *S. miltiorrhiza/Dalbergia odorifera* TCM mixture may have potential as an anti-anginal drug *(31)*.

Sodium Tanshinone II-A Sulfonate

Tu Chung (*Eucommia ulmoides*, Eucommiaceae)

The leaves can be used as a tea (Tu Chung tea), and the dried bark to supplement the liver and kidney, strengthen muscles and bones, and stabilize the fetus. The plant is said to be a longevity herb and exhibits antihypertensive effects.

Chemical Composition & Pharmacological Properties (32)

- Iridoids, including geniposidic acid, geniposide, asperulosidic acid, deacetyl asperulosidic acid, and asperuloside
- Phenols: pyrogallol, protocatechuic acid, and *p*-trans-coumaric acid
- Triterpenes and lignans

Geniposidic acid stimulates the parasympathetic nervous system through the muscarinic Ach receptor *(33)* and stimulates collagen synthesis in aged rats *(34)*.

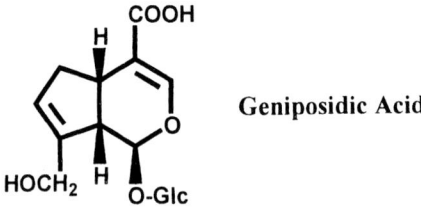

Geniposidic Acid

Wu Wei Tzu (*Schizsandra chinensis, Schisandraceae*)

Wu Wei Tzu is the dried fruit of *S. chinensis* (northern China) or *S. spenanthera* (southern China). Traditionally, it is used for dyspnea and cough, dry mouth and thirst, spontaneous diaphoresis, night sweats, insomnia, and amnesia.

Chemical Composition & Pharmacological Properties

The plant contains essential oils, lignans, and citric, malic, and tartaric acids. Dibenzocyclooctadiene lignans include schisandrin, deoxyschisandrin, pregomisin, and gomisins A-D, F-H, and J. Schisanhenol completely inhibits peroxidative damage of brain mitochondria and rat membrane *(35)*. Schisandrin B protects against hepatic oxidative damage in mice *(36)*. Gomisin A inhibits tumor promotion, probably due to its anti-inflammatory activity *(37)*, and gomisin G shows potent anti-HIV activity against HIV replication in H9 lymphocytes (EC$_{50}$ 0.006 µg/mL, TI 300) *(38)*.

Schizanhenol　　Schizandrin B　　Gomisin A　　Gomisin G

Astragalus or Huang-qi (*Astragalus membranaceüs, Leguminosae*)

Huang-qi is used in TCM prescriptions for ch'i (energy) deficiency and general weakness and specifically for shortness of breath and palpitation, collapse, spontaneous perspiration, night sweats, edema due to physical deficiency, chronic nephritis, pulmonary diseases, lingering diarrhea, rectal and uterine prolapse, nonfestering boils, and hard-to-heal sores and wounds. Immunostimulant, antioxidant, antiviral, and antitumor activities are found.

Chemical Composition & Pharmacological Properties

Astragalus contains flavonoids, polysaccharides, and triterpenoid saponins or astragalosides *(39)*. PG2, a polysaccharide fraction from this plant, was tested in animals and may be effective in counteracting cancer chemotherapy's severe side effects, including neutropenia, thrombocytopenia, and anemia *(40)*. In addition, astragalosides I, VI, and VII have shown antiviral suppression *(41)*.

Astragaloside I: R_1 = H, R_2 = R_3 = Ac
Astragaloside VI: R_2 = β-D-Glu, R_1 = R_3 = H
Astragaloside VII: R_1 = Glc, R_2 = R_3 = H

Kou Chi Tzu (*Lycium barbarum*, Solanaceae)

Traditionally, *L. barbarum* is used to supplement liver and kidney yin and to treat weakness, vertigo, excessive tearing, cough due to consumption, and diabetes. The dried red berries (Barbary wolfberry fruits) are similar to raisins and used to improve eyesight.

Chemical Composition & Pharmacological Properties

Polysaccharides and arabinogalactan-proteins are among the active components *(42,43)*. The glycan (LbGp4-OL) and, to a lesser extent, its glycoconjugate (LbGp4), enhanced splenocyte proliferation in normal mice, most likely by targeting B-lymphocyte cells *(44)*. Other active components include betaine, vitamins, and zeaxanthin.

Tung Chung Hsia Tsao

The parasitic fungus *Cordyceps sinensis* (Berk.) Sacc. (Hypocreaceae) attacks the caterpillar *Hepialus armoricanus* or Sphinx moth during its winter underground hibernation and slowly eats it away. In the summer, a rod-like fungal stroma grows out from mummified shell of the dead caterpillar. This stroma is harvested as Tung Chung Hsia Tsao (Winter Worm, Summer Grass). It is found naturally only in the highlands of the Himalayan region, Sichuan, Qinghai, Tibet, and Yunnan, but is now cultured as *Codyceps mycellia*. Tung Chung Hsia Tsao has been traditionally used to treat chronic cough, asthma and impotence, promote longevity, relieve exhaustion, and increase athletic prowess.

Chemical Composition & Pharmacological Properties

This product contains immunopotentiating polysaccharides, such as galactomannan, and antitumor polysaccharides, sterols, and adenosine derivatives *(45,46)*.

Coix Seeds (*Coix lachryma-jobi* var. *ma-yuen Stapf*)

The dried ripe kernels of *Coix lachryma-jobi* constitute the traditional Chinese medicine, coix seed, which is used to ease arthritis, control diarrhea, and eliminate edema by invigorating spleen function and promoting diuresis *(47)*. Coix seed is also found in the Chinese formulation Szu Shen Tan (Dioscorea Combination or Four Wonders Soup) along with *Dioscorea opposita*, *Nelumbo nucifera* (lotus seed), *Poria cocos* (Hoelen) and *Euryale ferax*. This Chinese prescription is well known in Taiwan for treating indigestion, especially in asthenic children. The formulation Lo Shih Shu or WTTC contains *Wisteria floribunda*, *Trapa bispinosa*, and *Terminalia chebula*, and Coix, and is formulated as a water-soluble ointment to treat gastric and rectal cancers, particularly to inhibit cancer cell growth and metastasis after cancer surgery. Kang Lai Te is a new anticancer drug from the active principles of coix seed. It is used as an emulsive i.v. injection and is effective in lung, liver, and bone cancer, particularly in reducing toxic side effects of chemotherapy.

Chemical Composition & Pharmacological Properties (48)

An acidic fraction composed of four free acids, palmitic, stearic, oleic and linoleic acids, shows antitumor activity; α-monolinolein inhibits tumorigenesis *(49)*; three glycans (coixans A, B, and C) show hypoglycemic activity *(50)*; and a benzoxazinone and a benzoxazolinone exhibit anti-inflammatory activity *(51)*.

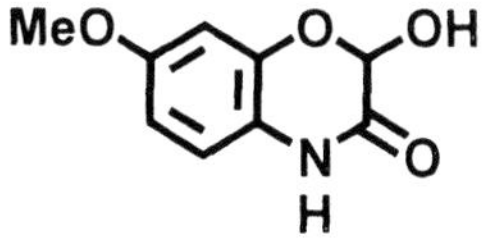

Benzoxazinone

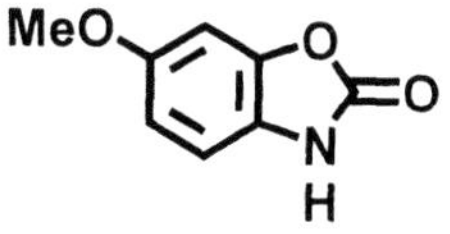

Benzoxazolinone (Coixol)

Ziziphus (*Ziziphus jujuba*, Rhamnaceae)

Dried ripe fruits of *Ziziphus* are used to tonify the spleen and stomach, moisten heart and lungs, nourish and pacify the spirit, smooth herbal action, and harmonize drugs. The roots have hypotensive effects, and the leaves can decrease the intake of sweets (as a taste-modifier and anti-obesity agent) *(52)*.

Chemical Composition & Pharmacological Properties

Among eight flavonoids in the seeds, swertisin and spinosin possess significant sedative activity *(53)*. The fresh leaves contain jujubasaponins, including ziziphin, which shows sweetness-inhibiting activity *(54, 55)*.

Swertisin: R = H
Spinosin: R = β-D-glc

Ziziphin

Middle Class Herbs

Tang Kuei or Dong Quai (*Angelica sinensis*, Umbelliferae)

The dried root of *Angelica sinensis* is used to activate blood circulation, regulate menstruation, relieve pain, and treat anemia, menstrual disorders, rheumatic arthralgia, and traumatic injuries.

Chemical Composition & Pharmacological Properties (56-59)

- Essential oils: ligustilide, *n*-butylidenephthalide – muscle-relaxant properties and antiproliferative effects on aortic smooth muscle cells
- Fatty acids: palmitic, linolic, stearic, and arachidonic acids

- Coumarins: bergapten, scopoletin, and umbelliferone
- Polysaccharides – immunostimulating and blood-tonifying
- High contents of vitamin B$_{12}$, folinic acid, and biotin – stimulating effect on hematopoiesis in bone marrow
- Tetramethylpyarazine & ferulic acid -- analgesic & antiinflammatory

Ligustilide   n-Butylidenephthalide   Tetramethylpyrazine   Ferulic acid

Chuan Chiung (*Cnidium officinale*, Umbelliferae)

The rhizome of *Ligusticum wallichii* (China) or *Cnidium officinale* (Japan) is used for headache, abdominal pain, arthralgia due to cold, tendon spasms, amenorrhea, menstrual disorders, and female genital inflammatory diseases. It invigorates blood circulation, promotes the flow of ch'i, and controls pain.

Chemical Composition & Pharmacological Properties

Ligustilide, butylidenephthalide, butylphthalide, senkyunolide, cnidilide *(60)*, and tetramethylpyrazine *(59)* are present. The herb increases myocardial contractility and coronary circulation, causes vasodilation, decreases heart rate and oxygen consumption, and lowers blood pressure *(61)*. Butylidenephthalide and ligustilide contribute to pentobarbital sleep effects in mice *(62)*. Among synthetic derivatives, BP-42 (4,5-dihydroxy-butylidenephthalide) showed the greatest antiproliferative effects in primary cultures of vascular mouse aorta smooth muscle cells, and may become a trial anti-atherosclerotic drug *(63)*.

4,5-Dihydroxy-butylidenephthalide

Ko Ken (*Pueralia lobata*, Leguminosae)

Kudzu vine root is one of the most important Chinese medical herbs. Ko Ken Tang (Pueraria Combination) is used to treat greater and sunlight Yang diseases, specifically with symptoms of fever, headache, and back/neck pain and stiffness. It is also used for cardiovascular disease, angina pectoris, and hypertension.

Chemical Composition & Pharmacological Properties

Ko Ken contains isoflavones, such as daidzein (7,4'-dihydroxyisoflavone), daidzin, and puerarin *(64-66)*. In addition to antiarrhythmic and immunostimulant activities, daidzein inhibits aldehyde dehydrogenase II and is uniquely used to treat alchoholism by decreasing alcohol craving and intake *(67)*. Genistein has estrogenic activity and inhibits oxidation of LDL *(68)*. Puerarin is a β-adrenergic blocker, inhibits platelet aggregation, and shows a cardioprotective effect *(69)*.

Daidzein: $R_1 = R_3 = R_4 = H$, $R_2 = OH$
Daidzin: $R_1 = R_3 = R_4 = H$, $R_2 = O\text{-}\beta\text{-D-glucose}$
Genistein: $R_1 = R_3 = H$, $R_2 = R_4 = OH$
Puerarin: $R_1 = O\text{-Glu}$, $R_2 = OH$, $R_3 = R_4 = H$

Shi Liu Pi (*Punica granatum*, Punicaceae)

Punica granatum (pomegranate) has been cultivated for centuries for its flavorful red fruit used in jams, jellies, and the drink "granadine". The root bark is used traditionally to purge intestinal parasites and the fruit husks are used as an antiseptic for gum, tonsil, and throat inflammation and infection.

Chemical Composition & Pharmacological Properties

The toxic alkaloid pelletierine is found in the root bark. Fruits with seeds contain polyphenolic compounds and estrogen. The latter is used for menopausal disorders and associated with fertility/contraception. The polyphenol ellagic acid shows anticarcinogenic effects *(70,71)* and antioxidant activity *(72)*. Ellagitannins, including punicalin, show anti-HIV activity *(73)*.

Ginger (*Zingiber officinale*, Zingiberaceae)

This spice or rhizome has botanical characteristics resembling those of turmeric. Ginger is known for antiemetic and antinausea properties and, as Kan Chiang, is used in Oriental traditional medicines for functional dyspepsia.

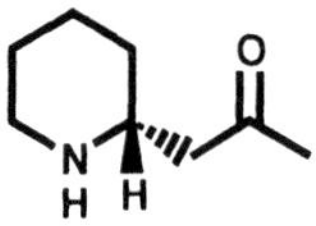

Pelletierine

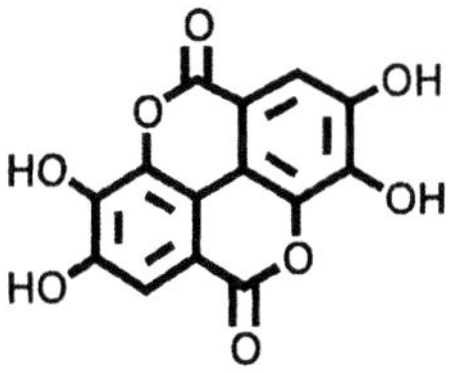

Ellagic Acid

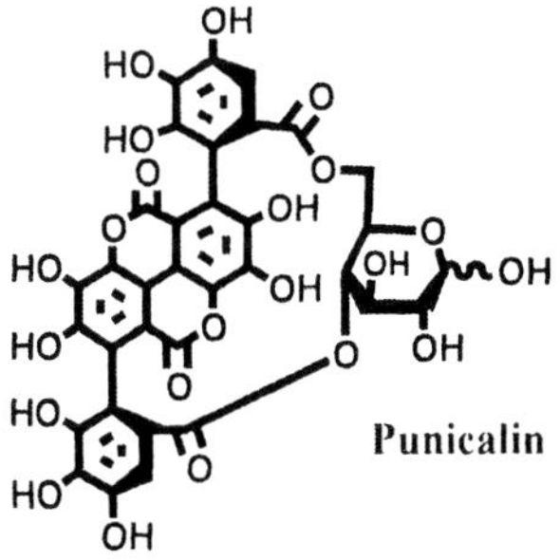

Punicalin

Chemical Composition & Pharmacological Properties

The major (30-70%) constituents of the essential oil are terpene hydrocarbons, including zingiberene, ar-curcumene, and α-bisabolene. The constituents responsible for ginger's pungent taste are 1-(3'-methoxy-4'-hydroxyphenyl)-5-hydroxyalkan-3-ones, also known as gingerols. [6]-Gingerol is a cholagogue, and [8]-gingerol is a hepatoprotective agent. Antioxidant activity has also been reported for *Z. officinale (74, 75)*.

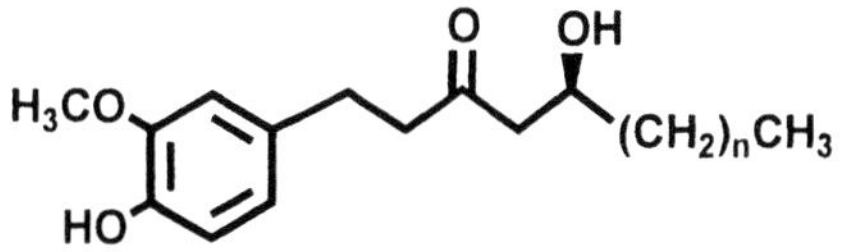

[6]-Gingerol, n = 4
[8]-Gingerol, n = 6

Green Tea (*Thea sinensis* or *Camellia sinensis*, Theaceae)

Green Tea is manufactured from the fresh leaves of *Thea sinensis*.

Chemical Composition and Pharmacological Properties

Green tea polyphenols (GTPs) [(-)-epicatechin (EC), (-)-epigallocatechin (EGC), (-)-epicatechin-3-gallate (ECG), and (-)-epigallocatechin-3-gallate (EGCG)] are the major bioactive constituents *(76)*. Their properties are:

- Cancer chemoprevention: GTPs inhibit promotion of carcinogenesis; EGCG inhibits tumor promotion (duodenal and skin cancers).
- Antibacterial activity: Inhibit the activity of bacterial exo-toxins.
- Inhibit influenza virus infection by blocking viral adsorption to cells *(77)*.
- Inhibit HIV RT: ECG, EGCG: IC_{50} = 10-20 ng/mL; but EC, EGC, and gallic acid were not active *(78)*.

(-)-Epicatechin (EC): R_1 = H, R_2 = OH
(-)-Epigallocatechin (EGC): R_1 = OH, R_2 = OH
(-)-Epicatechin-3-gallate (ECG): R_1 = H, R_2 =
(-)-Epigallocatechin-3-gallate (EGCG): R_1 = OH, R_2 =

The author's laboratory evaluated 38 tea polyphenols for anti-HIV activity in H9 lymphocytic cells. 8-C-Ascorbyl (-)-epigallocatechin and theasinensin-D had EC_{50} values of 3 and 8 μg/mL and therapeutic indexes of 9.5 and 5, respectively *(79)*.

8-C-Ascorbyl (-)-Epigallocatchin

Theasinensin-D

Ta Suan or Garlic (*Allium sativum*, Liliaceae)

Garlic is the bulb of *Allium sativum* L. and was first cultivated in ancient Egypt, Greece, Rome, India, and China for its safe and effective therapeutic benefits as well as culinary uses. It has been used to treat tumors, headaches, weakness and fatigue, wounds, sores, and infections. It lowers blood lipids and inhibits platelet aggregation.

Major Chemical Constituents of Garlic and Their Biological Activities

Fresh garlic contains sulfur-containing amino acids, including alliin, together with steroid saponins and polysaccharides. When the cells are crushed, allin is degraded and converted to allicin by alliinase. Allicin is converted by heat to many other compounds, including diallyl sulfide, which reduces tumor growth in animals and also shows anti-HIV effects. Additionally, allicin is a precursor of (*E*)-ajoene and (*Z*)-ajoene, which exhibit antithrombotic activity by inhibiting platelet aggregation, with the *Z* isomer being more active. Allicin itself shows antiprotozoal, antibacterial, antifungal and antiviral activities, exhibits hypoglycemic effects, and decreases blood cholesterol levels *(80,81)*. Alliin is also a precursor to allithiamine, a prodrug of Vitamin B_1 (VB_1). Allithiamine resists the activity of aneurinase (thiaminase) and is absorbed easily in the intestine. It is converted to VB_1 in the body and, thus, can be used as an active VB_1 *(82)*, although it does generate an unpleasant garlic odor.

Lower Class Herbs

Kansui (*Euphorbia kansui*, Euphorbiaceae)

Euphorbia kansui is widely distributed in Northwest China. Known as Kansui, its dried roots are used as an herbal remedy for ascites and cancer.

Chemical Composition & Pharmacological Properties

Ingenol derivatives, including kansuiphorins A, B, C, and D are found *(83, 84)*. Kansuiphorin A had *in vitro* IC$_{50}$ values ranging from 0.03–0.33 µg/mL against various leukemia, melanoma, and non-small cell lung, colon, and renal cancer cell lines, and kansuiphorins A and B were potent against P-388 leukemia in mice with T/C values of >176 and 177% at 0.1 and 0.5 mg/kg, respectively *(83)*.

Kansuiphorin A

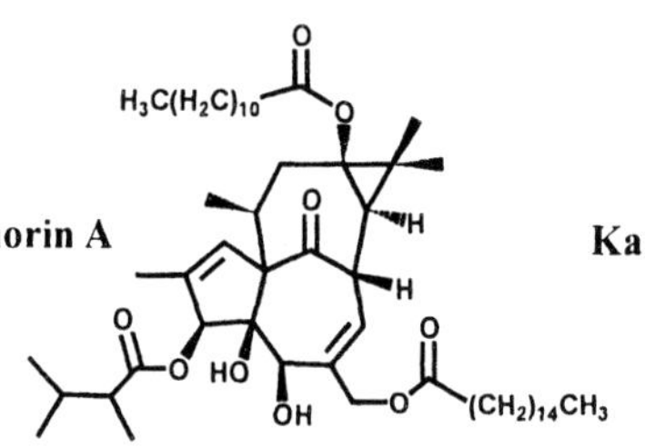

Kansuiphorin B

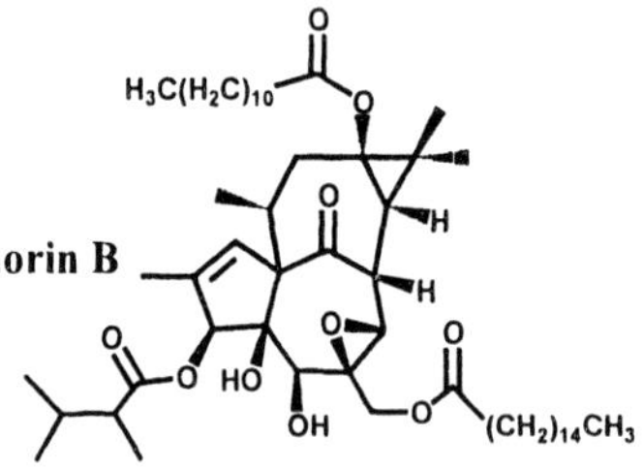

Hsia Ku Tsao (*Prunella vulgaris* L., Labiatae)

Hsia Ku Tsao (Spica Prunellae, Self-heal Spike) is the dried flowered fruit-spike of *P. vulgaris*. This herb is used for hypertension with headache, tinnitus, eye inflammation, and nocturnal eye pain. It also has hypotensive, antibacterial, and antitumor activities *(85)*.

Chemical Composition & Pharmacological Properties

Ursolic acid was isolated from the fruiting spikes using bioassay-directed fractionation. It showed significant cytotoxicity against P-388 and L-1210 leukemia cells and A-549 human lung carcinoma cells, and marginal cytotoxicity in KB, HCT-8 human colon, and MCF-7 mammary tumor cells *(86)*.

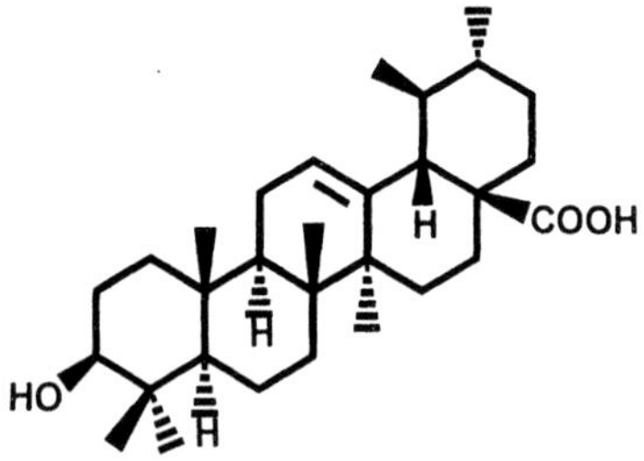

Ursolic Acid

Kuei Chiu (*Podophyllum emodi*, Berberidaceae)

Podophyllum emodi has long been used in China as an anticancer drug and to treat snakebites, periodontitis, skin disorders, coughs, and intestinal parasites. The dried roots are traditionally used as a contact cathartic.

Chemical Composition & Pharmacological Properties

The antimitotic lignan podophyllotoxin and other close derivatives are found in *P. emodi* and related species, including *P. peltatum* and *P. pleianthum*. Podophyllotoxin is a mitotic spindle poison. It inhibits the polymerization of tubulin and stops cell division at the start of metaphase *(87)*. It can be converted into the semisynthetic antineoplastic derivatives teniposide and etoposide, which are used to treat small cell lung cancer, testicular cancer, leukemias, lymphomas, and other cancers *(88-91)*. These semisynthetic derivatives are known as the epipodophyllotoxin series because they have the opposite stereochemistry and are glucosylated at the C-4 hydroxyl [with two of the glucose hydroxyl groups (at C-4" and C-6") blocked as either a thienylidene (teniposide) or ethylidene (etoposide)] and are demethylated at C-4'. Unlike podophyllotoxin, these derivatives do not affect microtubule assembly, but rather complex with the enzyme topoisomerase II (topo II), causing DNA double strand breakage and G2-phase arrest *(87)*. However, myelosuppression, drug resistance, and poor bioavailability are problems with etoposide's use *(92)*.

Etoposide: R = CH$_3$
Teniposide: R = 2-Thienyl

Podophyllotoxin

Using the principles of lead improvement, the author's laboratory has performed extensive structure-activity relationship, enzyme interaction, and computational studies to generate new compounds to overcome these limitations. In particular, several series of 4-alkylamino and 4-arylamino epipodophyllotoxin analogues were synthesized from the natural product podophyllotoxin *(93)*. New computational strategies have continued to play an important role in the rational design of improved etoposide analogs *(94, 95)*.

Compared with etoposide, several synthetic compounds have shown similar or increased % inhibition of DNA topo II activity and % protein-linked DNA breakage *(96)*, and even more notably, show increased cytotoxicity in etoposide-resistant cell lines. GL-331 *(97)*, which contains a *p*-nitroanilino moiety at the 4β position of etoposide, has emerged successfully from this preclinical development to proceed further along the drug development pathway.

GL-331

Like etoposide, GL-331 is a topo II inhibitor and causes cell cycle arrest. Formulated GL-331 has similar pharmacokinetic profiles to those of etoposide and shows desirable stability and biocompatability *(97)*. Genelabs Technologies, Inc. has patented GL-331 and its Phase I clinical trials as an anticancer drug were completed at the M.D. Anderson Cancer Center. Good antitumor efficacy was found in four tumor types (non-small and small cell lung, colon, and head/neck cancers) *(98, 99)*. The major toxicity was cytopenias, but side effects were minimal. GL-331's maximum tolerated dose (MTD) was 300 mg/m^2, while that for etoposide was 140 mg/m^2. Other advantages of GL-331 are greater activity both *in vitro* and *in vivo*, a shorter synthesis and, thus, easier manufacture, and evidence of activity in refractory tumors as it overcomes drug-resistance in many cancer cell lines (KB/VP-16, KB/VCR, P388/ADR, MCF-7/ADR, L1210/ADR, HL60/ADR, and HL60/VCR) *(98, 99)*.

❖ In summary, GL-331 is an exciting chemotherapeutic candidate and exemplifies successful preclinical drug development from Oriental herbal products.

Research on New Medicines from Herbal Products

An herb or herbal prescription is chosen as the source of a potential new drug based on folk or clinical experiences. The initial research (new lead discovery) focuses on isolation of a bioactive natural lead compound(s), as illustrated graphically below. After extraction of the herbal medicine, activity is verified by pharmacological testing. Bioactivity-directed fractionation and isolation (BDFI) of the active portions leads to the active principle(s). The newly discovered lead compound is then chemically modified to increase activity, decrease toxicity, or improve other pharmacological profiles. Compounds are evaluated through mechanism of action (e.g., enzymatic or antimitotic) and other appropriate assays. For example, preclinical screening in the National Cancer Institute's (NCI) *in vitro* human cell line panels and selected *in vivo* xenograft systems is used to select the most promising anticancer drug development targets. Efficacy and toxicity are evaluated, and production, formulation, and toxicological studies are performed prior to clinical trials.

Approaches to New Lead Discovery From Herbal Medicines Research

Clinical Experience of Herbal Medicines →

Extraction of a Target Herb or Prescription* →

Pharmacological Testing → Active Portions →

Bioactivity-Directed Isolation & Characterization →

Active Natural Products Leads

---

*Importantly, this same process should be applied to both single and formulated herbs leading to single active principles (single herbal-derived compounds), active fractions (herbal extracts), and effective and safe prescriptions (multiple herbal products).

By using these approaches, numerous drugs have been discovered from the active principles of TCM herbs, as described in the previous sections and below. Ephedrine and artemisinin exemplify drugs identified from single herbs, while indirubin demonstrates a drug discovered from an herbal formula.

Ephedras (*Ephedra spp.*, Ephedraceae)

Ephredras are the dried stems of *Ephedra sinica* Stapf., *E. intermedia* Schrenk et C.A. May., or *E. equisetina* Bge. In TCM, ephedra is the primary component of multi-herb formulas to treat asthma and bronchitis, cold and flu, cough and wheezing, fever, chills, lack of perspiration, headache, and nasal congestion.

Ephedras contain various alkaloids, primarily (-)-ephedrine, (+)-pseudoephedrine and the corresponding nor and N,N-dimethyl derivatives. Ephedrine is structurally similar to adrenaline and physiologically is an indirect sympathomimetic. Dietary supplements containing ephedras are widely used in weight reduction and energy enhancement *(100)*.

(-)-Ephedrine (+)-Pseudoephedrine

Qinghao (*Artemisia annua* L., Asteraceae)

Qinghao (Sweet Wormwood) is the dried aerial parts of *A. annua*. The herb has been used in China for over a thousand years to treat fever and malaria. Artemisinin (Qing Hao Su) was isolated as an active principle and directly kills malaria parasites in the erythrocytic stage with low toxicity to both animal and human organs. It was introduced clinically as a new type of antimalarial agent with safe and rapid action against chloroquine-resistant *Plasmodium falciparum* *(101)*. The novel endo-peroxide linkage is essential for the antimalarial activity. Among many synthetic derivatives, artemether is in clinical use in China and arteether is in Phase II clinical trials in the USA as an antimalarial drug *(102)*. The author's laboratory synthesized several antimalarial analogs related to artemisinin *(103-105)*.

Artemisinin

Artemether: R = CH$_3$
Arteether: R = CH$_2$CH$_3$

Indirubin (from Dang Gui Lu Hui Wan)

The prescription Dang Gui Lu Hui Wan is a traditional remedy for chronic myelocytic leukemia and contains *Angelica sinensis*, *Aloe vera*, *Gentiana scabra*, *Gardenia jasminoides*, *Scutellaria baicalensis*, *Phellodendron amurense*, *Coptis chinensis*, *Rheum palmatum*, *Aucklandia lappa*, and Indigo naturalis, which is a product made from leaves of *Baphicacanthus cusia*, *Indigofera tinctoria*, or *Isatis indigotica*.

The active ingredient of this prescription was identified as Indigo naturalis, and the antileukemic agent as indirubin *(106)*. The derivatives N,N'-dimethyl-indirubin and N-methylindirubin oxime were more potent than the parent compound against rat carcinosarcoma W256 and mouse leukemia L7212 *(107)*.

Indirubin: $R_1 = R_2 = H, X = O$
N,N'-Dimethylindirubin: $R_1 = R_2 = CH_3, X = O$
N-Methylindirubin Oxime: $R_1 = H, R_2 = CH_3, X = N\text{-}OH$

Modern Directions from an Ancient and Long-lasting Legacy

In "Shen Nung Pen Ts'ao Ching," the classical and longest surviving description of herbs, Shen Nung recorded 365 herbs, which together with many others, have been used in Chinese Herbal Medicine for over 4,000 years. According to TCM theories, herbs can affect the body in various positive ways, for instance, by strengthening and balancing the system, tonifying the organs, and optimizing the flow and use of energy (Chi; Qi). These properties, including antioxidant and antiaging, blood pressure-lowering, hypolipidemic, blood sugar-lowering, anti-allergic, and anti-arthritis effects, are ideal for dietary supplements to maintain health and for drugs to treat chronic health problems. Herbal products and their active principles can also treat acute illness. However, efficacy and safety of these products must be proven through rigorous study. Historically, only bioactive, pure, lead compounds have been the source of new pharmaceutical agents; however, in this century, three herbal-related sources should be explored as sources of dietary supplements and new drugs.

(1) Active Pure Compounds,
(2) Active Fractions, and
(3) Validated or Improved Effective and Safe Herbal Formulations

Conclusions

Oriental herbal products, which originate from TCM, use processed single or formulated herbal products as dietary supplements or as drugs to prevent, relieve, and cure many diseases. This ancient legacy will provide a strong base for continued development of modern high-quality dietary supplements and modern medicines in the 21st century. Herbs contain numerous bioactive compounds, and highly efficient BDFI, characterization, analog synthesis and mechanistic studies can develop such active principles as clinical candidates for world-class new drug development. New, effective, and safe world-class new medicines and dietary supplements will constantly be developed from bioactive lead compounds, active herbal fractions, and effective and safe TCM prescriptions.

Acknowledgement

This investigation is supported in part by grants CA-17625 and AI-33055 from the National Cancer Institute and the National Institute of Allergy and Infectious Disease, NIH, respectively, awarded to K. H. Lee.

References

1. Hosoya, E. In *Herbal Medicine:Kampo, Past and Present.* Takemi, T.; Hasegawa, M.; Kumagai, A.; Otsuka, Y. Eds.; Life Science Publishing Co.: Tokyo, 1985; pp 52-65.
2. Huang, K. C. In *The Pharmacology of Chinese Herbs.* CRC Press: Boca Raton, 1993; pp 3-7.
3. Hsu, H. Y. In *An Introduction to Chinese Medicine (Chung Kuo Yi Yao Gai Lun).* Committee on Chinese Medicine, Ed.; The Executive Yuan Press: Taipei, Taiwan, 1973; p 48.
4. Tsai, T. T.; Liao, S. N. In *The Application and Therapy of Ling Chih.* San Yun Press: Taichung, Taiwan, 1982.
5. Boik, J. In *Natural Compounds in Cancer Therapy.* Oregon Medical Press: Princeton, MN, 2001; pp 205-206.
6. Siao, M. S.; Lee, K. R.; Li, L. J.; Wang, C. T. In *Food Phytochemicals for Cancer Prevention.* Ho, C. T.; Osawa, T.; Huang, M. T.; Rosen, R. T. Eds.; ACS Symposium Series 547; American Chemical Society: Washington, DC, 1994; Vol. II, pp 342-354.
7. Bao, X.; Duan. J.; Fang, X.; Fang, J. *Carbohyd. Res.* **2001**, *336*, 127-140.
8. Min, B. S.; Gao, J. J.; Hattori, M.; Lee, H. K.; Kim, Y. H. *Planta Medica* **2001**, *67*, 811-814.
9. *American Cancer Society's Guide to Complementary and Alternative Cancer Methods.* American Cancer Society: Atlanta, GA, 2000; pp 314-315.
10. Sakagami, H.; Sugaya, K.; Utsumi, A.; Fujinaga, S.; Sato, T.; Takeda, M. *Anticancer Res.* **1993**, *13*, 671-675.
11. *American Cancer Society's Guide to Complementary and Alternative Cancer Methods.* American Cancer Society: Atlanta, GA, 2000; pp 345-346.
12. Chang, S. T. *Int. J. Med. Mushrooms* **1999**, *1*, 1-7.
13. Wasser, S. P.; Weis, A. L. *Int. J. Med. Mushrooms* **1999**, *1*, 31-62.
14. Lee, K.H.; Kozuka, M.; Tokuda, H. Unpublished Data.
15. Ohno, N.; Akanuma, A. M.; Miura, N. N.; Adachi, Y.; Motoi, M. *Pharm. Pharmacol. Lett.* **1999**, *11*, 87-90.
16. Shibata, S; Fujita, M.; Itokawa, H.; Tanka, O. *Chem. Pharm. Bull.* **1963**, *11*, 759.

17. Shibata, S. J. Korean Med. Sci. **2001**, *16*, 28-37.
18. Tanaka, O. In *Food Phytochemicals for Cancer Prevention.* Ho, C.T.; Osawa, T.; Huang, M.T.; Rosen, R.T. Eds.; ACS Symposium Series 547; American Chemical Society: Washington, DC, 1994; Vol. II, pp 342-354.
19. Yun, T. K.; Lee, Y. S.; Lee, Y. H.; Kim, S. I.; Yun, H. Y *J. Korean Med. Sci.* **2001**, *16*, -18.
20. Attele, A. S.; Wu, J. A.; Yuan, C. S. *Biochem. Pharmacol.* **1999**, *58*, 1685-1693.
21. Yun, T. K.; Choi, S. Y. *Internat. J. Epidemiol.* **1990**, *19*, 871-876.
22. Besso, H.; Kasai, R.; Wang, J. F.; Saruwatari, Y.; Fuwa, T.; Tanaka, O. *Chem. Pharm. Bull.* **1982**, *30*, 4534-4538.
23. Li, W.; Gu, C.; Zhang, H.; Awang, D. V. C.; Fitzloff, J. F.; Fong, H. H. S.; van Breemen, R.B. *Anal. Chem.* **2000**, *72*, 5417-5422.
24. Chan, T. W. D.; But, P. P. H.; Cheng, S. W.; Kwok, I. M. Y.; Lau, F. W.; Xu, H.X. *Anal. Chem.* **2000**, *72*, 1281-1287.
25. Dou, D. Q.; Hou, W. B.; Chen, Y. J. *Planta Medica* **1998**, *64*, 585.
26. Li, W. K.; Fitzloff, J. *J. Liquid Chromat. Related Technol.* **2002**, *25*, 29-41.
27. Wang, H. X.; Ng, T. B. *Biochem. Biophys. Res. Comm.* **2000**, *269*, 203-208.
28. Ma, W. G.; Mizutani, M.; Malterrud, K. E.; Lu, S. L.; Ducrey, F.; Tahara, S. *Phytochemistry* **1999**, *52*, 1133-1139.
29. Gan, F. Y.; Zheng G. Z. *Zhongguo Yaoxue Zazhi* **1992**, *27*, 138-142.
30. Kosuge, T.; et al. *Yakugaku Zasshi* **1981**, *101*, 629.
31. Sugiyama, A.; Zhu, B. M.; Takahara, A.; Satoh, Y.; Hashimoto, K. *Circulation J.* **2002**, *66*, 182-184.
32. Deyama, T.; Nishibe, S.; Nakazawa, Y. *Acta Pharmac. Sinica* **2001**, *22*, 1057-1070.
33. Nohara, T. Scientific study on Tochu-tea, The 3rd Symposium on Medicinal Foods, Abstract, **2000**, 67-76.
34. Li, Y. M.; Sato, T.; Metori, K.; Koike, K.; Che, Q. M.; Takahashi, S. *Biol. Pharm. Bull.* **1998**, *21*, 1306-1310.
35. Xue, J. Y.; Liu, G. T.; Wei, H. L.; Pan, Y. *Free Radical Biol. Med.* **1992**, *12*, 127-135.
36. Ip, S. P.; Yiu, H. Y.; Ko, K. M. *Mol. Cell. Biochem.* **2000**, *208*, 151-155.
37. Yasukawa, K.; Ikeya, Y.; Mitsuhashi, H.; Iwasaki, M.; Aburada, M.; Nakagawa, S.; Takeuchi, M.; Takido, M. *Oncology* **1992**, *49*, 68-71.
38. Chen, D. F.; Zhang, S. X.; Xie, L.; Xie, J. X.; Chen, K.; Kashiwada, Y.; Zhou, B. N.; Wang, F.; Cosentino, L. M.; Lee, K. H. *Bioorg. Med. Chem.* **1997**, *5*, 1715-1723.
39. Hirotani, M.; Zhou, Y.; Lui, H.; Furuya, T. *Phytochem.* **1994**, *36*, 665-670.
40. Wang, Y.P.; Li, X.Y.; Song, C.Q.; Hu, Z.B. *Acta Pharmacol. Sinica* **2002**, *23*, 263-266.
41. *Hong Kong Monograph of Chinese Materia Medica*, **2000**.
42. Peng, X. M.; Tian, G. *Carbohyd. Res.* **2001**, *331*, 95-99.

43. Qin, X. M.; Yamauchi, R.; Aizawa; Inakuma, T.; Kato, K. *Carbohyd. Res.* **2001**, *333*, 79-85.
44. Peng, X. M.; Huang, L. J.; Qi, C. H.; Zhang Y. X.; Tian, G. Y. *Chinese J. Chem.* **2001**, *19*, 1190-1197.
45. *Chinese Herbs*; SRA "Chinese Herbs" Editorial Committee, Ed.; Shanghai Science and Technology Publishing Company: Shanghai, 1999; Vol. 1, p 494.
46. Bok, J. W.; Lermer, L.; Chilton, J.; Klingeman, H. G.; Neil Towers, G. H. *Phytochemistry* **1999**, *51*, 891-898.
47. *A Coloured Atlas of the Chinese Materia Medica;* Pharmacopoeia Commission of the Ministry of Public Health, P.R. China; Guangdong Science & Technology Press: Guang Zhou, China, 1995; p 486.
48. Numata, M.; Yamamoto, A.; Moribayashi, A.; Yamada, H. *Planta Medica* **1994**, *60*, 356-359.
49. Tokuda, H.; Matsumoto, T.; Konoshima, T.; Kozuka, M.; Nishino, H.; Iwashima, A. *Planta Medica* **1990**, *56*, 653-564.
50. Takahashi, M.; Konno, C.; Hikino, H. *Planta Medica* **1986**, *52*, 64-65.
51. Otsuka, H.; Hirai, Y.; Nagao, T.; Yamasaki, K. *J. Nat. Prod.* **1988**, *51*, 74-79.
52. Suttisri R.; Lee I.R.; Kinghorn, A.D. *J. Ethnopharm.* **1995**, *47*, 9.
53. Cheng, G.; Bai, Y.; Zhao, Y.; Tao, J.; Liu, Y.; Tu, G.; Ma, L.; Liao, N.; Xu, X. *Tetrahedron* **2000**, *56*, 8915.
54. Kurihara Y.; Ookubo, K.; Tasaki H.; Kodama H.; Akiyama Y.; Yagi A.; Halpern B. *Tetrahedron* **1988**, *44*, 61.
55. Yoshikawa K.; Shimono N.; Arihara S. *Tetrahed. Lett.* **1991**, *32*, 7059.
56. Ye, Y. N.; Liu, E. S. L.; Shin, V. Y.; Koo, M. W. L.; Li, Y.; Matsui, H.; Cho, C. H. *Biochem. Pharmcol.* **2001**, *61*, 1439-1448.
57. Hsieh, M. T.; Wu, C. R.; Lin, L. W.; Hsieh, C. C.; Tsai, C. H. *Planta Med.* **2001**, *67*, 38-42.
58. Choy, Y. M.; Leung, K. N.; Cho, C. S.; Wong, C. K.; Pang, P. K. T. *Amer. J. Chin. Med.* **1994**, *22*, 137-145.
59. Ozaki, Y. *Chem. Pharm. Bull.* **1992**, *40*, 954-956.
60. Choi, H. S.; Kim, M. S. L.; Sawamura, M. *Flav. Frag. J.* **2002**, *17*, 49-53.
61. *New Medicine Development and Herbal Medicine Use II;* Itokawa, H., Ed.; CMC Press: Tokyo, 2001; p 266.
62. Matsumoto, K.; Kohno, S.; Ojima, K.; Tezuka, Y.; Kadota, S.; Watanabe, H. *Life Sci.* **1998**, *62*, 2073-2082.
63. Mimura, Y.; Kobayashi, S.; Naitoh, T.; Kimura, I.; Kimura, M. *Biol. Pharm. Bull.* **1995**, *18*, 1203-1206.
64. Murakami, T.; Nishikawa, Y.; Ando, T. *Chem. Pharm. Bull.* **1960**, *8*, 688.
65. Rong, H. J.; Stevens, J. F.; Deinzer, M. L.; DeCooman, L.; De Keukeleire, D. *Planta Medica* **1998**, *64*, 620-627.
66. Chen, G.; Zhang, J. X.; Ye, J. N. *J. Chromat. A* **2001**, *923*, 255-262.
67. Wing, M. K.; Vallee, B. L. *Phytochemistry* **1998**, *47*, 499-506.

68. Grynkiewicz, G.; Achmatowiz, O.; Pucko, W. *Herba Polonica* **2000**, *46*, 151-160.
69. Huang, K.C. In *The Pharmacology of Chinese Herbs*. CRC Press: Boca Raton, FL, 1999; p 99.
70. Constantinou, A.; Stoner, G. D.; Mehta, R.; Rao, K.; Runyan, C.; Moon, R. *Nutr. Cancer* **1995**, *23*, 121-130.
71. Kim, N. D.; Mehta, R.; Yu, W.; Neeman, I.; Livney, T.; Amichay, A.; Poirier, D.; Nicholls, P.; Kirby, A.; Jiang, W.; Mansel, R.; Ramachandran, C.; Rabi, T.; Kaplan, B.; Lansky, E. *Breast Cancer Res. Treat.* **2002**, *71*, 203-217
72. Noda, Y.; Kaneyuki, T.; Mori, A.; Packer, L. *J. Agric. Food Chem.* **2002**, *50*, 166-171.
73. Nonaka, G.; Nishioka, I.; Nishizawa, M.; Yamagishi, T.; Kashiwada, Y.; Dutschman, G. E.; Bodner, A. J.; Kilkuskie, R. E.; Cheng, Y. C.; Lee, K. H. *J. Nat. Prod.* **1990**, *53*, 587-595.
74. Nakatani, N. *Biofactors* **2000**, *13*, 141-146.
75. Andersen, L. F.; Moskaug, J. O.; Jacobs, D. R.; Blomhoff, R. *J. Nutrition* **2002**, *132*, 461-471.
76. Ho, C. T.; Ferraro, T.; Chen, Q.; Rosen, R. T.; Huang, M. T. In *Food Phytochemicals for Cancer Prevention*. Ho, C.T.; Osawa, T.; Huang, M. T.; Rosen, R.T. Eds., ACS Symposium Series 547; American Chemical Society: Washington, DC, 1994; Vol. II, pp 2-19.
77. Shimamura, T. In *Food Phytochemicals for Cancer Prevention*. Ho, C.T.; Osawa, T., Huang, M. T.; Rosen, R.T. Eds., ACS Symposium Series 547; American Chemical Society: Washington, DC, 1994; Vol. II, pp 101-104.
78. Nakane, H.; Hara, Y.; Ono, K. In *Food Phytochemicals for Cancer Prevention*. Ho, C.T.; Osawa, T., Huang, M. T.; Rosen, R. T. Eds., ACS Symposium Series 547; American Chemical Society: Washington, DC, 1994; Vol. II, pp 56-64.
79. Hashimoto, F.; Kashiwada, Y.; Nonaka, G.; Nishioka, I.; Nohara, T.; Cosentiono, L. M.; Lee, K. H. *Bioorg. Med. Chem. Lett.* **1995**, *6*, 695-700.
80. Harris, J. C.; Cottrell, S. L.; Plummer, S.; Lloyd, D. *Appl. Microbiol. Biotechnol.* **2001**, *57*, 282-286.
81. Reddy, B. S.; Rao, C. V. In *Food Phytochemicals for Cancer Prevention*. Ho, C. T.; Osawa, T., Huang, M. T.; Rosen, R.T. Eds., ACS Symposium Series 547; American Chemical Society: Washington, DC, 1994; Vol. II, pp 164-172.
82. Doyle, M. R.; Webster, M. J.; Erdmann, L. D. *Internat. J. Sport Nutri.* **1997**, *7*, 39-47.
83. Wu, T. S.; Lin, Y. M.; Haruna, M.; Pan, D. J.; Shingu, T.; Chen, Y. P.; Hsu, H. Y.; Nakano, T.; Lee, K. H. *J. Nat. Prod.* **1991**, *54*, 823-829.
84. Pan, D. J.; Hu, C. Q.; Chang, J. J.; Lee, T. T. Y.; Chen, Y. P.; Hsu, H. Y.; McPhail, D. R.; McPhail, A. T.; Lee, K. H. *Phytochemistry* **1991**, *30*, 1020-1023.

85. Zhu, Y.P. In *Chinese Materia Medica, Chemistry, Pharmacology and Applications.* Harwood Academic Publishers: The Netherlands, 1998; pp 120-122.
86. Lee, K. H.; Lin, Y. M.; Wu, T. S.; Zhang, D. C.; Yamagishi, T.; Hayashi, T.; Hall, I. H.; Chang, J. J.; Wu, R. Y.; Yang, T. H. *Planta Medica* **1988**, 308-311.
87. Cragg, G.; Suffness, M. *Pharmac. Ther.* **1988**, *37*, 425 and references cited therein.
88. Keller-Juslen, C.; Kuhn, M.; von Wartburg, A.; Stahelin, H. *J. Med. Chem.* **1971**, *14*, 936.
89. O'Dwyer, P.J.; Alonso, M.T.; Leyland-Jones, B.; Marsoni, S. *Cancer Treat. Rep.* **1984**, *68*, 1455.
90. VePesid Product Information Overview, Bristol Lab., 1983.
91. *Etoposide (VP-16) Current Status and New Developments*; Issell, B. F., Muggia, F. M., Carter, S. K., Eds.; Academic Press: Orlando, FL, p 1984.
92. van Maanen, J.M.; Retel, J.; deVries, J.; Pinedo, H.M. *J. Natl. Cancer Inst.* **1988**, *80*, 1526.
93. Lee, K. H.; Imakura, Y.; Haruna, M.; Beers, S. A.; Thurston, L. S.; Dai, H. J.; Chen, C. H.; Liu, S. Y.; Cheng, Y. C. *J. Nat. Prod.* **1989**, *52*, 606-613.
94. Cho, S.J.; Tropsha, A.; Suffness, M.; Cheng, Y.C.; Lee, K.H. *J. Med. Chem.* **1996**, *39*, 1383.
95. Xiao, Z.; Xiao, Y.D.; Feng, J.; Golbraikh, A.; Tropsha, A.; Lee, K.H. *J. Med. Chem.* **2002**, *45*, 2294.
96. Chang, J. Y.; Han, F. S.; Liu, S. Y.; Wang, H. K.; Lee, K. H.; Cheng, Y. C. *Cancer Res.* **1991**, *51*, 1755-1759.
97. Wang, Z.W.; Kuo, Y.H.; Schnur, D.; Bowen, J.P.; Liu, S.Y.; Han, F.S., Chang, J. Y.; Cheng, Y.C.; Lee, K.H. *J. Med. Chem.* **1990**, *33*, 2660.
98. Liu, S.Y.; Soikes, R.; Chen, J.; Lee, T.; Taylor, G.; Hwang, K.M.; Cheng, Y.C., presented at 84th AACR Annual Conference, Orlando, FL, May 19-22, 1993.
99. Fossella, F.V., University of Texas MD Anderson Cancer Center and Chen, J., Genelabs Technologies, Inc., personal communications (10/08/96).
100. Arditti, J.; Bourdon, J. H.; Spadari, M.; de Haro, L.; Richard, N.; Valli, M. *Acta Clinica Belgica* **2002**, *57*, 34-36.
101. Hsiao, P. K. In *Natural Products as Medicinal Agents*; Beal, J. L.; Reinhard, E., Eds.; Hippokrates Verlag: Stuttgart, Germany, 1981; pp351-394.
102. Zeng, Q.; Du, D.; Xie, D.; Wang, X.; Ran, C. *Chin. Trad. Herb. Drugs.* **1982**, *13*, 24-30.
103. Meshnick, S.R. In *Antimalarial Chemotherapy*, Rosenthal, P.J., Ed.; Human Press: Totowa, NJ, 2001, pp 191-201, and literature cited therein.

104. Avery, M. A.; MvLean, G.; Edwards, G.; Ager, A. In *Biologically Active Natural Products: Pharmaceuticals*; Cutler, S. J.; Cutler, H. G., Eds.; CRC Press LLC: Boca Raton, FL, 2000; pp 121-132.
105. Imakura, Y.; Yokoi, T.; Yamagishi, T.; Koyama, J.; Hu, H.; McPhail, D.R.; McPhail, A.T.; Lee, K.H. *J. Chem. Soc., Chem. Commun.*. **1988**, 372.
106. Imakura, Y.; Hachiya, K.; Ikemoto, T.; Yamashita, S.; Kihara, M.; Shingu, T.; Milhous, W.K.; Lee, K.H. *Heterocycles* 1990, *31,* 1011.
107. Imakura, Y.; Hachiya, K.; Ikemoto, T.; Kobayashi, S.; Yamashita, S.; Sakakibara, J.; Smith, F.T.; Lee, K.H. *Heterocycles* **1990**, *31,* 2125.

Chapter 2

Modernization of Traditional Chinese Herbal Medicine

Yuh-Chiang Shen, Wen-Fei Chiou, Guei-Jane Wang, Cheng-Jen Chou, and Chieh-Fu Chen

National Research Institute of Chinese Medicine, Taipei, Taiwan, Republic of China

Multiple drug composition and mechanism is the most unique character in Traditional Chinese Medicine (TCM). The mechanisms of action of TCM which produce biological effects are: 1). summation effect of one active component on different parts of the body, 2). summation effect of various active components acting on the same or a different part of the body, and 3). drug metabolism and interactions in the body. Compared with synthetical drugs, which usually contain a single chemical compound, TCM, or plants extract, contains many different chemicals. Plant extracts come from natural sources and it is well known that their quality are highly variable depending on the species and strains of the plant, geographical location, soil and culture conditions, growing conditions, climate and weather conditions, harvesting, part of the plant used, the solvent used to extract the plant, method of extraction (temperature, pressure, batch size, amount of solvent, filling height, velocity of flow), and storage. Even if all the above variables are well controlled, there are still some realistic problems, such as some active principles being unknown or only partially understood, the different modes of action being unclear, the lack of pharmacokinetic data, and the lack of stability and clinical studies under controlled conditions. It is our experience that plant extracts may be functionally variable even though the chemical profile seems rather uniform. Thus, to perform both chemical and pharmacological fingerprinting may be the only way to define the composition and functional uniformity of plant extracts in order to achieve a meaningful standardization.

Principles of Traditional Chinese Medicine (TCM)

Traditional Chinese medicine (TCM) postulates seven modes by which natural products can elicit pharmacologic effects, one when an agent is used alone, six when several are used concurrently. In the latter instance, the combined products in a prescription can interact with each other to augment or reduce potency or intrinsic activity with respect to beneficial or harmful effects. The six modes of drug interaction in Traditional Chinese Medicine can be interpreted by modern pharmacological terms (*1,2*):

Xiang xu (need each other), additive or synergistic enhancement of pharmacologic action by two or more substances with similar properties.

Xiang shi (use or reinforce each other), potentiation or synergism; enhancement of therapeutic action by substances with different properties.

Xiang wei (mutual respect or restraint), inhibition or reduction of pharmacologic effects by two or more substances with properties in common.

Xiang wu (multual dislike), inhibition or reduction of an effect of one drug by another with an opposing action.

Xiang sha (kill each other), the specific nullification of the effect of one compound by another agent by competitive antagonism, such as between agonist and antagonist compounds.

Xiang fan (oppose each other), incompatibility, not suitable for combination because severe adverse effects may result.

The composition of medical prescription according to status of 1. the principal, one or several drug(s) which produce therapeutic effect, 2. minister(s) which accelerate the therapeutic effect of the principal, 3. assistant(s) which assisting towards accomplishment of the principal, 4. convey, that directs or leads the principal drug to the proper channel (site of action or receptor).

Above are major principles guiding the formation of Chinese medical prescription, Fang-Ji. Thus, multiple drug composition and mechanism is the most unique character in TCM. The mechanisms of TCM which produce therapeutic effects are: 1). summation effect of one active component on different parts of the body, 2). summation effect of various active components acting on the same or a different part of the body, and 3). drug interactions in the body.

Comparison of Synthetic Drugs and TCM

Compared with synthetic drugs, an exact chemical compound, TCM, or plant extracts contain different chemicals. The side effect of a drug is strongly affected by its dosage. The greater the amount of a chemical that must be used to produce the therapeutic effect, the greater is the chance of a side effect. Traditionally, herbal medicine is almost never used as single chemical, but in a mixture form. If a mixture of chemicals containing a small percentage of the so-called active principle has activities equal to or greater than the single chemical alone, the chance of side effects will be less in the use of the extract than the single chemical. Therefore, a mixture is a much safer form of treatment than single chemicals. This provides a logical explanation as to why TCM and natural medicine have fewer side effects than pharmaceutical drugs. It also explains why, although 30-50% of pharmaceuticals are plant in origin, they have side effects for they lost the advantage of being in a mixture form as used traditionally. Hosoya and Yamamura studied the time course of antitussive action of various combinations of ephedra, in the original prescription (composed by *Ephedra herba*, *Armeniacae semen*, *Glycyrrhziae radix* and *Gypsum fibrosum*), and found that antitussive effect of the original prescription is best among six kinds of combination (*3*).

However, plant extracts come from natural sources and their quality are highly variable depending on the species and strains of the plant, geographical location, soil and culture conditions, growing conditions, climate and weather conditions, harvesting, part of the plant used, the solvent used to extract the plant, method of extraction in large quantity (temperature, pressure, batch size, amount of solvent, fill height, velocity of flow), and storage. The herbal drug *Evodia rutaecapar* (E.R.) is one that we studied most intensively. It was reported by Takagi that the content of synephrine was much higher in unripe fruit of E.R. then in ripe fruit (*4*). The alkaloids, flavonoids, acetophenones and other compounds not only vary from the fruit of different species of *Evodia* but also vary among different parts of the plant (*5*). Thus, lack of meaningful standardization is the universal problem in natural medicine development. There is not only product inconsistency among brands, but also inconsistency from batch to batch within the same brand. This then leads to the lack of consumer protection in the traditional functional food and nutraceutical industry. Thus, herbal preparations cannot be patented for composition and therapeutic use and cannot be used for clinical trials. Therefore, they cannot be developed into new drugs.

Even if all the above variables are well controlled, there are still some realistic problems. We found that the Rb_1 content of same species of American Ginseng is different (0.4-2.6%), and the immuno-stimulation effect is not related to the content of Rb_1 (Table 1).

Table I. Biological Effects of Different Sources of the Same Species of North American Ginseng

American Ginseng samples	Rb_1 content (% of raw material)	Immunostimulation (% of control)
A	0.4	1,623
B	1.6	5,271
C	2.6	5,679
D	1.9	10,481
E	1.1	4,936

The relaxant effect of *Panax ginsengs* on rabbit corpus cavernosum is not related to the major gensinosides (Rb_1, Rc, Re, Rg_1) (Figure 1).

Thus, some active principles are unknown or only partially understood. The three alkaloids isolated from *Evodia retaecapar*, dehydroevodiamine (DeHE), evodiamine (Evo) and rutecapine (Rut) have different modes of action on different target tissues, even though their structures show only minor differences (scheme 1A). The modes of action of these three compounds include inhibition

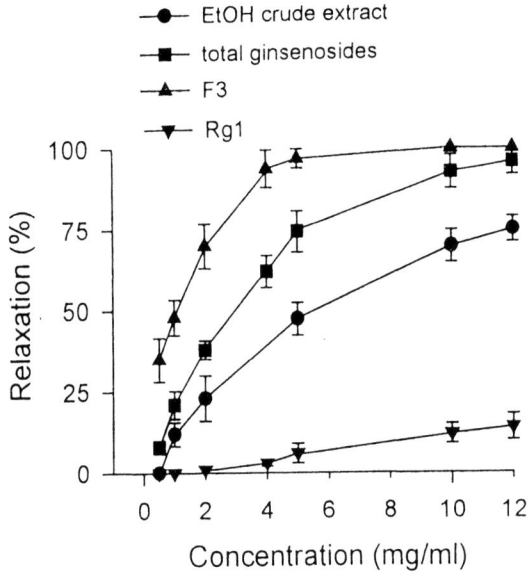

Figure 1. Relaxation effects of ethanol crude extract isolated from Korea red ginseng, total ginsenosides, partially purified fraction 3 (F3), and Rg_1 on phenylephrine-precontracted rabbit corpus cavernosum strips. Values are means ±S.E.M. from 7-9 experiments.

of the cardiac and vascular smooth muscles, activation of endothelial cells (*6, 7*), inhibition of macrophages (*8*) and a neuroprotective effect (*9*). However, there is little data on the effect(s) of other active components, and especially, their interaction. Thus, the different modes of action are still unclear. Under such conditions, the lack of pharmaco-kinetic data, and the lack of stability and clinical studies under good clinic practice are inevitable.

Quality Control of TCM

It is our experience that plant extracts may be functionally variable even though the chemical profile seems rather uniform. The guidance document entitled "Guidance for industry Botanical Drug Products" from the Food and Drug Administration of the Uunited States serves to delineate a possible path for researchers and industry to follow for botanical drug development and commercialization. This document further shows an appreciation of the valuable experience of historical use of botanical drugs and the difference between them and synthetic drugs. However, two points should be seriously considered.

1. Logically, chemical fingerprinting plus pharmacological fingerprinting are perhaps the best way, at this time, to provide assurance of product quality and consumer protection.
2. Batch-to-batch consistency should be in place at the earliest stage of drug development so that one can have confidence in the data from pre-clinical and clinical studies. Without the assurance of consistency in such preparations, data from all the studies are not comparable and rather meaningless.

TLC, HPLC, GC/MS, and LC/MS are good analytic instruments to identify chemical fingerprints.

For pharmacological fingerprinting, the pharmacodynamic and the pharmacokinetic evidences should be considered. All measurable changes of physiological, biochemical, and immunological parameters can be used to check the biological effect of plant extracts. If the pharmacokinetics of the major bioactive component(s) are known, we can design the ideal dosage regimen which keeps the blood concentration of active components safely between a therapeutically effective and a toxic concentration. Pharmacokinetic data, such as bioavailability, also can be generated for quality control of plant extracts. Microdialysis probes have been used in our Institute, for the on-line determination of drug concentration in the brain, bile duct, and blood for the pharmacokinetic studies of herbal bioactive component(s).

Anti-inflammatory Effect of *Radix Stephania tetrandra* (RST)

In this paper we discuss the work, done in our Institute, regarding the development of the Chinese herbal drug Fang-Ji (*Radix Stephania tetrandra*) as an anti-inflammatory functional food or as a pharmaceutical, an example both for pharmacological and chemical fingerprinting. The Chinese herbal drug, Fang-Ji, is used for the treatment of hypertension and some symptoms related to inflammation. Fang-Ji is the dry roots of Chinese herbal plants that come from two different Families including (1) Menispermaceae: *Stephania tetrandra* S. Moore, *Cocculus trilobus* (Thunb.) DC. and *Sinomenium acutum* (used in Japan) (2) Aristolochiaceae: *Aristolochia weslandi* and *Aristolochia heterophylla*. The main chemical constituents in the roots of *S. tetrandra* are the alkaloids tetrandrine (Figure 2B), fangchinoline (Figure 2B), cyclanoline (Figure 2C), oblongine (Figure 2C), menisine, and menisidine; those in *Cocculus trilobus* are trilobine (Figure 2D), isotrilobine, magnoflorine (Figure 2D), trilobamine, coclobine, menisarine, and normenisarine. *Aristolochia weslandi and Aristolochia heterophylla* contain aristolochic acid (Figure 2E) that is notorious for its nephrotoxic and carcinogenic effects.

Among these components, tetrandrine is the most well studied one as a calcium entry blocker of plant origin. Multiple pharmacological effects by tetrandrine have been reported, including an analgesic effect, anti-arrhythmic and anti-ischaemic effect, anti-hypertension action, and anti-tumor and anti-inflammatory functions. The article dealing with the cardiovascular pharmacology of tetrandrine, including anti-portal hypertension, was reviewed (*10*). With closely related but different chemical structures, we wondered if tetrandrine and other active components found in Fang-Ji displayed similar pharmacological effects. In this paper, we focus on the anti-inflammatory effects of the crude extract of *S. tetrandra*, tetrandrine, and its relative active compounds.

We had reported that tetrandrine can inhibit the Mac-1 (CD11b/CD18) up-regulation and adhesion by human neutrophils, possibly through impediment of calcium influx and reactive oxygen species (ROS) production (*11*). Further, we examined the anti-ischaemic effect by tetrandrine in an animal model and found that tetrandrine can ameliorate ischaemia-reperfusion injury of rat myocardium through inhibition of neutrophil priming and activation (*12*). As tetrandrine and fangchinoline are the active principles in *S. tetrandra*, we prepared various specially processed crude extracts from *radix S. tetrandra* (RST) and compared the bio-activity using an *in vitro* acute inflammatory cellular model. We found that the anti-inflammatory effect of RST extracts varied by different extracting methods with the ethanolic preparations being the most potent ones (Figure 3). The anti-inflammatory effects by ethanolic preparations (RST) were as potent as tetrandrine (Tet) and fangchinoline (Fan) in the prevention of phorbol-12-myristate-13-acetate (PMA)-induced ROS production (Figure 4) and adhesion (Figure 5) by human neutrophils. These effects of RST extracts, Tet and Fan

(A) Rutaecarpine
Dehydroevodiamine
Evodiamine

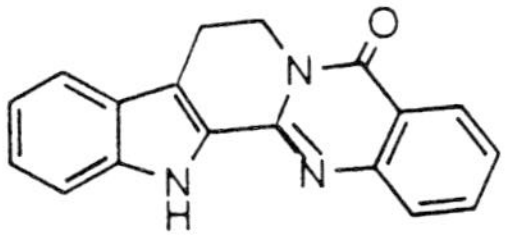

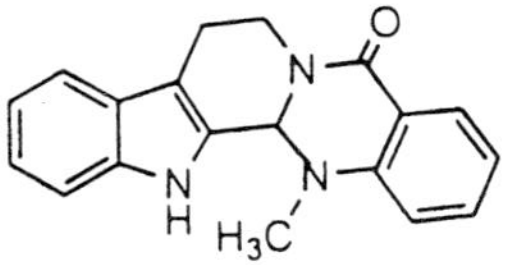

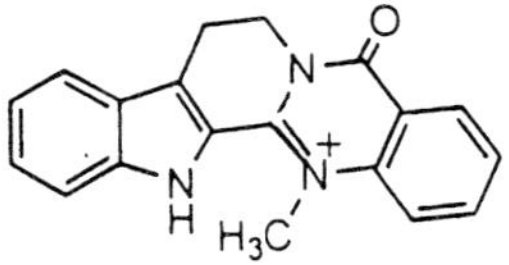

(B) Tetrandrine, R=CH₃; Fangchinilone, R=H

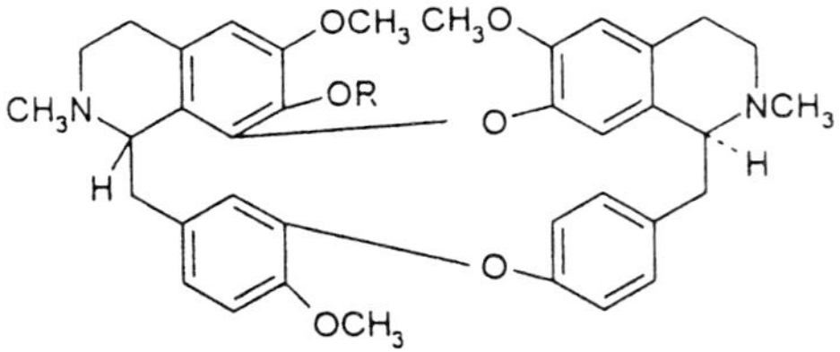

(C) Cyclanoline
Oblongine

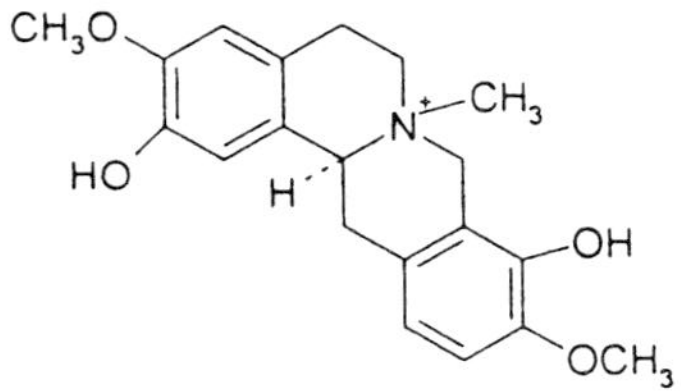

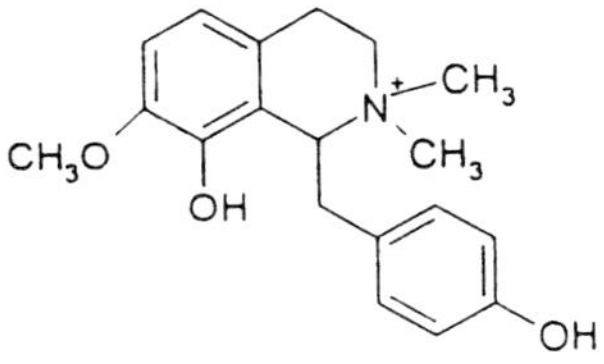

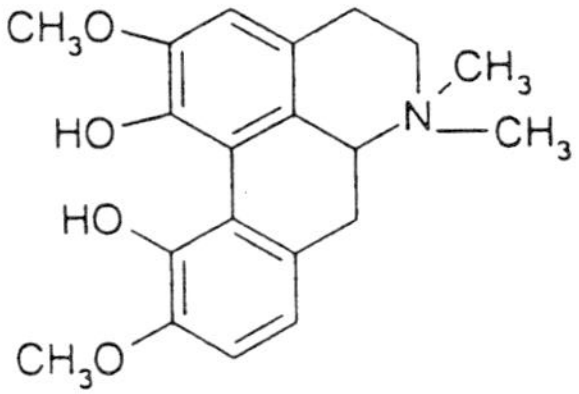

(D) Magnoflorine
Trilobine

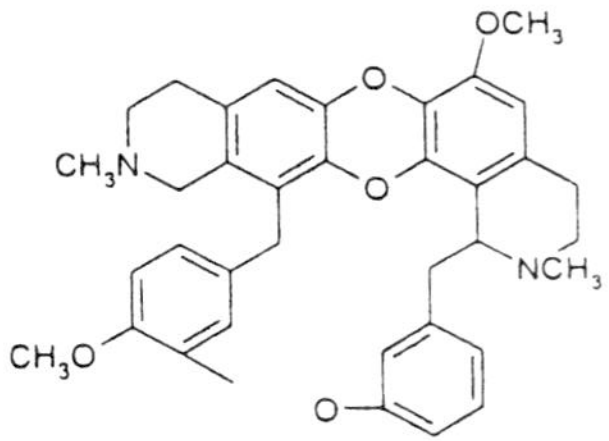

(E)

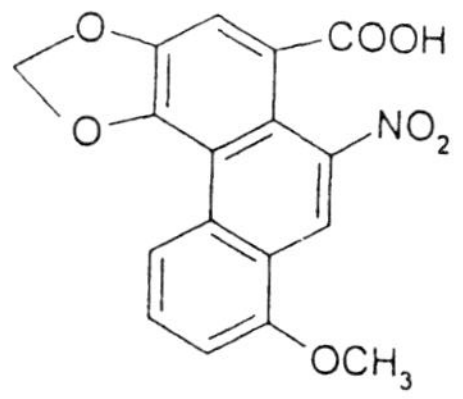

Aristolochic acid

Figure 2. Chemical Structure of alkaloids isolated form Evodia rutaecapar (A), Stephania tetrandra (B, C), Cocclus trilobus (D) and aristochic acid (E).

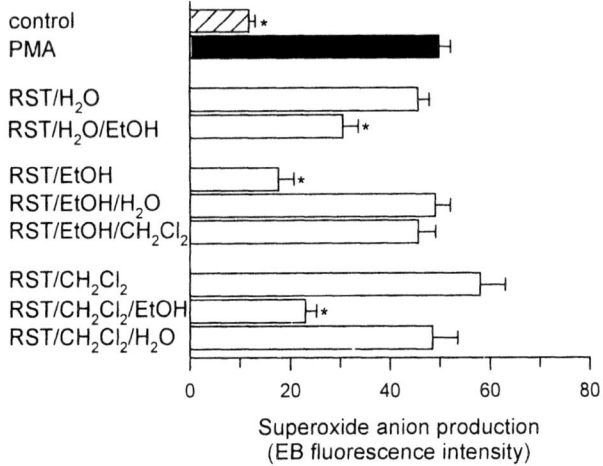

*Figure 3. Effects of various RST extracts on superoxide anion ($O_2^{\cdot -}$) production by PMA-stimulated human neutrophils. Samples were pretreated with various RST extracts (100 μg/mL) at 37 °C for 10 min. PMA (100 ng/mL)-induced $O_2^{\cdot -}$ (EB) production was measured by a flow cytometer (FACSCalibur^{TM}) 20 min after addition of PMA (100 ng/mL). *P < 0.05 as compared with samples treated with PMA alone. Values are means ±S.E.M. from 6 experiments. RST/H$_2$O, RST extracted by water only; RST/H$_2$O/EtOH, RST extracted by ethanol after water extraction; RST/EtOH, RST extracted by ethanol only; RST/EtOH/H$_2$O, RST extractd by water after ethanolic extraction; RST/EtOH/ CH$_2$Cl$_2$, RST extractd by CH$_2$Cl$_2$ after ethanolic extraction; RST/CH$_2$Cl$_2$, RST extracted by CH$_2$Cl$_2$ only; RST/CH$_2$Cl$_2$/EtOH, RST extracted by ethanol after CH$_2$Cl$_2$ extraction; RST/CH$_2$Cl$_2$/H$_2$O), RST extracted by water after CH$_2$Cl$_2$ extraction.*

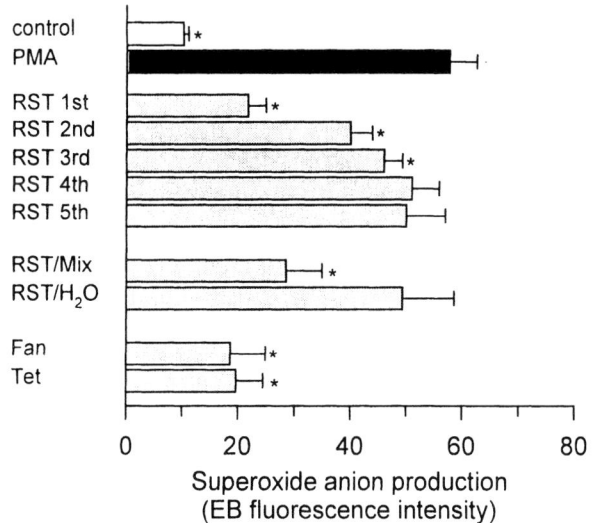

*Figure 4. Effects of RST extract from ethanolic fractions on superoxide anion ($O_2^{•-}$) production by PMA-stimulated human neutrophils. Samples were pretreated with various RST fractions (100 µg/mL) at 37 °C for 10 min. PMA (100 ng/ml)-induced $O_2^{•-}$ (EB) production was measured by a flow cytometer (FACSCalibur^{TM}) 30 min after addition of PMA (100 ng/mL). *P < 0.05 as compared with samples treated with PMA alone. Values are means ±S.E.M. from 6 experiments. RST 1st, RST extracted by first time of ethanol; RST/mix, combination of the first three time ethanolic fractions of RST; Tet, tetrandrine (100 µg/mL); Fan, fangchinoline (100 µg/mL).*

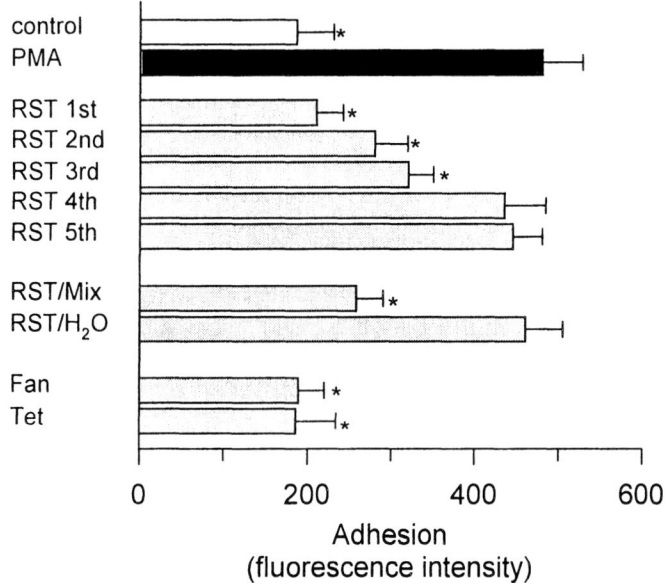

*Figure 5. Effects of RST extract from ethanolic fractions on PMA-activated adhesion by human neutrophils. Samples were pretreated with various RST fractions (100 μg/mL) at 37 ℃ for 10 min. PMA (100 ng/mL)-adhesion (BCECF fluorescence) was measured by a fluorescent plate reader (Cytofluor 2300, Millipore®) with excitation at 485 nm and emission at 530 nm 30 min after addition of PMA (100 ng/mL). *P < 0.05 as compared with samples treated with PMA alone. Values are means ±S.E.M. from 6 experiments. RST 1st, RST extracted by first time of ethanol; RST/mix, combination of the first three time ethanolic fractions of RST; Tet, tetrandrine (100 μg/mL); Fan, fangchinoline (100 μg/mL).*

were not due to cytotoxicity, because no significant cell death (less than 2.0%) was observed after experiments (about 4 hours) (Figure 6). It is interesting to note that combination of Tet and Fan enhanced cytotoxicity more than Tet, Fan, or RST used alone (Figure 6). To determine the active principles in the RST extracts, high-performance liquid chromatography (HPLC) was used, and we found that the major active components in the RST are (μg/ml) tetrandrine (0.95), fangchinoline (0.95), cyclanoline (0.95) and oblongine (1.69) (Figure 7) (*13*).

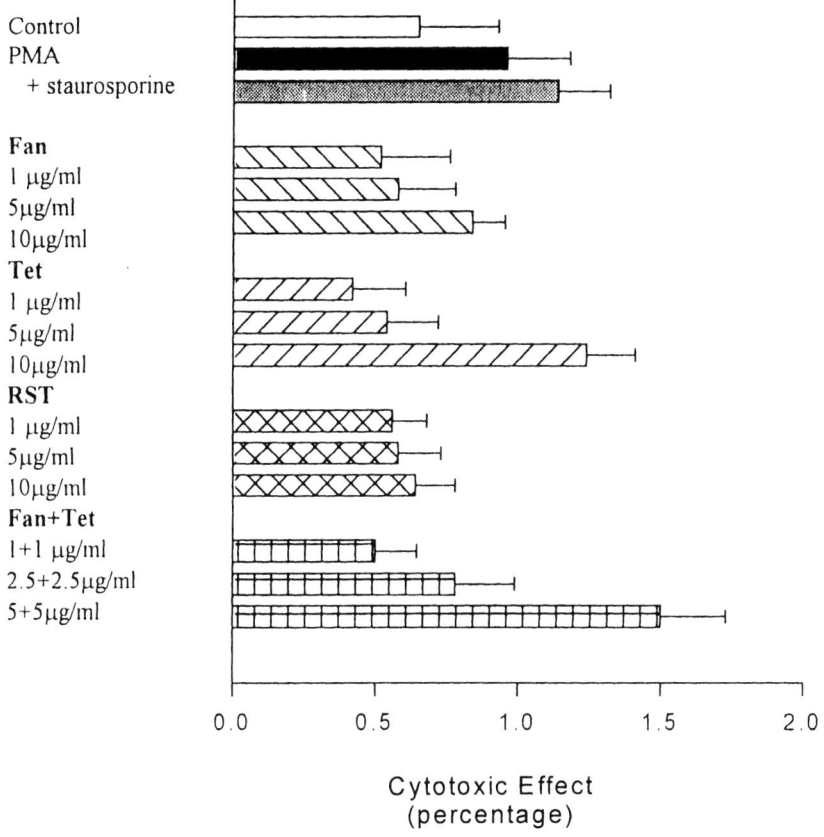

Figure 6. Summary on the cytotoxic effects of RST extract, tetrandrine, and fangchinoline on PMA-activated neutrophils. Cell viability was determined by counting the vital cell by trypan blue (0.4%) exclusion assay in the end of the experiments.

44

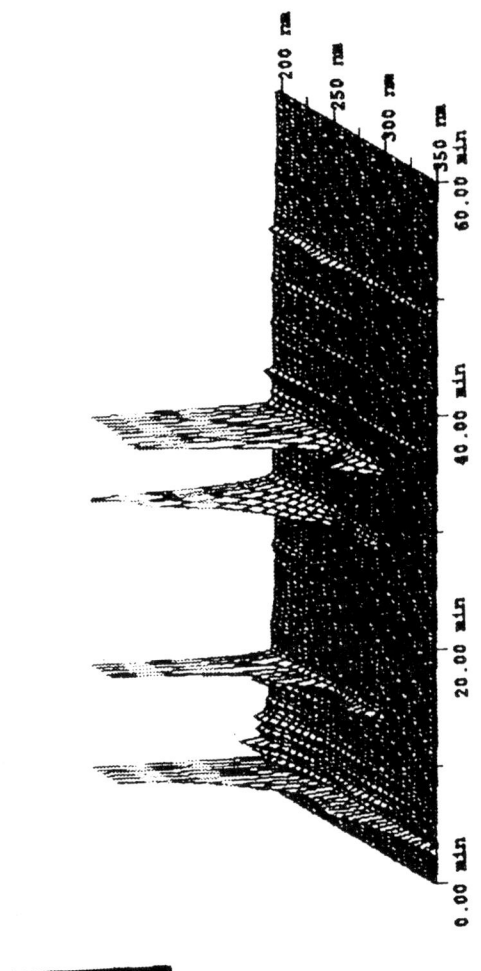

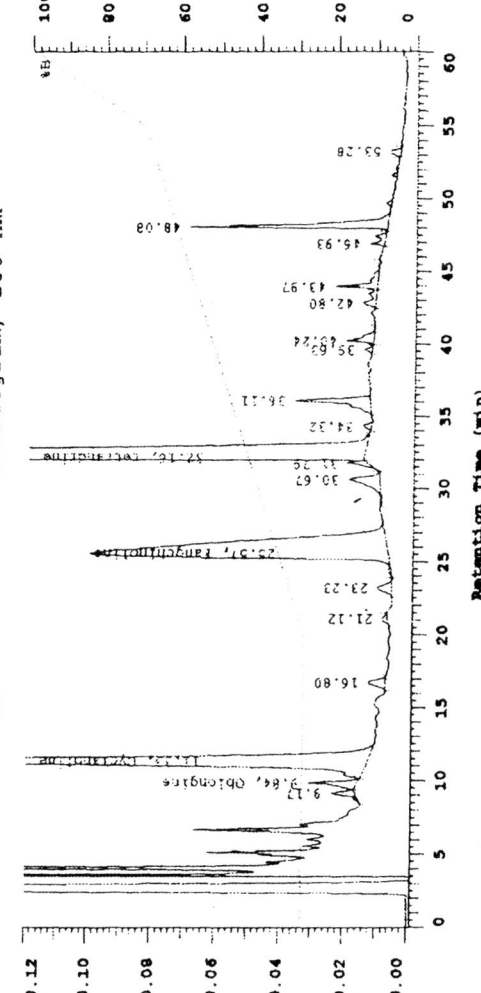

Figure 7. Chromatogram of ethanolic extract of Radix S. tetrandra. Upper panel, peak identities (from left to right): oblongine, cyclanoline, fangchinoline, and tetrandrine.

Cardiovascular Protective Effects of RST

We examined the cardiovascular protective effects of tetrandrine, fangchinoline and the specially processed crude extract from *radix S. tetrandra* (RST). We found that the RST extract with 9% and 6% in weight being tetrandrine and fangchinoline, respectively, have potent effect on the inhibition of the electrically-induced intracellular calcium transient and protein release during calcium paradox in a single rat ventricular myocyte (*14*). To compare the cardioprotective effects by RST extract and its active principles (tetrandrine and fangchinoline), a Langendorff isolated perfused rat heart preparation was used. We found that RST extract and tetrandrine (but not fangchinoline) produces equal potent cardioprotective and anti-arrhythmic effects as varapamil alone but circumventing the side effects of varapamil (*15*). We further studied the underlying anti-inflammatory mechanism(s) of specially processed crude extracts of *S. tetrandra* (SPRST) by human leukocyte and found that SPRST as well as its active principle tetrandrine and fangchinoline, can interfere with ROS production and calcium influx, through G protein modulation, to prevent Mac-1 dependent neutrophil adhesion (*16*). Since cyclanoline and oblongine are also effective in the inhibition of PMA-induced ROS production (our unpublished results) whether these components and/or other un-identified principles were involved in the anti-inflammatory effects of SPRST await further study.

References

1. Leong, W.E.; Liu, Y.Q.; Chen, C.F. *Prog. Drug. Res.* **1996**, *47*, 131-136.
2. Leong, W.E.; Chen, C.F. *J. Chin. Med.* **1998**, *9*, 229-237.
3. Hosoya, E.; Ya,a,ura, Y. In *Recent Advances in the Pharmacology of Kampo (Japanece Herbal) Medicines*, International Congress Series 854, Excerpta Medica: Amsterdam, 1988.
4. Takagi, S. *Shoyakugaku Zasshi* **1979**, *33*, 35-37.
5. Lin, L.C. Ph. D. Dissertation, Taipei Medical College, Taipei, Taiwan, R.O.C. 1994.
6. Wang, G.J.; Shan, J.; Pang, P.K.T.; Yang, M.C.M.; Chou, C.J.; Chen, C.F. *J. Pharmacol. Exp. Ther.* **1996**, *276*, 1016-1021.
7. Chiou, W.F.; Sung, Y.J.; Liao, J.F.; Shum, A.Y.C.; Chen, C.F. *J. Nat. Prod.* **1997**, *60*, 708-711.
8. Wang, G.J.; Wu, X.C.; Chen, C.F.; Lin, L.C.; Huang, Y.T.; Shan, J.; Pang, P.K.T. *J. Pharmacol. Exp. Ther.* **1999**, *289*, 1237-1244.
9. Wang, H.H.; Chou, C.J.; Liao, J.F.; Chen, C.F. *Eur. J. Pharmacol.* **2001**, *413*, 221-225,.

10. Huang, Y.T.; Hong, C.Y. *Cardiovasc. Drug Rev.* **1998**, *16*, 1-15.
11. Shen, Y.C.; Chen, C.F.; Wang, S.Y.; Sung, Y.J. *Mol. Pharmacol.* **1999**, *55*, 186-193.
12. Shen, Y.C.; Chen, C.F.; Sung, Y.J. *Brit. J. Pharmacol.* **1999**, *128*, 1593-1601.
13. Chou, C.J.; Lin, L.C.; Peng, C.; Shen, Y.C.; Chen, C.F. *J. Chin. Med.* **2002**, *13*, 39-48.
14. Wu, S.; Yu, X.C.; Shan, J.; Wong, T.M.; Chen, C.F.; Pang, P.K.T. *Life Sci.* **2001**, *68*, 2853-2861.
15. Yu, X.C.; Wu, S.; Wang, G.Y.; Shan, J.; Wong, T.M.; Chen, C.F.; Pang, P.K.T. *Life Sci.* **2001**, *68*, 2863-2872.
16. Shen, Y.C.; Chou, C.J.; Chiou, W.F.; Chen. C.F. *Mol. Pharmacol.* **2001**, *60*, 1083-1090.

Chapter 3

Targeting Inflammation Using Nutraceuticals

Mohamed M. Rafi, Prem N. Yadav, and Il-Kyung Maeng

Department of Food Science, Rutgers, The State University of New Jersey,
65 Dudley Road, New Brunswick, NJ 08901-8520

The use of herbal therapy or alternative medicine is becoming an increasingly attractive approach for the treatment of various inflammatory disorders. Significant research efforts in the laboratory and in the clinic are ongoing to understand the critical role of certain nutraceuticals in the regulation of inflammation. This article has reviewed few selected nutraceuticals and their possible mechanism of action in inflammation. The selected nutraceuticals includes, curcumin, resveratrol, epigallocatechin gallate (EGCG) and diarylheptanoids from ginger. In addition, we also discussed our recent work on a diarylheptanoid isolated from *Alpinia officinarum*, which have potential anti-inflammatory activities and inhibits the expression of COX-2 and iNOS mediated through NF-κB activation. This article summarizes possible targets in inflammation and will also provide insights for the development of new anti-inflammatory agents.

Introduction

Nutraceuticals have been touted for medical benefits, but little is known about their biological activity. The recent studies have repeatedly shown that many natural products marketed as nutraceuticals, deliver the health benefit in various disease conditions. A wide array of nutraceuticals present in edible and medicinal plants, have been reported to possess substantial anticarcinogenic and antiinflammatory activities. The majority of naturally occurring phenolics possess tremendous antioxidative and anti-inflammatory activities. Antiinflammatory properties of different nutraceuticals are mediated through the inhibition of production of cytokines (IL-1β, TNF-α, IL-6, IL-12, IFN-γ), nitric oxide (NO), prostaglandins and leukotrienes. The proinflammatory mediators NO and prostaglandins are produced by actions of cyclooxygenase-2 (COX-2) and inducible nitric oxide synthase (iNOS), respectively. These inflammatory mediators are soluble, diffusible molecules that act locally at the site of tissue damage and infection, and at more distant sites. The COX-2 and iNOS are important enzymes that mediate most of the inflammatory processes. Improper up-regulation of COX-2 and/or iNOS has been associated with pathophysiology of certain types of human cancers as well as inflammatory disorders (1,2). Since inflammation has been shown as one of the factors causing certain types of cancer, nutraceuticals with potent anti-inflammatory activities are thought to inhibit carcinogenesis. Examples are curcumin, gingerols diarylheptanoids, epigallocatechin gallate (EGCG), and resveratrol, which are known to inhibit various inflammatory mediators. Several studies have shown that eukaryotic transcription factor nuclear factor-kappa B (NF-κB) is involved in regulation of COX-2 and iNOS expression and these phytochemicals have been shown to inhibit COX-2 and iNOS expression by blocking improper NF-κB activation. Similarly, the inhibition of proinflammatory cytokine by these nutraceuticals has also been demonstrated in various inflammatory conditions (2,3). The most possible mechanisms underlying inhibition of NF-κB activation by aforementioned nutraceuticals is by inhibition of inhibitory kappa B kinase (IKK), which prevents the degradation of inhibitory-kappa B (IκB) and thereby hampers subsequent nuclear translocation of the functionally active subunit of NF-κB (4).

In this article, we have reviewed the literatures available for the mechanism of action and anti-inflammatory properties of selected nutraceuticals such as curcumin, resveratrol, green tea polyphenols and diarylheptanoids (chemical structures in Figure 1). In addition we have also shown the antiinflammatory properties of a diarylheptanoids from *Alpinia galanga* and different possible targets in inflammation.

Curcumin

Curcumin, an active ingredient of turmeric is commonly used as a spice to give the specific flavor and yellow color to curry (5). Turmeric has also been used for centuries as a traditional medicine to treat inflammatory disorders (6,7). Subsequent studies also demonstrated the anti-inflammatory properties of curcumin (8). Curcumin is known to have a variety of pharmacological effects,

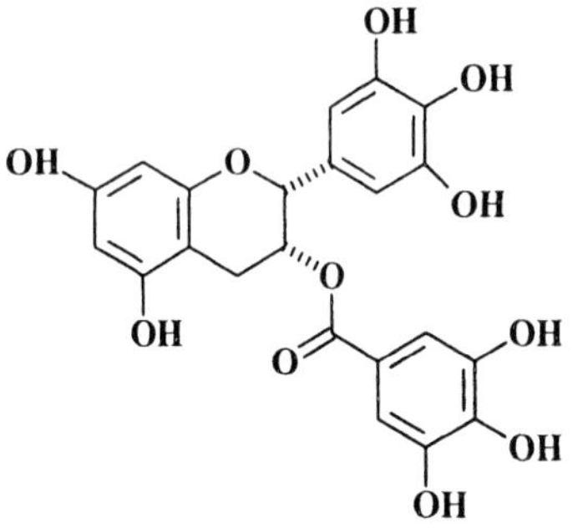

Epigallocatechin-3-*O*-gallate (EGCG)

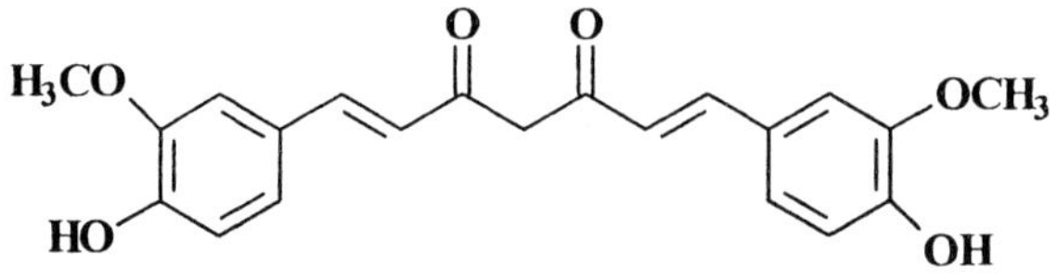

Curcumin

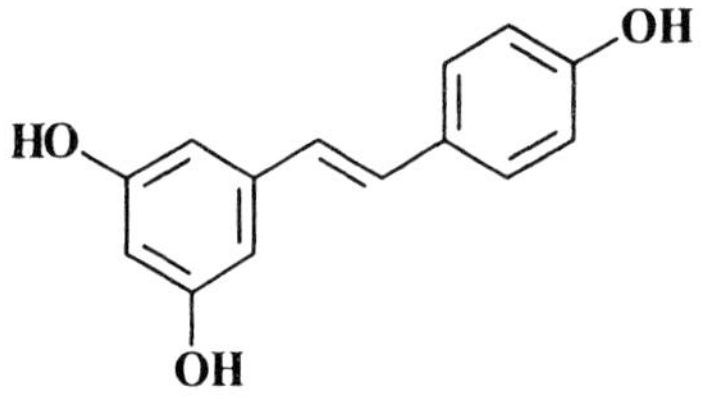

Resveratrol

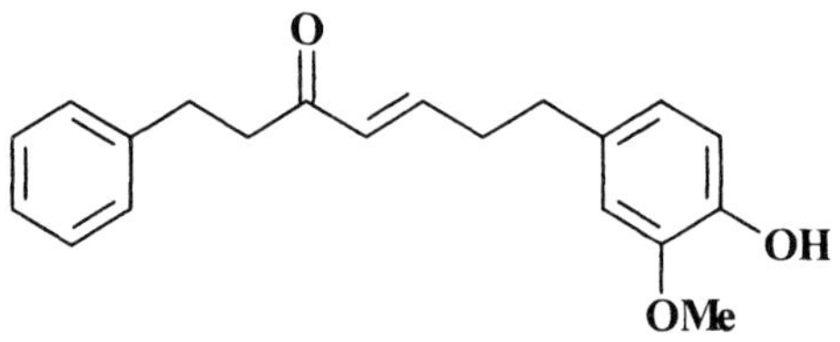

HMP (7-(4'-hydroxy-3'-methoxyphenyl)-1-phenylhept-4-en-3-one)

Figure 1. Chemical structure of some nutraceuticals.

including antitumor, anti-inflammatory, and anti-infectious activities (*2*). The pleiotropic effects of curcumin on various arms of immune system are well reported. Curcumin has been shown to selectively suppress the stimulation of T helper cells 1 (Th1) in various disease conditions (*9-11*). Treatment of mice with curcumin prevented trinitrobenzene sulfonic acid induced colonic inflammation (*3*). In this study, authors have clearly shown that curcumin treatment suppressed proinflammatory cytokine mRNA expression and the NF-κB activation in colonic mucosa. Subsequent study by Grandjean-Laquerriere et al (*12*), have shown that curcumin selectively upregulate IL-10 and inhibits the production of TNF-α, IL-6, in UVB induced keratinocytes. Moreover, Gukovsky et al, (*13*) have also demonstrated the *in vivo* anti-inflammatory effect of curcumin in pancreatic inflammation by decreasing the expression of IL-6, TNF-α and iNOS mediated through NF-κB and AP-1. It has also been reported that curcumin inhibits TNF-α-induced NF-κB activation in human myelomonoblastic leukemia cells and phorbol ester-induced c-Jun/AP-1 activation in mouse fibroblast cells (*14,15*). The molecular mechanism for NF-κB inhibition by curcumin is still not completely understood. However, inhibition of IKK leading to decreased degradation of IκB has been suggested as most plausible mechanism for action of curcumin (*16-19*). In Conclusion, all the evidences suggests that anti-inflammatory effect of curcumin is mediated through inhibition of transcriptional factors NF-κB and AP-1. The anti-inflammatory properties of curcumin have also been well demonstrated in animal models of atherosclerosis, Alzheimer's disease and arthritis (*20-23*). These properties of curcumin have been shown to be associated with its ability to inhibit the production of proinflammatory cytokines such as TNF-α, IL-1, IL-8, and inducible NO synthase (*1,24-26*). Although the exact mechanisms involved in the *in vivo* anti-inflammatory activity of curcumin is not fully defined, it prevents the activation of NF-κB, AP-1, and c-Jun kinase (*2,27*). Recent studies have also shown that curcumin inhibits IL-12 production from macrophages and thereby prevents the differentiation of Th1 cells in vitro (*28,29*). Natarajan and Bright have also demonstrated the beneficial effect of curcumin in Th1 cell-mediated inflammatory diseases of the central nervous system in experimental autoimmune encephalomyelitis (*11*). In addition, the anti-allergic effect of curcumin has also been demonstrated (*30*). Temkin et al (*30*) have clearly shown that curcumin significantly suppresses the recombinant human tryptase induced production of IL-6 and IL-8 from human peripheral blood eosinophils.

Resveratrol

Resveratrol (3,5,4'-trihydroxy-trans-stilbene), a natural polyphenolic phytoalexin, found in high to moderate quantities in various foods including grapes, peanuts and wine, has been shown to have anti-inflammatory properties. Recent *in vitro* and a limited number of *in vivo* studies have documented that physiological concentrations of resveratrol can modulate multiple molecular pathways thought to be associated with the development and progression of cardiovascular disease, cancer and inflammation (*31,32*). Resveratrol, a representative of hydroxystilbene, has received special attentions since it was known to have a central role in "French Paradox." The anti-inflammatory properties of Resveratrol are shown to be through inhibition of cyclooxygenase (*33*). It also inhibits arachidonate release (*32,34*), MAPK activation (*35,36*), protein kinase C (*37*), and degranulation of mast cells (*38*). Previous studies have shown that resveratrol inhibit the translocation of NF-κB into the nucleus and control the expression of iNOS (*39-42*). Moreover, Holmes-McNary et al (*42*) demonstrated that resveratrol blocks the TNFα induced activation of NF-κB, a transcription factor strongly associated with inflammatory diseases and oncogenesis. Manna et al (*43*) have also demonstrated that resveratrol inhibits TNF-α induced activation of mitogen-activated protein kinase (MAPK), c-Jun N-terminal kinase (JNK), reactive oxygen intermediate generation and lipid peroxidation and thereby blocks the activation of NF-κB and AP-1. Thus, it is apparent that NF-κB, AP-1 and associated kinases are the most vulnerable targets of resveratrol for its anti-inflammatory activities.

In addition, resveratrol possesses structural similarities with estrogenic compounds and have been suggested that it may exert some estrogenic activities through the estrogen receptors (*44-46*). Effect of estrogen on iNOS is somewhat controversial; some studies have shown that it inhibits nitric oxide production through a classic receptor-mediated pathway (*47*), while another study has shown that it does not have any effect (*48*). Recently, Cho et al (*49*) also confirmed that inhibition of NO by resveratrol is not mediated through estrogen receptor. Reactive oxygen intermediates (ROI) are important mediators of a variety of pathological processes, including inflammation and ischemia/reperfusion injury. Various proinflammatory cytokines and chemokines have been shown in human and animal ischemic brains, suggesting that hypoxia/reoxygenation may induce cytokine production through generation of ROIs. Wang et al (*50*) have shown the *in vitro* neuroprotective effect of resveratrol in glial cells. In this study, it is demonstrated that resveratrol significantly inhibited the IL-6 gene expression and protein secretion in mixed glial cultures under hypoxia/hypoglycemia followed by reoxygenation. In addition, the inhibition of TNF-α induced expression of adhesion molecules on endothelial cells and vascular leakage, by resveratrol is also demonstrated (*51*).

Green Tea Polyphenols

Tea is one of the most popular beverages in the world and various phytochemicals derived from Tea (*Camellia sinensis*) have drawn a great deal of interest due to beneficial effect on health. Certain epidemiological studies have suggested that regular tea consumption reduces the risk of cancer (52,53). Although tea consists of several components, interest has focused primarily on polyphenols, especially those found in green tea. The green tea polyphenols include (-)-epigallocatechin gallate (EGCG), (-)-epigallocatechin (EGC), (-)-epicatechin gallate (ECG) and (-)-epicatechin (EC). Of these, EGCG accounts for >40% of the total polyphenols (54). Following consumption, the polyphenols remain predominantly in their conjugated forms and are primarily excreted intact in the urine (55). These polyphenols have potent antioxidant properties including the scavenging of oxygen radicals and lipid radicals (54). Suganuma et al (56) reported that EGCG inhibits okadaic acid-induced TNFα production and inhibits its gene expression in BALB/3T3 cells. Green tea polyphenols also inhibit NO production and iNOS gene expression in isolated peritoneal macrophages by decreasing NF-κB activation (26,57). In a similar *in vivo* and *in vitro* study, Yang et al (58) have shown that anti-inflammatory mechanism of green tea polyphenols is mediated at least in part through down-regulation of TNF-α gene expression by blocking NF-κB activation. Exposure of skin to UV radiation can cause diverse biological effects, including induction of inflammation, alteration in cutaneous immune cells and impairment of contact hypersensitivity (CHS) responses. The beneficial effects of polyphenolic fractions from green tea in mouse models have been shown to protect against the carcinogenic effects of UVB radiation (52). This suggests that green tea, specifically polyphenols present therein, may be useful against inflammatory dermatoses and immunosuppression caused by solar radiations. Haqqi et al (59) have demonstrated that green tea polyphenolic fractions inhibits collagen induced arthritis by down regulating the expression of COX-2, IFN-γ, and TNF-α in arthritic models. Inflammatory cytokine, interleukin-1β (IL-1β) produced in an arthritic joint by activated synovial cells and infiltrating macrophages, is considered to be one of the most potent catabolic factors in joint diseases (60). IL-1β induces the production of several mediators of cartilage degradation, such as nitric oxide (NO) and matrix metalloproteinases and inhibits the concentration of tissue inhibitor of metalloproteinases in arthritic joints (61-62). Singh et al (63), have shown that EGCG inhibits IL-1β induced expression of iNOS and production of NO in human chondrocytes through the suppression NF-κB activation. These findings suggest that green tea polyphenols may be used as a therapeutic agent in inflammatory diseases.

Diarylheptanoids

Diarylheptanoids, which are structurally similar to curcumin are present in certain plants of ginger family (Zingiberaceae) and have been shown to have a strong anti-inflammatory properties. The diarylheptanoids, yakuchinone A (1-[4'-hydroxy-3'-methoxyphenyl]-7-phenyl-3-heptanone) and yakuchinone B (1-[4'-hydroxy-3'-methoxyphenyl]-7-phenylhept-1-en-3-one), which are present in *Alpinia oxyphylla* Miquel have been reported to be strong inhibitors of prostaglandin biosynthesis *in vitro* (64,65). It has been reported that phenolic diarylheptanoids such as yakuchinone B and demethyl-yakuchinone B derivative of yakuchinones, inhibit 5-lipoxygenase and cyclooxygenase activity (67,68). Yamazaki et al (69) have shown that a novel phenolic diarylheptanoid derivative, 1-(3,5-dimethoxy-4-hydroxyphenyl)-7-phenylhept-1-en-3-one (YPE-01), inhibits 5-lipoxygenase activity as well as it suppresses arachidonic acid- and 12-O-tetradecanoylphorbol 13-acetate (TPA)-induced ear edema in mice. However, it has been reported that nonphenolic diarylheptanoids also have anti-inflammatory effects in various experimental models of inflammation *in vivo* (70,71). Subsequent studies have revealed that yakuchinone A and yakuchinone B, can act as anti-tumor agents, as these compounds inhibits TPA-stimulated superoxide generation and TNF-α production in HL-60 cells (72,73).

Diarylheptanoids from *Alpinia officinarum* Hance (Zingiberaceae), structurally analogous to yakuchinone A and yakuchinone B also exhibit strong anti-inflammatory effects (74,75). Another structurally related curcuminoid, 7-(4'-hydroxy-3'-methoxyphenyl)-1-phenylhept-4-en-3-one (here abbreviated as HMP) from *Alpinia Officinarum* Hance has been shown to inhibit the prostaglandin synthase activity (68). Recently, we have also isolated this compound from the same plant by bioassay directed fractionation and studied its anti-inflammatory properties. We have shown that treatment of RAW 264.7 cells with HMP (6.25-25 μM) significantly inhibited lipopolysaccharide (LPS) stimulated nitric oxide (NO) production (Figure 2). This compound also inhibited the release of LPS induced proinflamatory cytokines, interleukin 1-β (IL1-β) and tumor necrosis factor α (TNF-α) from human PBMCs *in vitro* (Figure 3). In addition, western blotting and reverse transcription-polymerase chain reaction (RT-PCR) analysis demonstrated that HMP decreased LPS-induced inducible nitric oxide synthase (iNOS) and cyclooxygenase-2 (COX-2) protein and mRNA expression in RAW 264.7 cells (Figures 4 and 5). Furthermore, we showed that anti-inflammatory activity is mediated through transcription factor NF-κB (Figure 6). Our results suggest that HMP from *Alpinia officinarum* is a potent anti-inflammatory compound and could be useful for the treatment of inflammation (76).

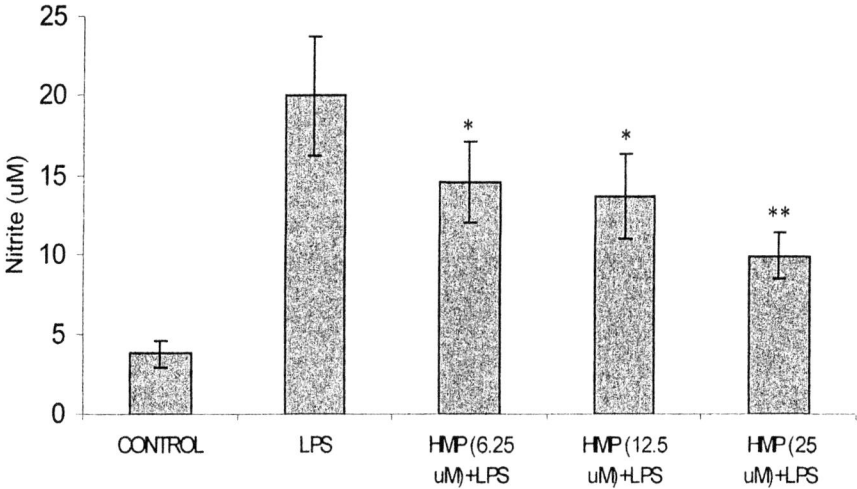

*Figure 2. Inhibition of nitric oxide production from RAW 264.7 by HMP. RAW 264.7 cells were treated with LPS (0.5 µg/mL) either alone or with different concentrations (25 µM-6.25 µM) of HMP for 24 hrs and supernatant were collected from each treatment group. The amount of nitrite was measured using Griess reagent. Each bar represents mean ± standard deviation (SD) of 4 replicates of one representative experiment of total five experiments. *Represents statistical significance of inhibition of LPS stimulated NO production by HMP as compared to LPS alone (*p<0.05; **p<0.01).*

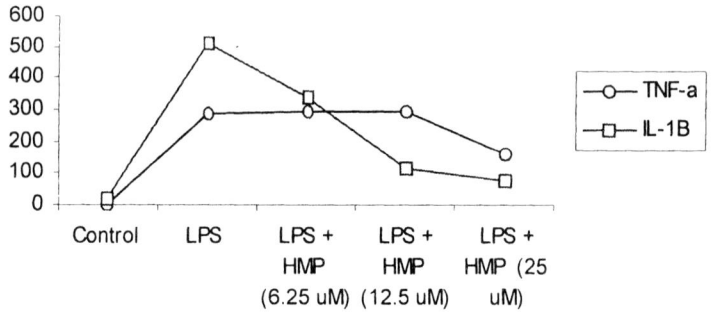

Figure 3. Inhibition of LPS stimulated IL-1β and TNF-α production from human PBMCS by HMP. The human PBMCs (0.5 X10⁶/mL) were cultured with LPS (10 ng/mL) either alone or with various concentrations of HMP (6.25-25 μM) for 18 hrs and amount of IL-1β and TNF-α in culture supernatant was quantitated in culture supernatant by ELISA.

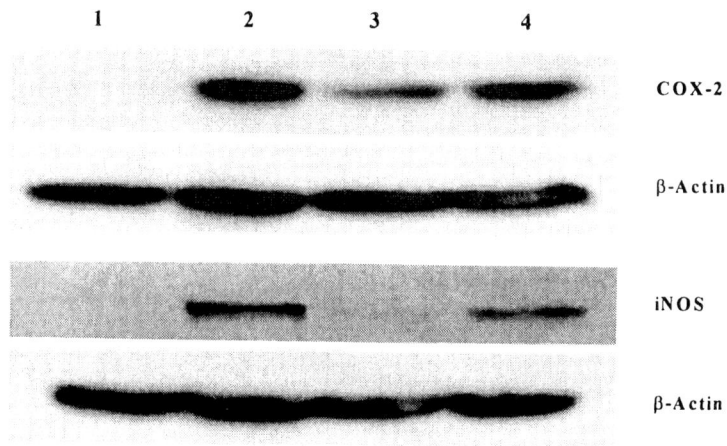

Figure 4. Inhibition of LPS induced iNOS and COX-2 protein expression by HMP. RAW 264.7 cells were treated with various concentrations of HMP and/or LPS (0.5 μg/mL) for 18 h. Total-cell lysate (40 μg) was resolved in 8-10% SDS-PAGE then transferred to PVDF membrane and detected with specific anti mouse iNOS, COX-2, β-actin monoclonal antibodies. Lane-1, Control (without any treatment); 2, LPS (0.5 μg/mL); 3, LPS and HMP 25 μM; 4, LPS and HMP-12.5 μM.

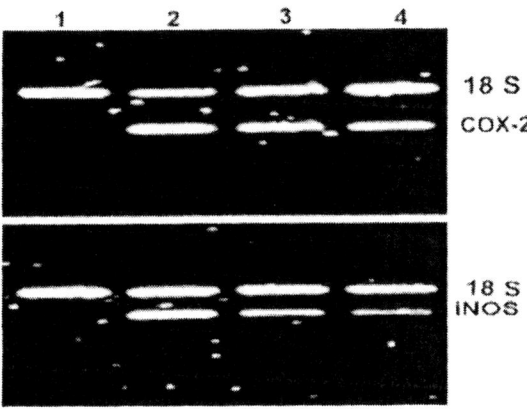

Figure 5. Effect of HMP on mRNA expression of iNOS and COX-2. Total RNA isolated from RAW 264.7 cells after 12 hrs of treatments with LPS either alone or with various concentration of HMP, was reverse transcribed to make cDNA and 2 µl (for iNOS) or 1 µl (for COX-2) of cDNA was amplified using gene specific primer (Ambion Inc). Equal volume (10 µl) of each PCR reaction was resolved in 2 % agarose gel. Lane 1, control; 2. LPS (0.5 µg/ml); 3, LPS+HMP (12.5 µM) and Lane, 4 LPS +HMP (25 µM).

Conclusion

Infection or injury triggers an immediate cascade of inflammatory signals. Various factors are released that recruit inflammatory leukocytes, such as neutrophils and macrophages. The infiltrating leukocytes are activated to express the iNOS and COX-2 leading to the production of nitric oxide and prostaglandin PGE_2, respectively. The infiltrated leucocytes at sites of inflammation also produce proinflammatory cytokines such as TNF-α, IL-1β, IL-12 etc, which altogether worsen the situation. Bremner et al have suggested that NF-κB is a master regulator and have a central role in the onset of inflammation (4). This transcription factor is composed of homo- or heterodimers of Rel-family proteins (p65, p50, p52 cRel, RelB), and the combination of Rel proteins might, in part, determine the pattern of gene expression. NF-κB is held in an inactive state in the cytosol by association with IκB. During inflammation, TNF-α and IL-β signaling in leukocytes activate IκB kinase (IKK), leading to the phosphorylation and degradation of IκB. Once, the NF-κB is separated from IκB, it gets translocated to the nucleus and induces expression of multiple genes, including iNOS and COX-2.

The anti-inflammatory effects of nutraceuticals (curcumin, resveratrol, EGCG and other diarylheptanoids) appear to be associated with their ability to inhibit NO, prosataglandins and cytokines. One of the most critical targets of these nutraceuticals involves NF-κB that regulates expression of a whole variety of genes, such as COX-2, iNOS and proinflammatory cytokines, responsible for inflammation (these targets are summarized in Figure 7).

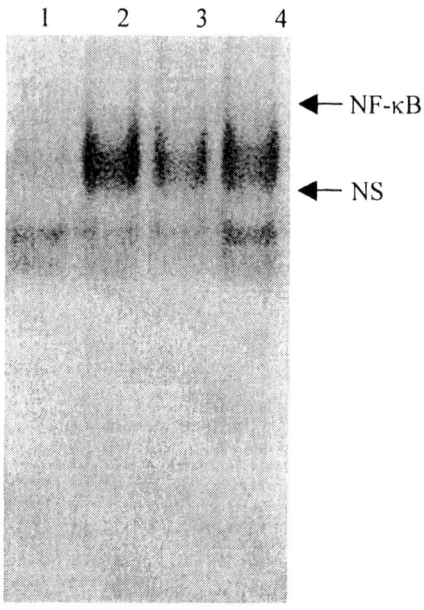

Figure 6. Inhibition of LPS induced NF-κB activation by HMP. Nuclear protein lysates of RAW 264.7 cells were prepared after 2 hrs of treatment with LPS (0.5 μg/ml) either alone or with different concentration of HMP. 6 μg of nuclear protein was used to DNA binding assay and total binding reaction was resolved in 5% nondenaturing polyacrylamide gel. The retarded bands were indicated with arrow. Data represent one of three similar results. Lane 1, control; 2, LPS (0.5 μg/ml); 3, LPS+HMP (25 μM); 4, HMP (12.5 μM). NS, non specific.

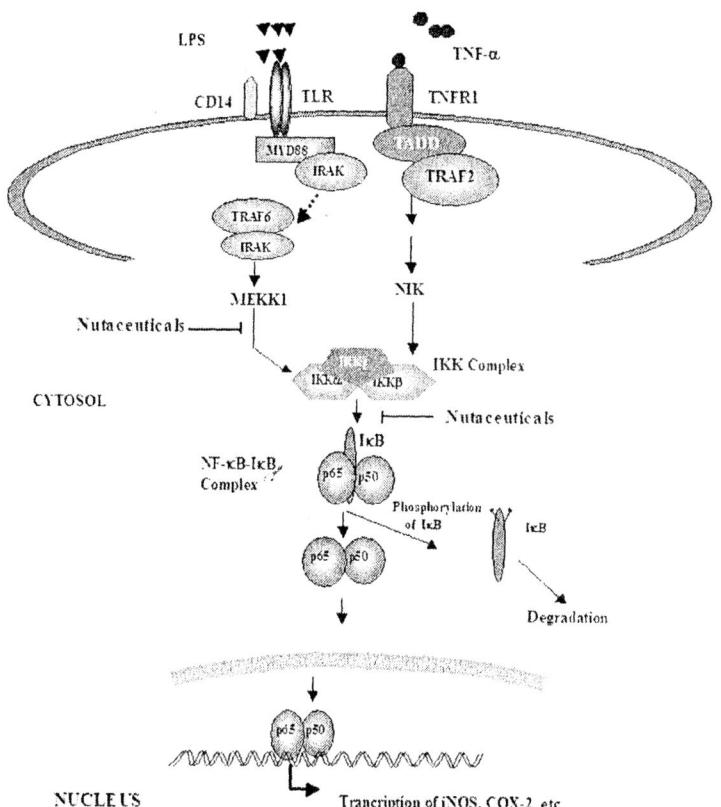

Figure 7. A schematic representation of signaling events for LPS and TNF-α and possible targets of nutraceuticals. The activation of NF-κB begins with stimulation of specific receptor families at the cell surface and recruitment of adaptor proteins, and leads to specific pathways of transduction controlled by various kinases. These pathways converge upon the IKK complex that, in turn, promotes the phosphorylation of IκB and ultimately degradation. As a consequence, the NF-κB inhibitory protein is removed and free NF-κB is rapidly translocated to the nucleus where it binds to specific promoter regions of various genes encoding, for example, iNOS, COX-2 and inflammatory cytokines. IKK, IκB kinase; LPS, lipopolysaccharide; MEKK1, mitogen-activated protein kinase/extracellular response kinase (MAPK/ERK) kinase kinase 1; MyD88, myeloid differentiation factor; IRAK, interleukin 1 receptor associated kinase; NIK, NF-κB-inducing kinase; TLR, Toll-like receptor; TRADD, TNF-receptor-associated death domain protein; TRAF, TNF-receptor-associated factors.

Acknowledgement

This work was supported by the grant from the New Jersey Agricultural Experiment Station, Cook College, Rutgers University.

References

1. Surh, Y. *Mutat. Res.* **1999**, *428*, 305-327.
2. Surh, Y.J.; Chun, K.S.; Cha, H.H.; Han, S.S.; Keum, Y.S.; Park, K.K.; Lee, S.S. *Mutat. Res.* **2001**, 480-481, 243-268.
3. Sugimoto, K.; Hanai, H.; Tozawa, K.; Aoshi, T.; Uchijima, M.; Nagata, T.; Koide, Y. *Gastroenterology* **2002**, *123*, 1912-1922.
4. Bremner P.; Heinrich M. *J. Pharm. Pharmacol.* **2002**, *54*, 453-472.
5. Govindaraja, V.S. *Crit Rev Food Sci Nutri* **1980**, *12*, 199.
6. Nadkarni, K.M. *Curcuma Long.* Popular Prskashan Publishing Co., Bombay, 1976.
7. Chopra, R.N.; Chopra, I.C.; Handa, K.L.; Kapur, L.D.Z. Indigenous drugs of India. In *Dhur* **1958**, 2nd Ed. pp. 325.
8. Ammon, H.P.T.; Walh M.A. Pharmacology of *Curcuma longa*. *Planta Med.* **1991**, *57*,1-7.
9. Kang, Y.J.; Lee, Y.S.; Lee, G.W.; Lee, D.H.; Ryu, J.C.; Yun-Choi, H.S.; Chang, K. *J. Exp. Pharmacol. Ther.* **1999**, *291*, 314–320.
10. Bhaumik, S.; Jyothi, M.D.; Khar A. *FEBS Lett.* **2000**, *483*, 78-82.
11. Natarajan, C.; Bright J.J. *J. Immunol.* **2002**, *168*, 6506-6513.
12. Grandjean-Laquerriere, A.; Gangloff, S.C.; Le Naour, R.; Trentesaux, C.; Hornebeck, W.; Guenounou, M. *Cytokine* **2002**, *18*, 168-177.
13. Gukovsky, I.; Reyes, C.N.; Vaquero, E.C.; Gukovskaya, A.S.; Pandol, S.J. *Am. J. Physiol. Gastrointest Liver Physiol.* **2003**, *284*, G85-95.
14. Singh, S.; Aggrawal, B.B. *J. Biol. Chem.* **1995**, *270*, 24995.
15. Huang, T.S.; Lee, S.C.; Lin, J.K. *Proc. Natl. Acad. Sci. USA* **1991**, *88*, 5292.
16. Kumar, A.; Dhawan, S.; Hardegen, N.J.; Aggrawal, B.B. *Biochem. Pharmacol.* **1998**, *55*, 775.
17. Pendurthi, U.R.; Williams, J.T.; Rao, L.V.M. *Arterioscler. Thromb. Vasc. Biol.* **1997**, *17*, 3406.
18. Bierhaus, A.; Zhang, Y.; Quehenberger, P.; Luther, T.; Haase, M.; Muller, M.; Mackman, N.; Ziegler, R.; Nawroth, P.P. *Thromb. Haemost.* **1997**, *77*, 772.
19. Jobin, C.; Bradham, C.A.; Russo, M.P.; Juma, B.; Narula, A.S.; Brenner, D.A.; Sartor, R.B. *J. Immunol.* **1999**, *163*, 3474-3483.

20. Mukhopadhyay, A.; Basu, N.; Ghatak, N.; Gujral, P.K. *Agents. Actions.* **1982**, *12*, 508-515.
21. Arora, R.B.; Kapoor, V.; Basu, N. *Indian J. Med. Res.* **1971**, *59*, 1289-1295.
22. Chandra, D.; Gupta, S.S. *Indian J. Med. Res.* **1972**, *60*, 138-142.
23. Ghatak, N.; Basu, N. *Indian J. Exp. Biol.* **1972**, *10*, 235-236.
24. Brouet, I.; Ohshima, H. *Biochem. Biophys. Res. Commun.* **1995**, *206*, 533-540.
25. Sreejayan, N.; Rao, M.N. *J. Pharm. Pharmacol.* **1994**, *46*, 1013.
26. Chan, M.M.; Huang, H.I.; Fenton, M.R.; Fong, D. *Biochem. Pharmacol.* **1998**, *55*, 1955-1962.
27. Kakar, S.S.; Roy, D. *Cancer Lett.* **1994**, *87*, 85-89.
28. Kang, BY.; Chung, S.W.; Chung, W.; Im, S.; Hwang, S.Y.; Kim, T.S. *Eur. J. Pharmacol.* **1999**, *384*, 191-195.
29. Kang, B.Y.; Song, Y.J.; Kim, K.M.; Choe, Y.K.; Hwang, S.Y.; Kim, T.S. *Br. J. Pharmacol.* **1999**, *128*, 380-384.
30. Temkin V.; Kantor B.; Weg, V.; Hartman, M.L.; Levi-Schaffer, F. *J. Immunol.* **2002**, *169*, 2662-2669.
31. Soleas G.J.; Diamandis, E.P.; Goldberg, D.M. *J. Clin. Lab. Anal.* **1997**, *11*, 287-313.
32. Kimura, Y.; Okuda, H.; Arichi, S. *Biochimica et Biophysica Acta* **1985**, *834*, 275–278.
33. Subbaramaiah, K.; Chung, W.J.; Michaluart, P.; Telang, N.; Tanabe, T.; Inoue, H.; Jang, M.; Pezzuto, J.M.; Dannenberg, A.J. *J. Biol. Chem.* **1998**, *273*, 21875–21882.
34. Kimura, Y.; Okuda, H.; Kubo, M. *J. Ethnopharmacol.* **1995**, *45*, 131–139.
35. El-Mowafy, A.M.; White, R.E. *FEBS Lett.* **1999**, *451*, 63–67.
36. Kawada, N.; Seki, S.; Inoue, M.; Kuroki, T. *Hepatology* **1998**, *27*, 1265–1274.
37. Kawada, J.R.; Ward, N.E.; Ioannides, C.G.; O'Brian, C.A. *Biochem.* **1999**, *38*, 13244–13251.
38. Cheong, H.; Ryu, S.Y.; Kim, K.M. *Planta Med.* **1999**, *65*, 266–268.
39. Tsai, S.H.; Lin-Shiau, S.Y.; Lin, J.K. *Br. J. Pharmacol.* **1999**, *126*, 673–680.
40. Matsuda, H.; Kageura, T.; Morikawa, T.; Toguchida, L.; Harima, S.; Yoshikawa, M. *Bioorg. Med. Chem. Lett.* **2000**, *10*, 323–327.
41. Wadsworth, T.L.; Koop, D.R. *Biochem. Pharmacol.* **1999**, *57*, 941–949.
42. Holmes-McNary, M.; Baldwin, A.S., Jr. *Cancer Res.* **2000**, *60*, 3477–3483.
43. Manna, S.K.; Mukhopadhyay, A.; Aggarwal, B.B. *J. Immuno.* **2000**, *164*, 6509–6519.
44. Chen, C.K.; Pace-Asciak, C.R. *General Pharmacology* **1996**, *27*, 363–366.

45. Mgbonyebi, O.P.; Russo, J.; Russo, I.H. *Int. J. Oncol.* **1998**, *12*, 865–869.
46. Gehm, B.D.; McAndrews, J.M.; Chien, P.Y.; Jameson, J.L. Proc. Nat. Acad. Sci. USA **1997**, *94*, 14138-14143.
47. Hayashi, T.; Yamada, K.; Esaki, T.; Mutoh, E.; Chaudhuri, G.; Iguchi, A. *J. Cardiovascular Pharmacol.* **1998**, *31*, 292–298.
48. Hunt, J.S.; Miller, L.; Platt, J.S. *Developmental Immunology* **1998**, *6*, 105–110.
49. Cho, D.I.; Koo, N.Y.; Chung, W.J.; Kim, T.S.; Ryu, S.Y.; Im, S.Y.; Kim, K.M. : *Life Sci.* **2002**, *71*, 2071-82.
50. Wang, M.J.; Huang, H.M.; Hsieh, S.J.; Jeng, K.C.; Kuo, J.S. *J Neuroimmunol,* **2001**, *112*, 28-34.
51. Bertelli, A.A.; Baccalini, R.; Battaglia, E.; Falchi, M.; Ferrero, M.E. *Therapie.* **2001**, *56*, 613-616.
52. Katiyar, SK.; Mukhtar, H. *World Rev. Nutr. Diet* **1996**, *79*, 154-184.
53. Stoner, G.D.; Mukhtar, H. *J. Cell Biochem. Suppl.* **1995**, *22*,169-180.
54. Salah, N.; Miller, N.J.; Paganga, G.; Tijburg, L.; Bolwell, G.P.; Rice-Evans, C. *Arch. Biochem. Biophys.* **1995**, *322*, 339-346.
55. Lee, M.J.; Wang, Z.Y.; Li, H.; Chen, L.; Sun, Y.; Gobbo, S.; Balentine, D.A.; Yang, C.S. *Cancer Epidemiol. Biomarkers Prev.* **1995**, *4*, 393-399.
56. Suganuma, M.; Okabe, S.; Sueoka, E.; Iida, N.; Komori, A.; Kim, S.J.; Fujiki, H. *Cancer Res.* **1996**, *56*, 3711-3715.
57. Lin, Y.L.; Lin, J.K. *Mol. Pharm.* **1997**, *52*, 465-472.
58. Yang, F.; de Villiers, W.J.; McClain, C.J.; Varilek, G.W. *J. Nutr.* **1998**, *128*, 2334-2340.
59. Haqqi, T.M.; Anthony, D.D.; Gupta, S.; Ahmad, N.; Lee, M.S.; Kumar, G.K.; Mukhtar, H. *Proc. Natl. Acad. Sci. U S A* **1999**, *96*, 4524-4529.
60. Van der Kraan, P.M.; van den Berg, W.B. *Curr. Opin. Clin. Nutr. Metab. Care* **2000**, *3*, 205-211.
61. Van de Loo, F.A.J.; Joosten, L.A.; van Lent, P.L.E.M.; Arntz, O.J.; van den Berg, W.B. *Arthritis Rheum.* **1995**, *38*, 164-172.
62. Wood, D.D.; Ihrie, E.J.; Dinarello, C.A.; Cohen, P.L. *Arthritis Rheum.* **1983**, *26*, 975-983.
63. Singh, R.; Ahmed, S.; Islam, N.; Goldberg, V.M.; Haqqi, T.M. *Arthritis Rheum.* **2002**, *46*, 2079-2086.
64. King, P.A.; Anderson, V.E.; Edwards, J.O.; Gustafson, G.; Plumb, R.C.; Suggs, J.W. *J. Am. Chem. Soc.* **1992**, *114*, 5430.
65. Yermilov, V.; Rubio, J.; Ohshima, H. *FEBS Lett.* **1995**, *376*, 207-210.
66. Flynny, D.L.; Rafferty, M.F. *Prostaglandins Leukotrienes Med.* **1986**, *22*, 357–360
67. Iwakami, S.; Shibuya, M.; Tseng, C.F.; Hanaoka, F.; Sankawa, U. *Chem. Pharm. Bull.* **1986**, *34*, 3960–3963.

68. Kiuchi, F.; Iwakami, S.; Shibuya, M.; Hanaoka, F.; Sankawa, U. *Chem. Pharm. Bull.* **1992**, *40*, 387–391.
69. Yamazaki, R.; Aiyama, R.; Matsuzaki, T.; Hashimoto, S.; Yokokura, T. *Inflammation Res.* **1998**, *47*, 182–186.
70. Claeson, P.; Panthong, A.; Tuchinda, P.; Reutrakul, V.; Kanjanapothi, D.; Taylor, W.C.; Santisuk, T. *Planta Med.* **1993**, *59*, 451–454
71. Claeson, P.; Pongprayoon, U.; Sematong, T.; Tuchinda, P.; Reutrakul, V.; Soontornsaratune, P.; Taylor, W.C. *Planta Med.* **1996**, *62*, 236–240
72. Adcock, I.M.; Brown, C.R.; Kwon, O.; Barnes, P.J.; *Biochem. Biophys. Res. Commun.* **1994**, *199*, 1518–1523.
73. Hur, G.M.; Ryu, Y.S.; Yun, H.Y.; Jeon, B.H.; Kim, Y.M.; Seok, J.H.; Lee, J.H. *Biochem. Biophys. Res. Commun.* **1999**, *261*, 917–922.
74. Lee, K.H.; Kim, D.G.; Shin, N.Y.; Song, W.K.; Kwon, H.; Chung, C.H.; Kang, M.S. *Biochem. J.* **1997**, *324*, 237–242.
75. Salgo, M.G.; Stone, K.; Squadrito, G.L.; Battista, J.R.; Pryor, W.A. *Biochem. Biophys. Res. Commun.* **1995**, *210*, 1025-1030.
76. Yadav, P.N.; Liu, Z.; Rafi, M.M. *J. Pharmacol. Exp. Ther.* **2003** (In press).

Biological Activity

Chapter 4

Carnosol from Rosemary Suppresses Inducible Nitric Oxide Synthase through Down-Regulating NF κB in Murine Macrophages

Jen-Kun Lin[1], Ai-Hsiang Lo[1], Chi-Tang Ho[2], and Shoei-Yn Lin-Shiau[3]

Institutes of [1]Biochemistry and [3]Toxicology, College of Medicine, National Taiwan University, Number 1, Section 1, Jen-Ai Road, Taipei, Taiwan, Republic of China
[2]Department of Food Science and Center for Advanced Food Technology, Rutgers, The State University of New Jersey, 65 Dudley Road, New Brunswick, NJ 08901–8520

Rosemary (*Rosmarinus officinalis* Labiatae) is a well-known and greatly valued herb that is native to southern Europe and is now cultivated throughout the world. Carnosol is a phytopolyphenol found in rosemary. In the present study, we investigated the antioxidant activity of carnosol and other compounds extracted from rosemary. Carnosol showed potent antioxidative activity in DPPH free radical scavenging. High levels of nitric oxide (NO) are produced by inducible NO synthase (iNOS) in inflammation and multiple stages of carcinogenesis. Treatment of mouse macrophage RAW 264.7 cell line with carnosol markedly reduced LPS-stimulated NO production in a concentration-related manner. But other tested compounds were less active. Western blot, RT-PCR, and Northern blot analyses demonstrated that carnosol decreased LPS-induced iNOS mRNA and protein expression. Carnosol treatment showed reduction of nuclear factor-κB (NF-κB) subunits translocation and NF-κB DNA binding activity in activated macrophages. Carnosol also showed inhibition of iNOS and NF-κB promoter activity in transient transfection assay. These activities were related to down regulation of inhibitor κB (IκB) kinase (IKK) activity by carnosol, thus inhibiting LPS-induced phosphorylation as well as degradation of IκBα. These results suggest that carnosol suppresses nitric oxide production and iNOS gene expression by inhibiting NF-κB activation and provides a possible mechanism for its anti-inflammatory and cancer chemopreventive actions.

Carnosol is the active ingredient isolated from rosemary (*Rosmarinus officinalis* Labiatae) originally found widely in southern Europe and now cultivated throughout the world. It has been used since antiquity to improve and strengthen the memory (*1*). Rosemary has a long-standing reputation as a tonic, invigorating herb, imparting a zest for life that is to some degree reflected in its distinctive aromatic taste (*1*). Its herb and oil are commonly used as spice and flavoring agents in food processing for its desirable flavor and high antioxidant activity (*2*). Rosemary was stated to act as a mild analgesic and antimicrobial agent in traditional herbal use (*3*). Rosemary contains flavonoids, phenols, volatile oil and terpenoids (*3,4*). The antioxidant activity of an extract from rosemary leaves is comparable to known antioxidants, such as butylated hydroxyanisole (BHA) and butylated hydroxytoluene (BHT), without the cytotoxic and carcinogenic risk of synthetic antioxidants (*5,6*). Among the antioxidant compounds in rosemary leaves, about 90% of the antioxidant activity can be attributed to carnosol and carnosic acid.

Nitric oxide (NO) is a short-lived, small molecule, free radical produced from L-arginine (*7*) in a reaction catalyzed by NO synthase (NOS). NO has many biological functions. It is involved in vasodilatation, neurotransmission, tissue homeostasis, wound repair, inflammation and cytotoxicity (*8-10*). Physiologically low concentrations of NO produced by iNOS possess beneficial roles in much of the antimicrobial activity of macrophages against pathogens (*8,9,11,12*). However, over production of NO and its derivatives, such as peroxynitrite and nitrogen dioxide, are found to be mutagenic *in vivo*, and also provoke the pathogenesis of septic shock and diverse autoimmune disorders (*13-17*). Furthermore, NO and its oxidized forms (NOx) have also been shown to be carcinogenic (*18,19*). Excessive amounts of reactive nitrogen species (RNS) increase oxidative stress in the body (*20*). RNS actively participate in the metabolic activation of procarcinogens; oxidative deterioration of lipids, proteins and DNA; altering membrane fluidity; cellular homeostasis and gene expression.

RNS are considered as major biodeterminants in the process of diseases and carcinogenesis (20,21). Therefore, inhibiting high-output NO production by blocking iNOS expression or its activity may be a useful strategy for treatment of NO-related disorders. It is worthy to note that the recent emphasis on the role of NO in pathological conditions has led to the discovery of new therapeutic agents. Antioxidants such as (-)-epigallocatechin-3-gallate (EGCG) (22), resveratrol (23), and naturally occurring flavonoids including apigenin and kaempferol (24) have been reported to suppress NO production and their inhibition mechanisms are based on their ability to inhibit the activation of NF-κB. Although the broad spectrum of research on rosemary and its extracts has concentrated on its action as a cancer chemopreventive agent, the action mechanisms are not yet well established. Thus, here we compare the antioxidative properties of rosemary phytochemicals and examine their effects on NO generation, iNOS expression, and NF-κB activation in LPS-stimulated RAW 264.7 cells. The present data suggest that carnosol could protect against endotoxin-induced inflammation by blocking NF-κB activation, thereby suppressing iNOS expression.

Materials and Methods

Chemicals and Reagents

Carnosol and ursolic acid were isolated from rosemary as described (5). Carnosic acid and rosmarinic acid were isolated from the ground dried leaves of rosemary, sequentially, by hexane and n-butanol extractions. The final products were purified by column chromatography on silica gel (2). LPS (lipopolysaccharide, *Escherichia coli* 026:B6), sulfanilamide and naphthyl-ethylenediamine dihydrochloride were purchased from Sigma Chemical Co. (St. Louis, MO). Isotopes were obtained from Amersham (Arlington Heights, IL). RT-PCR reagents were purchased from Promega (Madison, WI). Polynucleotide kinase was purchased from Pharmacia (Piscataway, NJ).

The mouse iNOS promoter plasmid was generously provided by Dr. Charles J. Lowenstein, Johns Hopkins University (9). The plasmid construct contains a 1.75 kb *Hinc* III restriction fragment upstream of the macrophage NOS gene which was cloned into the GeneLight luciferase vector system (Promega). The pNFkB-Luc plasmid was purchased from Stratagene Corp. (La Jolla, CA).

HPLC Separation of Rosemary Components

Rosemary extracts were obtained as described (5). In brief, 10 g dried ground leaves of rosemary were extracted with 50 mL methanol at 60°C, twice, and filtered through 3M filter paper. The filtrate was bleached with activated charcoal. The final filtrate was concentrated to 10 mL with a rotary vacuum evaporator. After adding 15 mL water, the precipitate was collected and dried. After dissolving the precipitate in methanol, a 20 µL sample was filtered through a 4.5 µm filter disk and subjected to HPLC analysis. The components of rosemary extracts were separated using reverse HPLC system with an Osmosil C18 packed column (5 µm, 4.6 x 250 mm, Nacalai Tesque, Inc., Kyoto, Japan). The mobile phase was composed of MeOH: Water: Phosphoric acid = 80:19.98: 0.02 (v/v/v) and ran by an isocratic elution at a flow rate of 1 ml/min (Waters 600E system controller). The constituent peak signal was detected at 285 nm with a Waters 484 tunable absorbance detector system and the data was plotted and integrated using a Waters 745 data module.

Measurement of DPPH Free Radicals Scavenge Activity

α,α-Diphenyl-β-picrylhydrazyl (DPPH) was purchased from Sigma. 100 µM DPPH (in 60% absolute alcohol) was mixed with vehicle (DMSO) only or with different concentrations of tested compounds. Absorption at 517 nm was measured using a Hitachi U3201 UV-visible spectrophotometer. The decrease of absorbance was shown when tested compounds possessed free radicals scavenging activity (25). DMSO did not interfere with the reaction.

Cell Culture

The mouse monocyte-macrophage cell line RAW 264.7 (ATCC TIB-71) was cultured in DMEM containing 10% heat inactivated fetal bovine serum (GIBCO Life Technologies, Grand Island, NY). Except for transient transfection assay, cells were plated in dishes at a density of 4×10^6/mL for 18-24 h before activation by LPS (1 µg/mL). Tested compounds were co-treated with LPS and the final DMSO (as vehicle) concentration was less than 0.2 % (v/v).

Estimation of Nitrite

The nitrite accumulated in the culture medium was measured as an indicator of NO production according to the Griess reaction (26).

Western Blots for iNOS and IκBα

Total cellular protein (for iNOS, α-tubulin, IκBα) was prepared using lysis buffer containing 10 % glycerol, 1% Triton X-100, 1 mM sodium orthovanadate, 1 mM EGTA, 5 mM EDTA, 10 mM NaF, 1 mM sodium pyrophosphate, 20 mM Tris-HCl pH 7.9, 100 μM β-glycerophosphate, 137 mM NaCl, 1 mM phenylmethylsulfonyl fluoride (PMSF), 10 μg/mL aprotinin and 10 μg/mL leupeptin. Cytosolic fraction (for PARP, α-tubulin, c-Rel, p65 and p50) and nuclear fraction (for PARP, α-tubulin, c-Rel, p65, and p50) were prepared according to a modified procedure of Xie et al (*30*). Cells were suspended in hypotonic buffer containing 10 mM HEPES pH7.6, 10 mM KCl, 0.1 mM EDTA, 1 mM dithiothreitol (DTT), 0.5 mM PMSF, 1.5 mM MgCl$_2$, 1 mM PMSF, 1 μg/ml aprotinin and 1 μg/mL leupeptin. The nuclei were pelleted by centrifugation at 3000 *g* for 5 min. The supernatants containing cytosolic proteins were collected. Nuclei lysis was performed with hypertonic buffer containing 30 mM HEPES, 1.5 mM MgCl$_2$, 450 mM KCl, 0.3 mM EDTA, 10 % glycerol, 1 mM DTT, 1 mM PMSF, 1 μg/mL aprotinin and 1 μg/mL leupeptin. The supernatants containing nuclear proteins were obtained by centrifugation at 12,000 *g* for 20 min. The nuclear proteins were also retained at –70 °C for use in the DNA biding assay. 30-50 μg proteins were separated on sodium dodecyl sulfate-polyacrylamide gel (SDS-PAGE) (8 % for iNOS and α-tubulin, 10% for PARP, α-tubulin, c-Rel, p65, p50 and IκBα) and electro-transferred to polyvinylidene difluoride (PVDF) membrane (Immobilon^P, Millipore, Bedford, MA). The membrane was preincubated in phosphate buffered saline (PBS) containing 0.01 % Tween-20, 1% bovine serum albumin (BSA) and 0.2% NaN$_3$ overnight at 4 °C, and then incubated with anti-iNOS monoclonal antibody (Transduction Laboratories, Lexingtons, KY), anti-α-tubulin monoclonal antibody (Oncogene Science, Cambridge, U.K.), anti-phospho (Ser32)-specific IκBα, antibody (New England Biolabs, Bevery, MA) anti-PARP (UBI, Lake Placid, NY) or anti-IκBα, -c-Rel, -p65, -p50 polyclonal antibodies (Santa Cruz Biochemicals, Santa Cruz, CA). After incubation with horseradish peroxidase- or alkaline phosphatase-conjugated anti-rabbit/goat/mouse IgG antibody, the immunoreactive bands were visualized with enhanced chemiluminescence reagents (ECL, Amersham), or with colorigenic substrates: nitro blue tetrazolium (NBT) and 5-bromo-4-chloro-3-indolyl-phosphate (BCIP) as suggested by the manufacturer (Sigma Chemical Co.). Data were quantified using a densitometer (IS-100 Digital Imaging System).

RT-PCR and Northern Blotting

Cells were washed with ice-cold PBS and total RNA was isolated by acid guanidinium thiocyanate-phenol-chloroform extraction (27). 5 μg total RNA was reverse-transcribed (RT) into cDNA using Moloney murine leukemia virus (M-MLV) reverse transcriptase and oligo (dT)$_{18}$ primer by incubating the reaction mixture (15 μL) at 37 °C for 90 min. The polymerase chain reaction (PCR) was performed as described previously (22). A final volume of 25 μL contained dNTPs (each at 200 μM), 1X reaction buffer, 0.4 μM each primer (iNOS; forward: 5'-CCCTTCCGAAGTTTCTGGCAGCAGC-3' and reverse: -5' - GGCTGTCAGAGAGCCTCGTGGCTTTGG-3'; G3PDH: 5'-TGAAGGTCGGTGTGAACGGATTTGGC-3' and 5'-CATGTAGGCCATGAGGTCCACCAC-3'), 3 μL of RT product, and 50 units/ml Super Taq DNA polymerase. After an initial denaturation for 5 min at 95 °C, 30 cycles of amplification (95 °C for 30 s, 65 °C for 45 s, and 72 °C for 2 min) were performed followed by a 10 min extension at 72 °C. 5 μL of each PCR product was electrophoresed on a 2% agarose gel and visualized by ethidium bromide staining (497-bp iNOS fragment and 983-bp G3PDH fragment). For Northern blot analysis, total RNA (20 μg) was denatured with formaldehyde/formamide and incubated at 65 °C for 15 min, size fractioned on 1.2% formaldehyde-containing agarose gel, and transferred to Hybond-N nylon membrane (Amersham Corp, Arlington Heights, IL) in 20X standard sodium citrate (SSC) buffer. The blotted membrane was hybridized with [α-^{32}P]dCTP labeled (Random Primer labeling kit, Amersham) iNOS fragment at 42 °C overnight. After a stringent wash, the membrane was dried and autoradiographed with Kodak X-ray film (Rochester, NY). Amplification of glyceradehyde-3-phosphate dehydrogenase (G3PDH) and ethidium bromide staining of ribosomal RNA (18S and 28S) served as a control for sample loading and integrity. Data were quantified using a densitometer (IS-100 Digital Imaging System), and represented one of three similar experimental results.

Immunoprecipitation and IKK Kinase Assay

Immunoprecipatation was performed on the whole cell lysates (300 μg) with anti-IKKα/β antibody and protein A/G plus agarose beads (Santa Cruz Biotechnology) for 18 h at 4 °C in IP buffer containing 1 % Triton X-100, 0.5 % NP-40, 150 mM NaCl, 10 mM Tris pH7.4, 1 mM EDTA, 1 mM EGTA, 0.2 mM sodium orthovanadate, 0.2 mM PMSF, 10 μg/mL aprotinin and 1 μg/mL leupeptin, with gentle rotation. Immunoprecipitated IKK was washed three times with IP buffer and two times with kinase buffer containing 50 mM HEPES pH7.4, 10 mM MgCl$_2$, 2.5 mM EDTA, 10 mM β-glycerophosphate, 1 mM NaF, 1 mM DTT, 10 mM p-nitrophenyl phosphate (PNPP), 300 μM sodium orthovanadate, 1 mM benzamidine, 10 μg/ml aprotinin and 1 μg/mL leupeptin. IKK kinase activity was assayed by phosphorylation of GST-IκBα fusion protein (1 μg in each sample, Santa Cruz) in a final volume of 40 μl kinase buffer with 5 μCi [γ-^{32}P] dATP (5000 Ci/mmol, Amersham) for 30 min at 30 °C. Reaction was terminated by the addition of 10 μl 5X Laemmli's loading buffer and heated at 100 °C for 10 min. The sample was resolved by 10% SDS-PAGE, dried and visualized by autoradiography. Data were quantified using a densitometer (IS-100 Digital Imaging System) and represented one of three similar experimental results.

Electrophoretic Mobility Shift Assay(EMSA)

Nuclear extract (5 μg) was mixed with the double-stranded NF-κB oligonucleotide, 5'-ATGTGAGGGGACTTTCCCAGGC-3' end-labeled by [γ-^{32}P] dATP (underlining indicates a κB consensus sequence or a binding site for NF-κB/c-Rel homodimeric and heterodimeric complexes) and salmon sperm DNA in binding buffer containing 100 mM KCl, 30 mM HEPES, 1.5 mM MgCl$_2$, 0.3 mM EDTA, 1 mM DTT, 1 mM PMSF, 1 μg/mL aprotinin, 1 μg/mL leupeptin and 10% glycerol) at room temperature for 20 min. DNA/protein complex was separated from free probe on a 4.5% non-denaturing TBE polyacrylamide gel in 0.5X Tris/borate/EDTA buffer (TBE: 44.5 mM Tris, 44.5 mM boric acid and 1 mM EDTA). The specificity of binding was examined by competition with the unlabeled oligonucleotide. The gel was dried and subjected to autoradiography. Data were quantified using a densitometer (IS-100 Digital Imaging System) and represented one of three similar experimental results.

Transient Transfection and Luciferase Assay

When the RAW 264.7 cells were confluent, the medium was replaced with serum-free Opti-MEM (Gibco BRL). Then the cells were transfected with pNF-κB-Luc, or pGL2-iNOS plasmid reporter gene using LipofectAMINE™ (Gibco BRL). After an additional 24 h incubation, this medium was replaced with complete medium. After 24 h, the cells were trypsinized and equal numbers of cells were plated in 12-well tissue culture plates for 12 h. Cells were then treated with LPS (1 μg/mL) alone or with carnosol (20 μM) for 3 h. Each well was washed twice with cold phosphate-buffered saline and harvested in 150 μL of lysis buffer (0.5 M HEPES pH 7.8, 1 % Triton X-100, 1 mM CaCl$_2$ and 1 mM MgCl$_2$). Aliquots of 100 μl of cell lysate were used to assay luciferase activity with the Luc-Lite™-M luciferase reporter gene assay kit (Packard Instrument Co., Meriden, CT). Luminescence was measured in a TopCount® Microplate Scintillation and Luminescence Counter (Packard 9912V1; Meriden, CT) in single photon counting mode for 0.05 min/well, following 5 min adaptation in the dark. Luciferase activities were normalized to protein concentrations. Data obtained represented one of three similar experimental results.

Results

Composition of Rosemary Phytochemicals

The major rosemary phytochemicals including carnosol, carnosic acid, ursolic acid and rosmarinic acid (Figure 1) were separated by HPLC as described in Materials and Methods and illustrated in Figure 2. The retention time of these compounds were carnosic acid, 19.5 min; carnosol, 7.7 min; ursolic acid, 13 min and rosmarinic acid, 2-12 min. It seemed that rosmarinic acid was not resolved well in this system. The relative amounts of these compounds in rosemary extracts were similar to previous findings (5) as 16.5-19.2% ursolic acid; 3.8-4.6% carnosol; 0.1-0.5% carnosic acid and trace amount of rosmarinic acid, respectively.

Effects of Rosemary Phytochemicals on the Viability of RAW 264.7 Cells

Cells were treated with rosemary phytochemicals (0-20 μM) for 24 h. Cell viability was assayed using ATP-Lite™-M non-radioactive cell proliferation assay kit and luminescence intensity measured by Topcount Microplate Scintillation and Luminescence Counter. The results indicated that ursolic acid showed strong inhibitory effect on the viability of RAW 264.7 cells; carnosol and carnosic acid showed moderately activity, while rosmarinic acid was less active than carnosol and carnosic acid (Figure 3).

Antioxidative Properties of Rosemary Phytochemicals

The antioxidant activity of the four phytochemicals isolated from rosemary were assayed in a DPPH free radical scavenge system. Carnosic acid, carnosol and rosmarinic acid showed concentration-dependent scavenging ability . These phytochemicals were more potent than vitamin C and vitamin E. The 50% DPPH

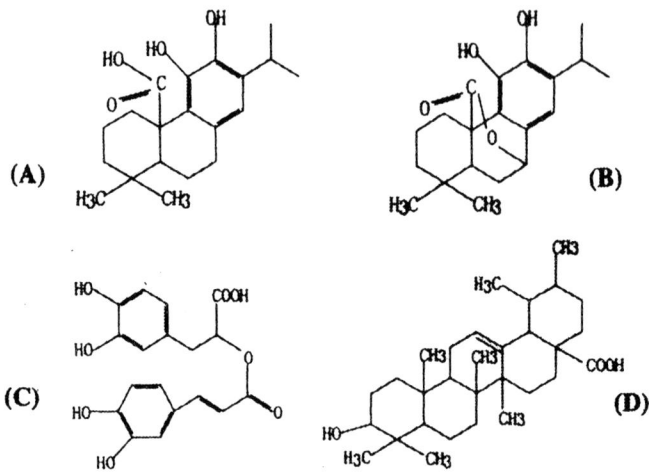

Figure 1. Structure of rosemary phytochemicals. (A) carnosic acid; (B) carnosol; (C) rosmarinic acid; (D) ursolic acid.

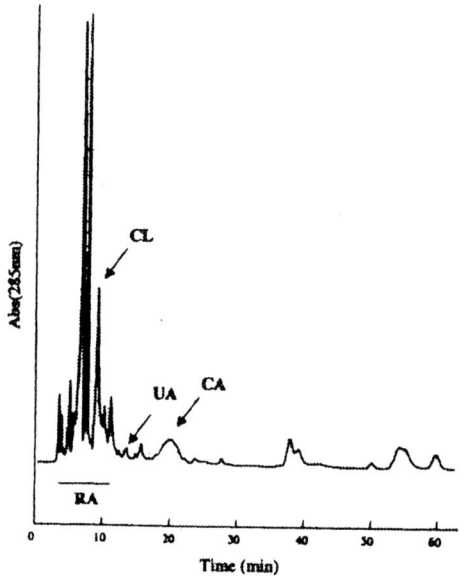

Figure 2. High performance liquid chromatographic separation of rosemary phytochemicals. The chromatographic conditions were as described in the Materials and Methods. The abbreviations are : CL, carnosol; UA, ursolic acid; CA, carnosic acid; RA, rosmarinic acid.

scavenge concentrations are carnosic acid = 0.6 μM, carnosol = 0.59 μM, rosmarinic acid=0.49 μM and ursolic acid > 10 μM as approximation values coming from one of the three separate experiments. Rosmarinic acid was the most potent antioxidant over the same concentration range and ursolic acid showed no effects on DPPH scavenging.

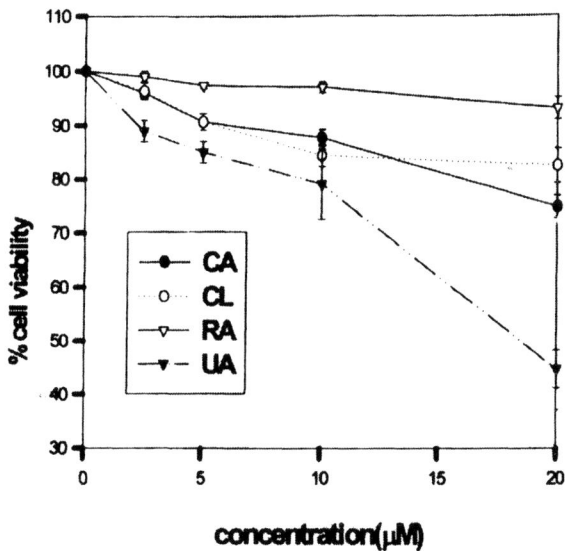

Figure 3. Effects of rosemary phytochemicals on viability of RAW 264.7 cells. Cells were treated with rosemary phytochemicals for 24 h. Cell viability was assayed using ATP-Lite-M nonradioactive cell proliferation assay kit and luminescence intensity measured by Topcount Microplate Scintillation and lumininescence counter. The percentage of cell viability was normalized to control. Data represent means ± SE triplicate tests.

Effects of Rosemary Phytochemicals on LPS-induced NO Production

We used nitrite production as indicator of NO released in LPS-activated macrophage to investigate the anti-inflammatory effects of the four rosemary phytochemicals. The concentration-response relationships of nitrite formation (Figure 4) were determined 24 h after treatment of LPS (1 μg/mL) only or along with the tested compounds. Carnosol inhibited nitrite production by > 50 % at 10 μM. Carnosol was found to reduce NO generation in a concentration-dependent manner. Carnosic acid and rosmarinic acid showed less inhibitory effect over the same concentration range. Cell viability assay verified that the inhibition of carnosol was not due to general cellular toxicity. The inhibitory ability of ursolic acid might be due to its cytotoxic effect. The phytochemicals dissolved in DMSO did not interfere with the Griess reaction.

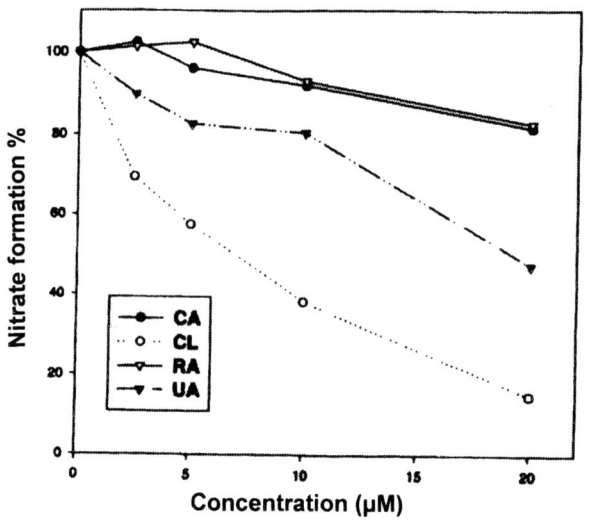

Figure 4. Effects of rosemary phytochemicals on LPS-induced nitrite formation in RAW 264.7 cells. Cells were treated with or without indicated concentrations of tested compounds [carnosic acid (CA), carnosol (CL), rosmarinic acid (RA), ursolic acid (UC)] and LPS (1 μg/mL) for 24 h. Amounts of nitrite released to culture medium was determined by Griess reagent and read at OD 550 nm with a sodium nitrite standard curve. Data represent means ± SE of triplicate tests.

Suppression of iNOS Protein and mRNA Expression by Carnosol

As shown in Figure 4, among the compounds tested, carnosol was the most potent in inhibiting the production of nitrite. After stimulating with LPS, the level of iNOS mRNA continually increased between 1-12 h following the increased iNOS protein level. Inhibition of iNOS protein by carnosol was detected in a concentration-dependent manner after 18 h treatment. The inhibitory concentration for iNOS protein expression was similar to that for reduction of nitrite formation. In order to investigate whether the suppression of iNOS protein by carnosol was due to reduced iNOS mRNA expression, a RT-

PCR analysis for total mRNA samples extracted from RAW 264.7 cells after 12 h treatment was carried out. The amplification of cDNA with primers specific for mouse iNOS and G3PDH (as control gene) is shown in Figure 5. The results indicated that lower levels of iNOS mRNA were expressed in the presence of carnosol in LPS-activated macrophages. Similar results were obtained from Northern blot analysis of iNOS mRNA in cell extracts (Figure 6). These data suggested carnosol modulated iNOS expression at the transcriptional level. Consistent with previous findings, carnosol almost completely suppressed iNOS gene expression at 20 µM (Figure 6) thus inhibiting the production of NO in LPS-stimulated RAW 264.7 cells (Figure 4). To further investigate the importance of LPS and carnosol in modulating expression of iNOS, transient transfection was performed using mouse iNOS luciferase promoter constructs. LPS-induced iNOS promoter activity was inhibited by carnosol. Carnosol might block LPS-mediated signal transduction.

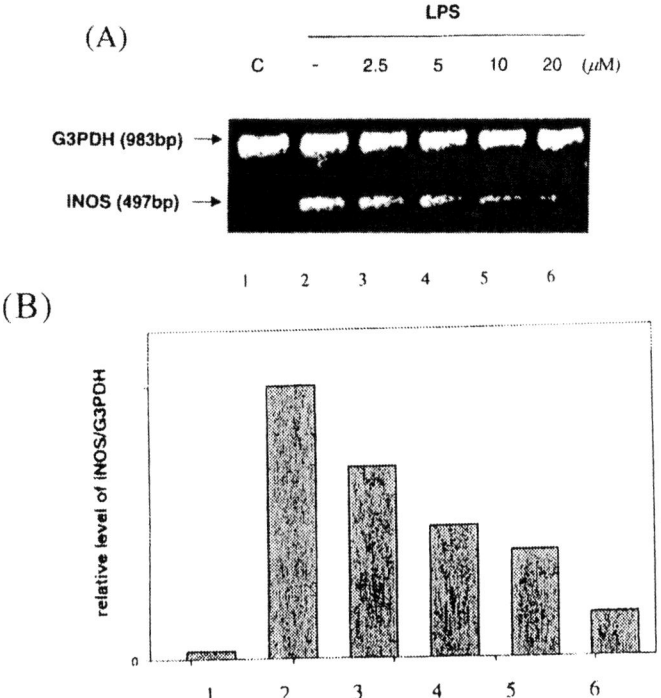

Figure 5. RT-PCR analysis of iNOS mRNA. (A) Cells were treated with different concentrations of carnosol and LPS (1 µg/mL) for 12 h. 5 µg of total RNA was subjected to reverse transcription-polymerase chain reaction (RT-PCR) and PCR product was resolved in 1% agarose gel. C, control. (B) Data was quantified using a densitometer (IS-100 Digital Imaging System), and the relative level observed with LPS alone is set at 1.

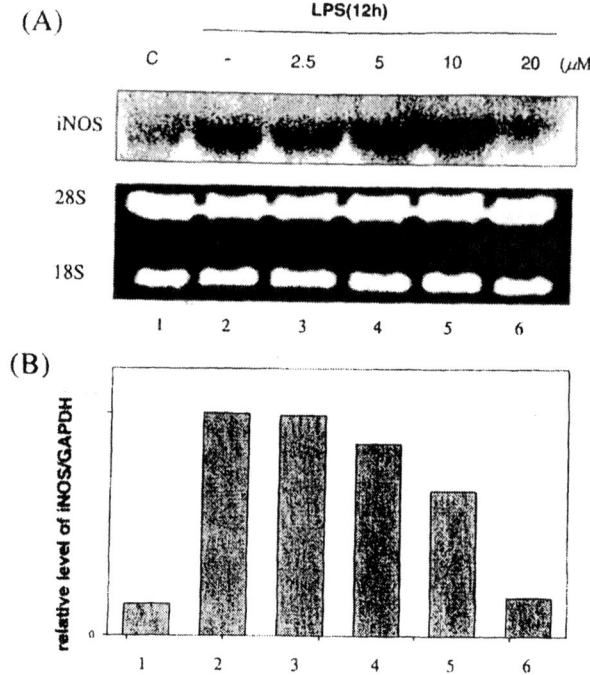

Figure 6. Northern blot analysis of iNOS mRNA. (A) Total RNA was extracted from treated cells and separated on 1.2% formaldehyde-containing agarose gel then transferred onto nylon membrane. Upper panel: the blots were hybridized by ^{32}P-labelled iNOS probe as described in Materials and Methods. Lower panel: ethidium bromide staining of ribosomal RNA (18S and 28S) was shown as control. C, control. (B) Data was quantified using a densitometer (IS-100 Digital Imaging System) and represents one of three similar results. The relative level observed with LPS alone is set at 1.

Suppression of Nuclear Contents of NF-κB/Rel Family Members by Carnosol

It has been demonstrated that the transcription factor NF-κB is involved in the activation of iNOS by LPS induction. We tested if carnosol perturbed the translocation of NF-κB/Rel family subunits (c-Rel, p65, and p50) into the nucleus in LPS-stimulated RAW 264.7 cells. Nucleus and cytosolic extracts were prepared and subjected to immunoblot analysis. The amounts of nuclear c-

Rel, p65, and p50 were increased at 30 min after LPS treatment. However, co-incubation with LPS plus carnosol resulted in the reduction of nuclear contents of c-Rel and p65 proteins (28).

Suppressing Effect of Carnosol on LPS-induced NF-κB Activation

To investigate if carnosol specifically inhibited activation of NF-κB, gel mobility shift assay was performed to analyze NF-κB DNA binding activity. As shown in a previous study (28), the induction of specific NF-κB DNA binding activity by LPS was inhibited by carnosol. The specificity of binding was examined by competition with the addition of excess consensus unlabeled oligonucleotide. In an additional study, transient transfection with a NF-κB-dependent luciferase reporter plasmid was done to confirm whether carnosol inhibited NF-κB binding activity in LPS-induced macrophages. The results indicated that carnosol inhibited LPS-induced NF-κB transcriptional activity.

Suppression of Carnosol on LPS-induced Phosphorylation and Degradation of IκBα

Since the activation of NF-κB is correlated with the hyperphosphorylation of IκBα and its subsequent degradation, we examined the phosphorylated and protein levels of IκBα by immunoblot analysis. After 12 min activation by LPS, the serine-phosphorylated IκBα protein was detected by anti-Ser32-phospho-specific IκBα antibody as illustrated in Figure 7A. Carnosol inhibited LPS-induced IκBα phosphorylation. Time course experiment showed that treatment of LPS caused degradation of IκBα after 20-40 min (Figure 7B) and IκBα recovered to basal level after 50 min. As shown in Figure 7C, the amount of IκBα was decreased by the treatment of LPS for 30 min, and the treatment with carnosol effectively sustained the IκBα protein content. The pattern of inhibition on IκBα phosphorylation by carnosol was parallel to the pattern of inhibition on its degradation. These results suggest that inhibition of NO production by carnosol (Figure 4) occurs via blocking phosphorylation, as well as degradation of IκBα protein (Figure 7); thus preventing the translocation and activation of NF-κB in the nucleus.

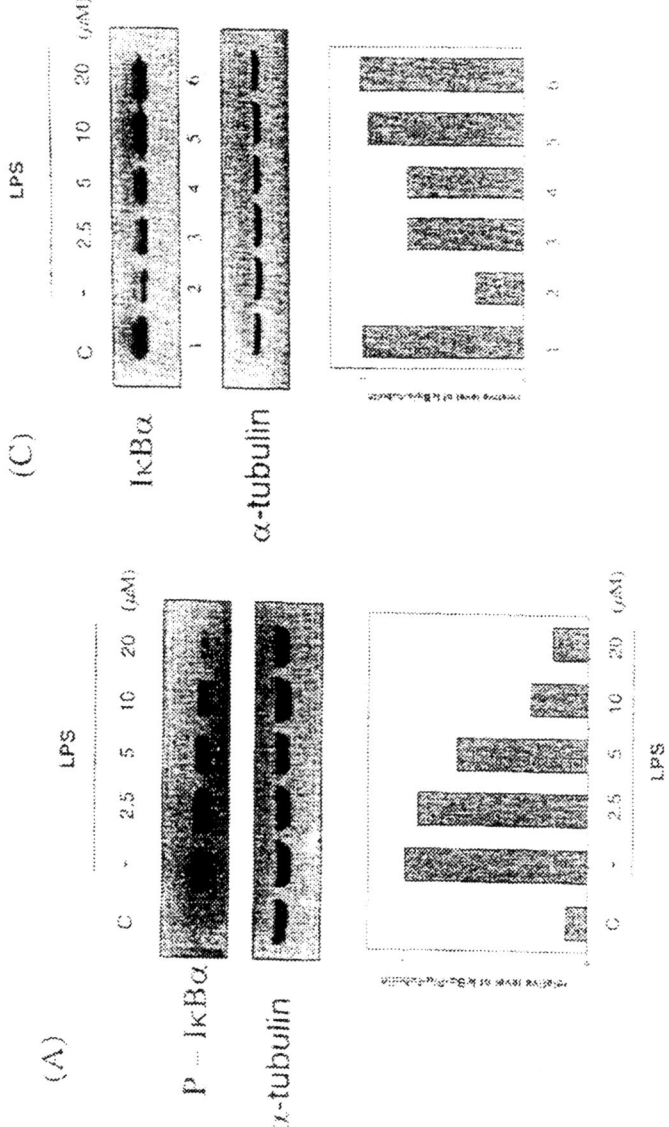

(B)

Figure 7. The inhibition of LPS-mediated IκBα phosphorylation and degradation by carnosol in RAW264.7 cells. Cells were treated with LPS (1 μg/mL) without or with carnosol and incubated for 12 min or 30 min. Total cell lysates were prepared for Western blot analysis for the content of IκBα protein. (A) After 12 min activation, the phosphorylated IκBα was detected by anti-phospho(ser32)-specific IκBα antibody. (B) Time course experiment if IκBα degradation in LPS-stimulated RAW264.7 cells. (C) The content of IκBα protein was detected after 30 min activation by anti- IκBα antibody. Data were quantified using a densitometer (IS-100 Digital Imaging System), and represents one of three similar results. The relative level observed with LPS is set at 1 in (A); the relative level observed with control is set at 1 in (C). C, control.

Effects of Carnosol on Activation of IKK

To further confirm the inhibition of IκBα phosphorylation and degradation by carnosol and the correlation between phosphorylation of IκBα and IκB kinase (IKK) activity, we assayed IKK activity in cell culture. When we assayed the immunoprecipitated IKK activity by phosphorylation of GST-IκBα, IKK activity was inhibited by carnosol in LPS-induced RAW 264.7 cells. Carnosol had no effect on the level of IKK protein (28). This result suggested that carnosol inhibited iNOS expression via down regulated IKK kinase activity thus preventing NF-κB activation.

Discussion

Like many other antioxidants, rosemary and its polyphenols possess not only antioxidative activities, but also antitumorigenic activities. Huang et al. (5) investigated the inhibitory effects of rosemary extract, carnosol and ursolic acid on tumor formation in mouse skin. They found that topical application of rosemary inhibits B(a)P- and DMBA-induced initiation of tumor and TPA-induced tumor promotion in DMBA-initiated mice. Carnosol and ursolic acid were found to be strong inhibitors of TPA-induced inflammation, ornithine decarboxylase activity and tumor promotion in mouse skin (5). It was suggested that carnosol acted like other nonsteroidal phenolic anti-inflammatory agents, such as curcumin, which inhibited the metabolism of arachidonic acid.

In the present study, carnosol, carnosic acid and rosmarinic acid showed effective DPPH free radicals scavenging activity. Carnosic acid, carnosol, rosmarinic acid and ursolic acid effectively inhibited pUC19 DNA strand break induced by Fenton reaction (28). The reaction of iron (II) with hydrogen peroxide is generally considered to yield the hydroxyl radicals (29). These compounds may react with and/or scavenge the hydroxyl radicals produced by Fenton reaction.

We had reported that many antioxidants possessed anti-inflammatory effects (22-24). These findings provided a significant molecular basis for the mode of actions of dietary biochemical active compounds in preventing cancer and inflammation. In this study, we demonstrated that carnosol inhibited LPS-induced NO production in mouse macrophage RAW 264.7 cells with an IC$_{50}$ of 9.4 μM. Carnosic acid, rosmarinic acid and ursolic acid had less effect although they possessed different extents of antioxidant or DNA protection activity. The antioxidant activity of a given phytopolyphenol is related, but not really consistent with, its effect on the LPS-induced NO production. To identify the direct scavenging effect of carnosol, we determined the amount of NO remaining in the supernatant shortly after the addition of carnosol to the supernatant of LPS-stimulated RAW 264.7 cells; however, the amount of NO pre-existed in the supernatant was not changed by carnosol (data not shown). From these results, we assumed that the inhibition of NO production in LPS-stimulated RAW 264.7 cells by carnosol occurred via modulation of iNOS.

Many compounds modulate LPS-stimulated iNOS at the transcription level. To determine the inhibitory effect of carnosol on the LPS-induced iNOS protein and mRNA expression, Western blot, RT-PCR and Northern blot analysis were performed. Carnosol concentration-dependently inhibited LPS-induced iNOS mRNA and protein expression. LPS positively regulated NF-κB/Rel for iNOS and other immune or inflammatory gene expression (*30*). Among those transcription factors, NF-κB is necessary to confer inducibility by LPS in mouse macrophages (*31*). Kleinert and Euchenhofer (*32*) also reported that in 3T3 cells, at least three different signal transduction pathways could stimulate iNOS mRNA expression. All these pathways seem to converge in the activation of the essential transcription factor NF-κB though a marked intercell variability exists (*33-35*). NF-κB is composed mainly of two proteins: p65 and p50. In unstimulated cells, NF-κB exists in the cytosol in a quiescent form bound to its inhibitory protein, IκB (*36*). After stimulating the cells with various agents, IκB becomes phosphorylated and goes to subsequent proteolytic degradation. Releasing IκB from NF-κB allows the activation and nuclear accumulation of NF-κB subunits (*37*). To identify the mechanisms involved in the effects of carnosol on iNOS expression, we investigated the transcriptional regulation of NF-κB. EMSA and promoter activity studies showed that carnosol treatment inhibited activation of these κB-binding complexes. Immunoblot analysis also showed that the inhibition of NF-kB activity by carnosol might be the results of the inhibition of IκBα phosphorylation and degradation and subsequent reduction of the translocation of NF-κB subunits. p50 forms heterodimers with p65 and c-Rel, the Rel family protein. Here we found that carnosol reversed the nuclear translocation of each subunit with different ratios. This may be due to the different degradation rates of subunits, and this interesting phenomenon deserves further investigation. We also identified that inhibition of IKK activity was important for the anti-inflammatory action of carnosol.

Phosphorylation plays important roles in activating protein tyrosine kinase, MAPK and protein kinase C in mediating LPS signaling. Reactive oxygen species have been proposed to be involved in the activation of NF-κB via regulation of various redox-sensitive protein kinase or tyrosine kinase (*38-40*). Antioxidants, thiols and iron chelators can specifically prevent activation of NF-κB (*41*). Furthermore, many antioxidants or phytopolyphenols such as pyrrolidine dithiocarbamate (*42*), N-acetyl-L-cysteine (*43*) and EGCG (*22*) inhibit NO production via regulating NF-κB activity. Recently, aspirin was also reported to possess antioxidative and DNA protection properties, as well as inhibit NF-κB (*44*). Carnosol possessed high antioxidant activity in previous studies (*45*), as well as our own studies. Our findings showed that carnosol can block activation of NF-κB by interfering with the signal-induced phosphorylation of IκB.

The NO concentration in tumors seems to be an important factor in carcinogenesis, and a number of studies have shown that high NO levels are tumoricidal, whereas low NO levels are tumorigenic. Some cancers are less susceptible to the toxic effects of NO, and in these types NO can actually stimulate their growth by promoting angiogenesis and tumor cell proliferation (*46*). Most experimental solid tumors were shown to have increased expression of iNOS, and with NO being both angiogenic and genotoxic, its increased production may contribute to carcinogenesis. Additionally, NO generated in tumors increases vascular permeability, which enhances nutritional supply and sustains the rapid growth of tumor cells (*46*). Tumoricidal or tumorigenic actions of NO are highly dependent on the conditions in the tumor microenvironment. A number of experimental and clinical examples confirm the duality of the cellular effects of NO and may ultimately determine its biological role at any given time (*46*). Inflammatory responses induced by various pathogens can accelerate tissue damage and mutagenesis, and NO may effectively sustain the growth of the transformed cells.

In the present study, we have demonstrated the suppression of iNOS expression through inhibiting of IKK activity and down-regulation of NFκB activation by rosemary phytochemicals. These findings have provided a possible molecular mechanism for their anti-inflammation and cancer chemoprevention.

Acknowledgements

This study was supported by the National Science Council NSC 89-2320-B-002-245 and NSC 89-2320-B-002-223; by the National Health Research Institute NHRI-EX 90-8913BL; by the National Research Institute of Chinese Medicine, NRICM-90102; and by the ministry of Education, ME89B-FA01-1-4.

References

1. Chevallier, A. *The Encyclopedia of Medicinal Plants*. Dorling Kindersley Limited, London, 1996, p. 125.
2. Ho, C.-T.; Ferraro, T.; Chen, Q.; Rosen, R.T.; Huang, M.T. In *Food Phytochemicals for Cancer Prevention II*. Ho, C.-T.; Osawa, T.; Huang, M.T.; Rosen, R.T. (Eds.),. Am. Chem. Soc. Symp. Ser., *547*, American Chemical Society, Washington, DC, 1994, pp. 2-19.
3. Newall, C.A. *Herbal Medicines-A Guide For Health Care Professionals*. The Pharmaceutical Press, London. 1996.
4. Collin, M.A.; Charles, H. P. *Food Microbiol*. **1987**, *4*, 311-315.
5. Huang, M.T.; Ho, C.-T.; Wang, Z.Y.; Ferraro, T.; Lou, Y.R.; Stauber, K.; Ma, W.; Georgiadis, C.; Laskin, J.D.; Conney, A.H. *Cancer Res*. **1994**, *54*, 701-708.

6. Ito, N.; Fukushima, S.; Hagiwara, A.; Shibata, M.; Ogiso, T. *J. Natl. Cancer Inst.* **1983**, *70*, 343-352.
7. Nathan, C. *FASEB J.* **1992**, *6*, 3051-3064.
8. Moncada, S.; Palmer, R.M.; Higgs, E. A. *Pharmacol. Rev.* **1991**,*43*, 109-142.
9. Lowenstein, C.J.; Alley, E.W.; Raval, P.; Snowman, A.M.; Snyder, S.H.; Russell, S.W.; Murphy, W. J. *Proc. Natl. Acad. Sci.U.S.A.* **1993**, *90*, 9730-9734.
10. Lowenstein, C.J.; Snyder, S.H. *Cell* **1992**, *70*, 705-707.
11. Cook, H.T.; Cattell, V. *Clin. Sci. (Colch.)* **1996**, *91*, 375-384.
12. Hibbs, J.B., Jr.; Taintor, R.R.; Vavrin, Z. *Science* **1987**, *235*, 473-476.
13. Liu, R.H.; Hotchkiss, J.H. *Mutat. Res.* **1995**, *339*, 73-89.
14. Wink, D.A.; Kasprzak, K.S.; Maragos, C.M.; Elespuru, R.K.; Misra, M.; Dunams, T.M.; Cebula, T.A.; Koch, W.H.; Andrews, A.W.; Allen, J.S. *Science* **1991**, *254*, 1001-1003.
15. Nguyen, T.; Brunson, D.; Crespi, C.L.; Penman, B.W.; Wishnok, J.S.; Tannenbaum, S.R. *Proc. Natl. Acad. Sci.U.S.A.* **1992**, *89*, 3030-3034.
16. Kilbourn, R.G.; Gross, S.S.; Jubran, A.; Adams, J.; Griffith, O.W.; Levi, R.; Lodato, R.F. *Proc. Natl. Acad. Sci.U.S.A.* **1990**, *87*, 3629-3632.
17. Miller, M.J.; Sadowska-Krowicka, H.; Chotinaruemol, S.; Kakkis, J.L.; Clark, D.A. *J. Pharmacol. Exp. Ther.* **1993**, *264*, 11-16.
18. Halliwell, B. *Lancet* **1994**, *344*, 721-724.
19. Ohshima, H.; Bartsch, H. *Mutat. Res.* **1994**, *305*, 253-264.
20. Kehrer, J.P. *Crit. Rev. Toxicol.* **1993**, *23*, 21-48.
21. Bagchi, D.; Bagchi, M.; Stohs, S.J.; Das, D.K.; Ray, S.D.; Kuszynski, C.A.; Joshi, S.S.; Pruess, H.G. *Toxicology* **2000**, *148*, 187-197.
22. Lin, Y.L.; Lin, J.K. *Mol. Pharmacol.* **1997**, *52*, 465-472.
23. Tsai, S.H.; Lin-Shiau, S.Y.; Lin, J.K. *Br. J. Pharmacol.* **1999**, *126*, 673-680.
24. Liang, Y.C.; Huang, Y.T.; Tsai, S.H.; Lin-Shiau, S.Y.; Chen, C.F.; Lin, J.K. *Carcinogenesis* **1999**, *20*, 1945-1952.
25. Ratty, A.K.; Sunflamoto, J.; Das, N.P. *Biochem. Pharmacol.* **1988**, *37*, 989-995.
26. Kim, H.; Lee, H.S.; Chang, K.T.; Ko, T.H.; Baek, K.J.; Kwon, N.S. *.J. Immunol.* **1995**, *154*, 4741-4748.
27. Chomczynski, P.; Sacchi, N. *Anal. Biochem.* **1987**, *162*, 156-159.
28. Lo, A.H.; Liang, Y.C.; Lin-Shiau, S.Y.; Ho, C.-T.; Lin, J.K. *Carcinogenesis*, **2002**, 23, 983-991.
29. Koppenol, W.H. In *Free Radical Damage and Its Control.* Rice-Evans, C.A.; Burdon, R.H.E. (Eds.) Elsevier, Amsterdam, 1994, pp 3-24.
30. Jeon, Y.J.; Kim, Y.K.; Lee, M.; Park, S.M.; Han, S.B.; Kim, H.M. *J. Pharmacol. Exp. The*, **2000**, *294*, 548-554.

31. Xie, Q.W.; Kashiwabara, Y.; Nathan, C. *J. Biol. Chem.* **1994**, *269*, 4705-4708.
32. Kleinert, H.; Euchenhofer, C.; Ihrig-Biedert, I.; Forstermann, U. *J. Biol. Chem.* **1996**, *271*, 6039-6044.
33. Vincenti, M.P.; Burrell, T.A.; Taffet, S.M. *J. Cell Physiol.* **1992**, *150*, 204-213.
34. Feuillard, J.; Gouy, H.; Bismuth, G.; Lee, L.M.; Debre, P.; Korner, M. *Cytokine* **1991**, *3,* 257-265.
35. Muroi, M.; Suzuki, T. *Cell Signal.* **1993**, *5,* 289-298.
36. Baeuerle, P.A.; Baltimore, D. *Science* **1988**, *242*, 540-546.
37. Baeuerle, P.A; Baltimore, D. *Cell* **1996**, *87*, 13-20.
38. Kang, J.L.; Go, Y.H.; Hur, K.C.; Castranova, V. *J. Toxicol. Environ. Health A.* **2000**, *60,* 27-46.
39. Kang, J.L.; Pack, I.S.; Lee, H.S.; Castranova, V. *Toxicology* **2000**, *151*, 81-89.
40. Kang, J.L.; Pack, I.S.; Hong, S.M.; Lee, H.S.; Castranova,V. *Toxicol. Appl. Pharmacol.* **2000**, *169,* 59-65.
41. Mulsch, A.; Schray-Utz,B.; Mordvintcev, P.I.; Hauschildt, S.; Busse, R. *FEBS Lett.* **1993**, *321*, 215-218.
42. Schini-Kerth, V.; Bara, A.; Mulsch, A.; Busse, R. *Eur. J. Pharmacol.* **1994**, *265*, 83-87.
43. Pahan, K.; Sheikh, F.G.; Namboodiri, A.M.; Singh, I. *Free Radic. Biol. Med.* **1998**, *24*, 39-48.
44. Shi, X.; Ding, M.; Dong, Z.; Chen, F.; Ye, J.; Wang, S.; Leonard, S.S.; Castranova,V.; Vallyathan,V. *Mol. Cell Biochem.* **1999**, *199*, 93-102.
45. Aruoma, O.I.; Halliwell, B.; Aeschbach, R.; Loligers, J. *Xenobiotica* **1992**, *22,* 257-268.
46. Buga, G.M.; Ignarro, L.J. In :*Nitric Oxide: Biology and Pathobiology.* Ignarro, L. J. (Ed.), Academic Press, San Diego, **2000**, pp 895-920.

Chapter 5

Hypolipidemic Effect and Antiatherogenic Potential of Pu-Erh Tea

Lucy Sun Hwang[1], Lan-Chi Lin[1], Nien-Tsu Chen[1], Huei-Chiuan Liuchang[1], and Ming-Shi Shiao[2]

[1]Graduate Institute of Food Science and Technology, National Taiwan University, Taipei, Taiwan, Republic of China
[2]Department of Medical Research and Education, Veterans General Hospital, Taipei, Taiwan, Republic of China

Pu-Erh tea is a fermented tea produced in Yunnan area, a southwestern part of China. The manufacture of Pu-Erh tea involves natural fermentation and prolonged storage at ambient temperature. Lovastatin was identified in batches of Pu-Erh tea. This study demonstrated that uptake of Pu-Erh tea reduced plasma cholesterol and triacylglycerol in cholesterol-fed hamsters. Results also suggested that the cholesterol-lowering effect of Pu-Erh tea was caused by a combination of lovastatin and tea polyphenols. Although the content of EGCG was low, Pu-Erh tea yet exhibited strong antioxidant activities that scavenged DPPH radical and inhibited LDL oxidation *in vitro* and *ex vivo*. This study indicates that Pu-Erh tea drinking may reduce the risk factors in atherosclerosis-related ischemic heart disease.

© 2003 American Chemical Society

Elevation of plasma cholesterol, particularly low-density lipoprotein cholesterol (LDL-C), is positively correlated with coronary heart disease (CHD), a major vascular disease predominantly causing by atherosclerosis (1,2). Recent studies have indicated that LDL oxidation, endothelial dysfunction, and inflammation play important roles in the molecular pathogenesis of atherosclerosis (3). Oxidized LDL (OxLDL) appears in the circulation and tends to infiltrate into the aortic endothelium (4). Antioxidants, which inhibit LDL oxidative modification, may reduce early atherogenesis and slow down the disease progression to an advanced stage (5).

Tea (*Camellia sinensis*) is a very popular and widely consumed beverage in the world. The biological activities and pharmacological functions of tea natural components, particularly the polyphenols, have attracted great attention for years (6,7). Tea catechins decrease micellar solubility and intestinal absorption of cholesterol in rats (8). The cholesterol-lowering effect of (-)-epigallocatechin gallate (EGCG) (Figure 1), a major catechin in tea, on experimental hypercholesterolemia in rats has been reported. It indicates that the anti-hypercholesterolemic effect of EGCG is primarily due to the inhibition of absorption of exogenous cholesterol from the digestive tract and partly due to the enhanced elimination of endogenous cholesterol (9). Feeding of green and black teas reduce lipid peroxidation in rat liver and kidney (10). Epicatechin isomers from jasmine green tea and tea polyphenols inhibit free radical-induced cell lysis and oxidative damage to red blood cells (11,12). Green and black teas reduce LDL oxidizability and atherosclerosis in cholesterol-fed rabbits (13). EGCG and theaflavins from tea inhibit Cu^{2+} induced LDL oxidation, a process involving cholesteryl ester degradation and apoB-100 fragmentation (14). The relative potency of five common tea polyphenols (flavan-3-ol derivatives) on Cu^{2+} mediated oxidative modification of LDL follows the trend: EGCG>ECG>EC>C>EGC (15). The potency to inhibit LDL oxidation *in vitro* by green tea extract has also been compared with those of vitamin C and E. The antioxidant effect of green tea is not due to metal chelation (16). *Ex vivo* antioxidant effects of green tea and black tea have been demonstrated in human (17,18). Green tea is more potent in inhibiting LDL oxidation *in vitro* and *ex vivo*. At least one study has indicated that black tea consumption does not protect LDL from oxidation (19).

Pu-Erh tea is a fermented tea produced in Yunnan area, a southwestern part of China. The manufacture of Pu-Erh tea involves natural fermentation and prolonged storage at an ambient temperature and high moisture environment. This tea has gained popularity recently. It is speculated that the ingredients in Pu-Erh tea can be originated from the tea leaves and their transformed products by endogenous enzymes. Besides, metabolites derived from the microorganisms, such as *Aspergillus*, and biotransformation products originated from tea may contribute to the biological activities of Pu-Erh tea. A previous study has indicated that Pu-Erh tea reduces plasma cholesterol and triacylglycerol in female Wistar rats (20). The findings suggest that Pu-Erh tea may ameliorate

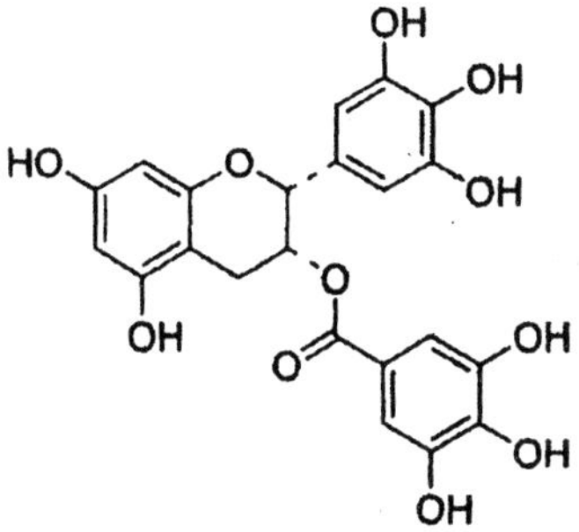

(-)-Epigallocatechin-3-gallate (EGCG)

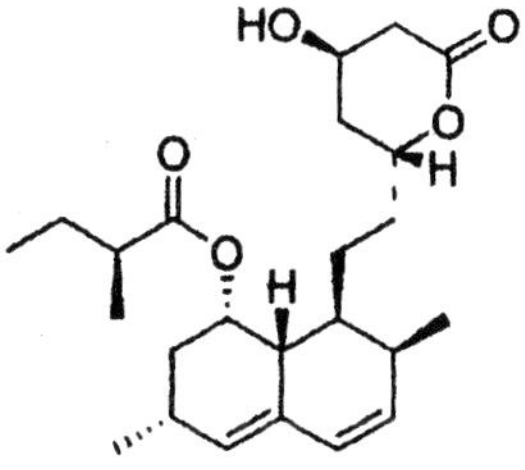

Lovastatin

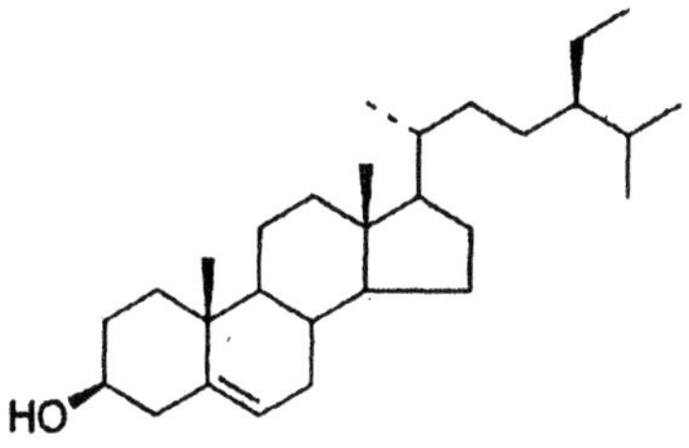

β-Sitosterol

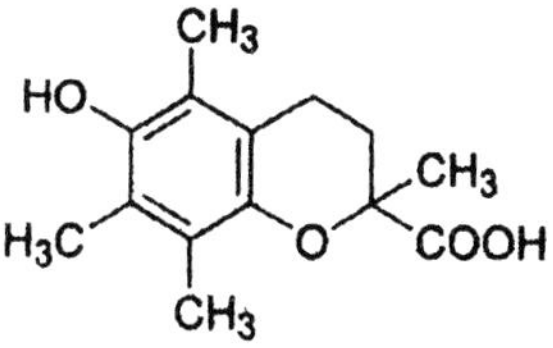

Trolox

Figure 1. Structures of (-)-epigallocatechin-3-gallate (EGCG), lovastatin, β-sitosterol, and trolox.

hyperlipidemia and atherosclerosis. The potential of Pu-Erh tea to inhibit LDL oxidation has not been elucidated and compared with other teas. The content of catechins in Pu-Erh tea is significantly lower than those of green and black teas. It is less likely to attribute the beneficial effect to catechins. We therefore studied the effects of Pu-Erh tea on lipid metabolism in hamsters and the antioxidant potential to reduce LDL oxidation *in vitro* and *ex vivo*.

Materials and Methods

Materials

Cholesterol, β-sitosterol, 1,1-Diphenyl-2-picrylhydrazyl (DPPH), 1,1,3,3-tetramethoxypropane (TMP), bovine serum albumin (BSA), α-tocopherol, and retinyl acetate were purchased from Sigma (St Louis, MO).

Pu-Erh tea was obtained from Yunnan, China. It was ground into powder (2.5 kg) and extracted with boiling water (1:10; w/v). The aqueous extract was filtered, concentrated, and lyophilized. The dry powder (designated as PET) was kept at -20 °C before use. The extraction yield was 29.2% (w/w).

Human hepatoma cells (Hep G2) were obtained from the cell bank, Veterans General Hospital-Taipei. Male, Golden Syrian hamsters (n=40) (body weight 116±5 g) were obtained from the Animal Center, National Science Council, Taiwan. Rodent chow (Purina 5001) was purchased from Purina.

Inhibition of Cholesterol Synthesis in Hep G2 Cells

The potential of PET to inhibit cholesterol biosynthesis was determined by the inhibition of the incorporation of [2-^{3}H]acetate and R-[2-^{14}C]mevalonate into cholesterol (21). Human hepatoma cells (Hep G2) were cultured in DMEM supplemented with 10% fetal calf serum (FCS) in 12-well culture dishes (3x10^5 cell/well) at 37 °C for 48 hours. After transferring to fresh medium, cells were treated with PET and labeled precursors ([2-^{3}H]acetate 2.5 µCi; R-[2-^{14}C]mevalonate, 1.3 µCi). Lovastatin, in hydroxy acid form (5 and 10 µM), was used as a positive control. Incorporation experiment was carried out at 37 °C for 2 hours. After removal of medium and extensive washing, cells were harvested and crude total lipids were collected and saponified. Unlabeled cholesteryl oleate was used as a carrier in saponification. Cholesterol was recovered by extraction with *n*-hexane. Radioactivity was measured by using a liquid scintillation counter.

Identification of Lovastatin in Pu-Erh Tea

PET that exhibited inhibitory activity in cholesterol biosynthesis in cultured Hep G2 cells was again prepared in a larger scale. The powder was extracted with ethyl acetate (1:10, w/v). The ethyl acetate layer was concentrated and partially purified by silica gel column chromatography and preparative TLC (silica gel plates, 20x20-cm, 2-mm thickness; *n*-hexane: ethyl acetate=1:1, v/v). The silica gel, in the Rf range corresponding to lovastatin, was collected and extracted with ethyl acetate. Further purification was carried out by semi-preparative reversed-phase HPLC (C_{18}, 8.0x250-mm) (*22*). Final identification of lovastatin in PET was carried out by mass spectrometry (Micromass Platform System; Manchester, UK).

DPPH Radical Scavenging

The radical scavenging activity of PET was determined by using the stable radical DPPH (1,1-diphenyl-2-picrylhydrazyl). In a final volume of 300 µL, the reaction mixture contained 167 µM DPPH and PET (the powder was initially dissolved in water and finally diluted in 10% ethanol). After 15-min incubation in an incubator shaker, the absorption at 517 nm was taken and the value was corrected against a blank without DPPH (*23*). The radical scavenging activities of catechins, probucol, and trolox were also determined for comparison.

Inhibition of Cu^{2+}-induced LDL Oxidation

Human sera were obtained from healthy adult donors after overnight fasting. LDL was isolated by ultracentrifugation with the density adjusted by NaBr (ρ 1.006-1.063). LDL fraction was dialyzed with PBS at 4 °C in darkness for 24 hr. LDL oxidation was induced by Cu^{2+} (10 µM). LDL oxidation, determined by conjugated diene formation, was monitored by the increase of UV absorption at 234 nm (*24,25*). Antioxidant activity was determined by the capability to inhibit conjugated diene formation and prolong the lag phase (T_{lag}, min). A concentration dependent curve was obtained for the determination of IC_{50} value. Probucol and trolox were used as positive controls.

Animal Study

Hamsters were acclimatized for 2 weeks in the 12-hour light dark controlled animal house before randomly assigned to five groups. Animals in each group (n=8) were fed with one of the five diets for 28 days. N, normal diet (Purina 5001); HC, high-cholesterol diet (normal diet plus 1% cholesterol, w/w); Sitosterol, HC plus 1.0% β-sitosterol (w/w); PET-1 (HC plus 1.0% PET, w/w), PET-2 (HC plus 2.0% PET, w/w), respectively. At the end of feeding period, animals were sacrificed after overnight fasting. Feces, collected in the last two days of feeding, were dried, ground, and saponified with an ethanolic KOH solution. During the 28-day feeding period, we adhered to the guideline for care and use of laboratory animals.

Human Study

Sixteen healthy male adults (mean age, 23.2±2.2 yr.; BMI, 21.9±1.2 kg/m^2) with normal plasma lipids (TC < 200 mg/dL, TG < 200 mg/dL, and plasma glucose < 126 mg/dL) were recruited with written consent (26,27). Volunteers were recommended to maintain their dietary habit and physical activity for at least 7 days before the experiment. Pu-Erh tea was prepared by extracting with boiling water (50 g Pu-Erh tea/1000 mL water) for 5 minutes. To reduce the bitter and mellow herbal taste, the extract was diluted 2.5 folds before providing to the volunteers. Volunteers were asked to drink 1000-mL tea extract (equivalent to the extract of 20 g Pu-Erh tea) each day for 7 days. Plasma samples were obtained on day 0 and day 7 after overnight fasting. Several volunteers are regular tea drinkers. During the 7-day period, Pu-Erh tea extract was the only beverage consumed except drinking water. No participant withdrew from this study. Plasma samples, before and after the 7-day drink period, were subjected to lipid and lipoprotein analysis.

General Lipid Analysis

Serum cholesterol and TG levels were determined by enzymatic methods (28,29). Serum free fatty acid (FFA) level was determined by a colorimetric method (30). Human LDL was obtained by ultracentrifugation. The density range between 1.006-1.063 g/mL was collected as LDL. To determine the oxidative susceptibility in Cu^{2+}-induced oxidation, LDL was dialyzed in phosphate-buffered saline (PBS) (10 mM, pH 7.4) and subjected to oxidation without prolonged storage (25,31). Hepatic cholesterol contents of hamsters were determined by a modified colorimetric assay after saponification and extraction (32). To avoid interference from plant sterols in the colorimetric assay, fecal cholesterol content was determined by reversed-phase HPLC (mobile phase: methanol : acetonitrile = 56:44, v/v).

Oxidative Susceptibility of Human LDL *ex vivo*

Oxidation of LDL was initiated by adding with 10 µM Cu^{2+} in PBS. The time course of LDL oxidation at ambient temperature was monitored by the formation of conjugated dienes, which was determined by the increase of UV absorption 234 nm. Lag phase (T_{lag}, min) was defined as the intercept of the tangent drawn to the steepest segment of the propagation phase to the horizontal axis (*24*).

α-Tocopherol Content in Human LDL

A 200-µl aliquot of LDL was added with an equal volume of ethanol. The mixture was immediately extracted with 1.0 mL n-hexane containing BHT (0.4 mg/mL) in darkness. The hexane layer was dried by a stream of nitrogen and the residue was dissolved in 100 µL mobile phase (acetonitrile : tetrahydrofuran = 70/30, v/v). α-Tocopherol in the mixture was determined by reversed-phase HPLC (*33*). The detector wavelength was set at 292 nm. Retinyl acetate was used as an internal standard.

Statistical Analysis

Results were expressed as mean±SEM. Statistical analyses were obtained by using unpaired t test or analysis of variance (ANOVA). A p value less than 0.05 was considered as statistically significant.

Results

Inhibition of Cholesterol Synthesis in Hep G2 Cells

Treatment of PET inhibited cholesterol synthesis in cultured Hep G2 cells (Table I). The potential of PET to inhibit cholesterol synthesis was demonstrated by the incorporation of [2-^{3}H]acetate into cholesterol. PET inhibited cholesterol biosynthesis in a dose-dependent manner (Table I). The conversion of R-[2-^{14}C]mevalonate to cholesterol was not affected. Double-labeling experiment showed that PET inhibited cholesterol biosynthesis (40 µg/mL of PET had 56% inhibition) at the pre-mevalonate stage without affecting the post-mevalonate steps. In the assay system, fetal calf serum was maintained in DMEM. The IC_{50}

value of PET was 36 µg/mL. As a positive control, lovastatin, an HMG-CoA reductase inhibitor, inhibited cholesterol biosynthesis in cultured Hep G2 cells. The EGCG content in PET (0.59 mg/g tea), which was significantly lower than that in green tea (20.1 mg/g tea), was unable to inhibit cholesterol synthesis in Hep G2 cells to an equivalent extent (data not shown).

Lovastatin in Pu-Erh Tea

Lovastatin ($C_{24}H_{36}O_5$; molecular weight 404.55) was identified in some batches of Pu-Erh tea. The identification was based on HPLC and mass spectrometry (APCI) after partial purification (Figure 2). The molecular ion of lovastatin (m/z) was clearly identified. Coupling with the inhibitory assay by using Hep G2 cells, we have examined several preparations of Pu-Erh tea. Results also showed that not all batches of PET Pu-Erh tea contained lovastatin (data not shown). The content of lovastatin in different supplies of Pu-Erh tea varied greatly from undetectable level to 0.86 mg/g dry weight of PET (aqueous extracts of Pu-Erh tea). For PET preparations exhibiting inhibitory activities in cholesterol biosynthesis, the inhibition was caused by lovastatin predominantly. Batches of Pu-Erh tea, which contained no or less lovastatin, also were less potent to inhibit cholesterol biosynthesis in cultured Hep G2 cells.

DPPH Radical Scavenging

The IC_{50} value of PET to scavenge DPPH radical was 8.3 µg/mL. The potency of PET was compared with those of tea catechins, trolox, and probucol (on equal weight basis). The DPPH radical scavenging activities were in the order: EGCG > trolox > green tea > Pu-Erh tea > black tea. As to the aqueous extract, the potency of Pu-Erh tea to scavenge DPPH radical was close to that of green tea (IC_{50} value 7.5 µg/mL) and better than black tea (IC_{50} value 14.3 µg/mL).

Inhibition of LDL Oxidation *in vitro*

PET exhibited strong antioxidant activity to inhibit LDL oxidation *in vitro*. The antioxidant activities, based on the prolongation of lag phase in Cu^{2+}-induced LDL oxidation, were in the following order: trolox > PET > green tea > probucol > black tea. The IC_{50} values of the aqueous extracts of teas to inhibit Cu^{2+}-induced LDL oxidation were PET (1.4 µg/mL), green tea (1.8 µg/mL), and

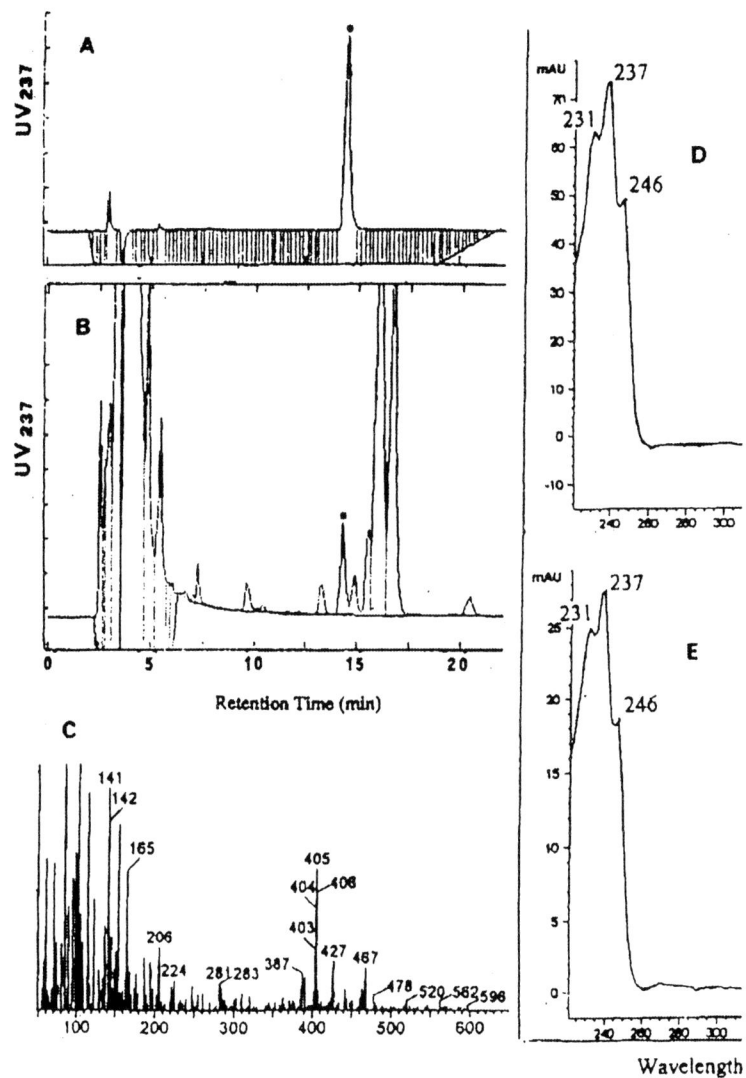

Figure 2. Identification of lovastatin in Pu-Erh tea. After partial purification by preparative TLC and semi-preparative reversed-phase HPLC, lovastatin in an enriched fraction was identified by mass spectrometry. A, authentic lovastatin in reversed phase HPLC; B, peak corresponding to lovastatin in a preparativeTLC purified fraction; C, mass spectrum (m/z) of a semi-preparative HPLC enriched fraction containing lovastatin. Peak corresponding to lovastatin is marked by an asterisk. D, UV spectrum of lovastatin; E, UV spectrum of lovastatin collected from reversed-phase HPLC.

Table I. Inhibition of Cholesterol Biosynthesis in Hep G2 Cells by PET

	Conc. (μg/mL)	[2-^{3}H]Acetate (10^3 dpm/10^5 cells/h)	R-[2-^{14}C]Mevalonate (10^3 dpm/10^5 cells/h)
Control (n=3)	-	5.21±0.07	0.41±0.01
Lovastatin (n=4)	2.0	1.82±0.02 (-65%)	0.38±0.02 (-6.8%)
	4.0	0.84±0.06 (-84%)	0.38±0.02 (-7.8%)
PET (n=4)	20	3.97±0.17 (-24%)	0.39±0.01 (-5.1%)
	40	2.29±0.09 (-56%)	0.39±0.02 (-5.3%)
	80	1.58±0.04 (-70%)	0.38±0.02 (-6.5%)

NOTE: Human hepatoma cells (Hep G2) were cultured in DMEM. Cells were treated with [2-^{3}H]acetate (2.5 μCi) and R-[2-^{14}C]mevalonate (1.3 μCi) for 2 h. Lovastatin, in hydroxy acid form, was used as a positive control. Data are expressed as mean±SD (10^3 dpm/10^3 cells/h). The percentages of inhibition in treated groups, as compared with control, are shown in parentheses.

black tea (3.3 μg/mL). The aqueous extracts of green tea and PET had approximately equal antioxidant activities to inhibit LDL oxidation *in vitro*.

Animal Study

Hamsters fed with a high cholesterol diet (1%, w/w) were chosen as the animal model. There was no significant difference in body weight or adipose tissue weight among five animal groups after a 28-day feeding period (Table II). High cholesterol-diet feeding significantly increased the liver weight (5.3 g vs. 9.3 g, p<0.05). β-Sitosterol treatment group (Sitosterol) and two PET treatment groups (PET-1 and PET-2) exhibited reduced liver weights, as compared with that of high cholesterol diet group (HC) (p<0.05). Hepatic cholesterol contents were also significantly reduced in two PET-treated groups. β-Sitosterol treatment also reduced hepatic cholesterol content (Table II).

Table II. Body, Liver and Adipose Tissue Weights of Hamsters

	N (n=8)	HC (n=8)	Sitosterol (n=8)	PET-1 (n=8)	PET-2 (n=8)
Body wt (g)					
Initial	115±5 [a]	117±5 [a]	115±4 [a]	117±6 [a]	116±4 [a]
Final	142±8 [b]	148±11 [b]	146±10 [b]	142±12 [b]	141±8 [b]
Gain	28±4 [c]	31±9 [c]	31±7 [c]	24±7 [c]	25±7 [c]
Liver wt (g)	5.3±0.6 [d]	9.3±0.8 [f]	8.2±0.7 [e]	8.1±0.9 [e]	8.4±0.6 [e]
Liver/Body wt ratio (g/100g)	3.8±0.6 [g]	6.3±0.3 [h]	5.6±0.3 [h]	5.8±0.4 [h]	5.9±0.2 [h]
Adipose wt (g)	3.7±0.5 [i]	3.8±0.8 [i]	3.9±0.6 [i]	3.3±0.7 [i]	3.5±0.6 [i]
Adipose/Body wt ratio(g/100g)	2.6±0.3 [j]	2.6±0.4 [j]	2.6±0.3 [j]	2.0±0.8 [j]	2.5±0.3 [j]

NOTE: Animals were fed for 28 days. N, normal diet; HC, high cholesterol diet (1.0% cholesterol); Sitosterol, HC plus 1.0% β-sitosterol; PET-1, HC plus 1.0% Pu-Erh tea aqueous extract; PET-2, HC plus 2.0% Pu-Erh tea aqueous extract. Results are shown as mean ±S.D. Data with same superscript in the same row were not significantly different (p>0.05).

Serum cholesterol level in the HC group was significantly higher (748 mg/dL) than that in the control diet group (121 mg/dL) (P<0.05). PET treatment significantly decreased serum cholesterol (-34.8% in PET-1 group, p<0.05; −39.0% in PET-2 group, p<0.05) (Table III). As a positive control, β-Sitosterol treatment (Sitosterol group) reduced serum cholesterol by 27.4% (p<0.05). Serum triacylglycerol (TG) level was significantly increased after 28-days high cholesterol diet feeding (HC group). β-Sitosterol feeding did not affect the TG level. The TG levels in two PET treatment groups were reduced by 30.7% (PET-1 group) and 41.6% (PET-2 group) (p<0.05). Fasting serum free fatty acid (FFA) level is a marker of lipolysis, a biochemical event inversely correlated with insulin sensitivity, in the adipose tissue. PET treatment reduced serum FFA in the high treatment group (PET-2) only. β-Sitosterol had no FFA lowering effect.

Fecal cholesterol, as a marker to negatively correlate intestinal cholesterol absorption, in Sitosterol group was significantly increased (p<0.05). High dose of PET treatment (PET-2) also enhanced fecal cholesterol secretion. The relative potency in the induction of fecal cholesterol secretion in PET-2 was 44% of that in Sitosterol group (Table IV).

Table III. Levels of Liver Cholesterol, Serum Cholesterol, TG and FFA in Hamsters

	N (n=8)	HC (n=8)	Sitosterol (n=8)	PET-1 (n=8)	PET-2 (n=8)
Liver Chol (mg/g)	1.5±0.1 [a]	26.8±4.6 [c]	21.6±0.9 [b]	21.6±1.2 [b]	21.7±0.9 [b]
Serum Chol (mg/dL)	121±20 [d]	748±289 [f]	543±188 [e]	488±147 [e]	466±70 [e]
TG (mg/dL)	574±132 [g]	1003±351 [h]	915±314 [h]	695±227 [g]	586±111 [g]
FFA (μM)	566±119 [i]	578±185 [j]	634±113 [k]	514±148 [i]	499±119 [i]

NOTE: Same as Table II.

Table IV. Fecal Cholesterol Contents of Hamsters

	N (n=8)	HC (n=8)	Sitosterol (n=8)	PET-1 (n=8)	PET-2 (n=8)
Chol content (mg/g)	2.5±0.5 [a]	18.8±0.5 [b]	22.9±0.5 [d]	19.8±0.4 [b,c]	20.6±0.4 [c]
Change (mg/g)	---	---	+4.1	+1.0	+1.8
% of change	---	---	+100	+24	+44

NOTE: Same as Table II.

Human Study

During the 7-day drink study on young, healthy male adults (n=16), no change of body weight was observed and no observable adverse effect was detected. Plasma levels of TC, TG, FFA, LDL-cholesterol, and HDL-cholesterol were not changed (Table V). Susceptibility of LDL to oxidation *ex vivo* was determined by the lag phase (T_{lag}) in Cu^{2+}-induced oxidation (24). The lag phase was increased significantly after Pu-Erh tea drinking (before 114±31 min; after 135±30 min) ($p<0.05$). The α-tocopherol content in LDL was determined by reversed-phase HPLC. Data were expressed as number of α-tocopherol per LDL particle. After 7-day Pu-Erh tea drink, the α-tocopherol content in LDL was significantly increased (before 8.7±1.6; after 10.1±2.1) ($p<0.05$) (Table V).

Table V. Human Study

	Before (n=16)	After (n=16)	Change
Serum Cholesterol (mg/dL)	182±23	180±25	NS
TG (mg/dL)	111±26	108±22	NS
FFA (mM)	407±208	388±214	NS
LDL-Chol (mg/dL)	98±17	97±18	NS
HDL-Chol	49±10	50±8	NS
Lag phase LDL oxidation (min)	114±31	134±30	p<0.05
α-Tocopherol per LDL	8.7±1.6	10.1±2.1	p<0.05

NOTE: NS, not significant.

Discussion

Teas are generally categorized as fermented and non-fermented teas according to the extent of fermentation in the processing. Pu-Erh tea is unique since it is a naturally fermented tea following by prolonged storage. The ingredients in Pu-Erh tea may derive from tea leaves and the biotransformed products by the endogenous enzymes during the processing. Metabolites, either that derived from the fermentation microorganisms or the biotransformation products originally from tea leaves, may contribute to the biological activities. The fungus *Aspergillus* has been previously identified in Pu-Erh tea (20). We also found that *Monascus* species and the lovastatin-producing strain of *Aspergillus terreus* could grow on PET-supplemented culture medium and produce lovastatin (data not shown).

Previous studies from other laboratories have demonstrated that the hypocholesterolemic effect of tea is due to tea catechins by decreasing the micellar solubility and intestinal absorption of dietary cholesterol (8). EGCG is the major catechin in tea leaves to inhibit the absorption of exogenous cholesterol from digestive tract. Moderate enhancement in the elimination of endogenous cholesterol by tea catechins has been reported (9). A previous study has demonstrated that Pu-Erh tea reduces plasma cholesterol and triacylglycerol in female Wistar rats (*20*). Since the catechin content in Pu-Erh tea is significantly lower than that of green tea, the lipid-lowering effect in experimental animals can not be attributed to EGCG in Pu-Erh tea.

This study demonstrated that the Pu-Erh tea contained lovastatin in a low, but yet detectable amount. The aqueous extract of Pu-Erh tea (PET) inhibited cholesterol biosynthesis in cultured human hepatoma cells (Hep G2). PET did not affect post-mevalonate events in the cholesterol pathway, since the incorporation of labeled mevalonate into cholesterol was not affected. Direct evidence to support the occurrence of lovastatin in PET was based on extensive purification and identification of lovastatin in its lactone form by mass spectrometry. To enrich lovastatin, PET was solvent extracted to recover lovastatin in lactone form. The content of lovastatin in Pu-Erh tea varied greatly among different batches. The situation is not unexpected, since the preparation procedure of Pu-Erh tea involves natural fermentation. The growth of *Aspergillus* and production of lovastatin in Pu-Erh tea during the fermentation and storage is not under control.

Green and black teas have been reported to inhibit human LDL oxidation *in vitro* and *ex vivo* (*13,15-18*). However, at least one study has indicated that black tea is not effective to protect LDL from oxidative modification (*19*). Current study demonstrated that the aqueous extracts of Pu-Erh tea (PET) scavenged DPPH radical and inhibited Cu^{2+}-induced LDL oxidation *in vitro*. The antioxidant potency of PET was similar to that of green and black tea extracts but the content of EGCG in Pu-Erh tea was lower than that of the green tea. It is suggested that the antioxidant activity of PET is due to the oligomeric phenolic compounds and theaflavins.

Hamsters, fed with a high cholesterol diet, were chosen as the animal model in this study. Hamsters are considered as a better animal model than rats for the study of lipid and lipoprotein metabolism (*34,35*). Since lovastatin content varied and EGCG content was low in PET, they were not chosen as a positive control. We chose β-sitosterol, a cholesterol-lowering phytosterol affecting intestinal lipid absorption, as a positive control. Dietary phytosterols, including β-sitosterol and sitostanol, reciprocally influence cholesterol absorption and biosynthesis in hamsters (*36*). The dose of PET used in animal study is higher than that for human consumption. However, the dose range used in PET-1 and PET-2 groups are still reasonable. Since Pu-Erh tea is a fermented tea that develops a strong and distinct herbal taste, we decided not to add PET to the drinking water in the animal study. Treatment of PET in the diet to hamsters reduced serum total cholesterol, TG, and FFA. Fecal cholesterol secretion was also increased. We speculated that the cholesterol-lowering effect of PET is a compounding effect of lovastatin and inhibitor to reduce intestinal cholesterol absorption. The later is most likely the polyphenols including theaflavins in Pu-Erh tea.

The human study is not a double blind, placebo-controlled clinical trial. The end point is simply a surrogate marker, namely oxidative susceptibility of LDL *ex vivo*. As in green and black teas drinking (*18,19,27*), Pu-Erh tea has a mild but significant effect in protecting human LDL from oxidation. The α-tocopherol content in LDL was elevated. Pu-Erh tea did not contain α-tocopherol to a significant amount. It is concluded that Pu-Erh tea drinking may indirectly protect LDL from oxidative modification by scavenging free radicals in the circulation. Consequently, α-tocopherol in LDL is protected.

In conclusion, uptake of Pu-Erh tea lowers plasma cholesterol and triacylglycerol in cholesterol-fed hamsters. The lipid lowering effects of Pu-Erh tea is likely caused by a combination of lovastatin and tea polyphenols. As a naturally fermented tea with lower EGCG and other catechins, Pu-Erh tea still exhibits strong antioxidant activities to scavenge DPPH radical and inhibit LDL oxidation *in vitro* and *ex vivo*. This study suggests that Pu-Erh tea drinking may reduce the risk factors in atherosclerosis-related ischemic heart disease.

Acknowledgements

The authors thank Ms. H.-R. Huang for technical assistance. This study was supported by the National Science Council (M.-S. Shiao), the Department of Health (L.S. Hwang), and Veterans General Hospital-Taipei, Taiwan, ROC.

References

1. Cleeman, J. I. *J. Am. Med. Assoc.* **2001**, *285*, 2486-2497.
2. Glass, C. K.; Witztum, J. *Cell* **2001**, *104*, 503-516.
3. Savla, U. *Nature Medicine* **2002**, *8*, 1207-1262.
4. Steinberg, D. *Circulation* **1997**, *95*, 1062-1071.
5. Parthasarathy S.; Santanam, N.; Ramachandran, S.; Meilhac, O. *J. Lipid Res.* **1999**, *40*, 2143-2157.
6. Ninomiya, M.; Unten, L.; M. Kim, M. *Chemical and physicochemical properties of green tea polyphenols.* in Chemistry and Applications of Green Tea; Yamamoto, Y.; Juneja, L. J.; Chu, D. C.; Kim, M. Ed, CRC Press, Boca Raton, **1997**. pp 23-35.
7. Yang, C. S.; Maliakal, P.; Meng, X. *Annu. Rev. Pharmacol. Toxicol.* **2002**, *42*, 25-54.
8. Ikeda, I.; Imasato, Y.; Sasaki, S.; Nakayama, M.; Nagao, H.; Takeo, T.; Yayabe, F.; Sugano, M. *Biochim. Biophys. Acta* **1992**, *1127*, 141-146.
9. Chisaka, T.; Matsuda, H.; Kubomura, Y.; Mochizuki, M.; Yamahara, J.; Fujimura, H. *Chem. Pharm. Bull.* **1988**, *36*, 227-233.

10. Sano, M.; Takahashi, Y.; Yoshino, Y.; Shimoi, K.; Nakamura, Y.; Tomita, I.; Oguni, I.; Konomoto, H. *Biol. Pharm. Bull.* **1995**, *18*, 1006-1008.
11. Zhang, A.; Zhu, Q. Y.; Luk, Y. S.; Ho, K. Y.; Fung, K. P.; Chen, Z.-Y. *Life Sci.* **1997**, *61*, 383-394.
12. Grinberg, L. N.; Newmark, H.; Kitrossky, N.; Rahamim, E.; Chevion, M.; Rachmilewitz, E. A. *Biochem. Pharm.* **1997**, *54*, 973-978.
13. Tijburg, L. B. M.; Wiseman, S. A.; Meijer, G. W.; Weststrate, J. A. *Atherosclerosis* **1997**, *135*, 37-47.
14. Miura, S.; Watanabe, J.; Sano, M.; Tomita, T.; Osawa, T.; Hara, Y.; Tomita, I. *Biol. Pharm. Bull.* **1995**, *18*, 1-4.
15. Miura, S.; Watanabe, J.; Tomita, T.; Sano, M.; Tomita, I. *Biol. Pharm. Bull.* **1994**, *17*, 1567-1572.
16. Luo, M.; Kannar, K.; Wahlqvist, M. L.; O'Brien. R. C. *Lancet* **1997**, *349*, 360-361.
17. Ishikawa, T.; Suzukawa, M.; Ito, T.; Yoshida, H.; Ayaori, M.; Nishiwaki, M.; Yonemura, A.; Hara, Y.; Nakamura, H. *Am. J. Clin. Nutr.* **1997**, *66*, 261-266.
18. Serafini, M.; Ghiselli, A.; Ferro-Luzzi, A. *Eur. J. Clin. Nutr.* **1996**, *50*, 28-32.
19. MaAnlis, G. T.; McEneny, J.; Pearce, J.; Young, I. S. *Eur. J. Clin. Nutr.* **1998**, *52*, 202-206.
20. Sano, M.; Takenaka, Y.; Kojima, R.; Saito, S.-I.; Tomita, I.; Katou, M.; Shibuya, S. *Chem. Pharm. Bull.* **1986**, *34*, 221-228.
21. Cheng, H.-C.; Yang, C.-M.; Shiao, M.-S. *Hepatology* **1993**, *17*, 280-286.
22. Shen, P. M.; Shiao, M.-S.; Chung, H.-R.; Lee, K. R.; Chao, Y.-S.; Hunt, V. J. *Chin. Chem. Soc.* **1996**, *43*, 451-457.
23. Blois, M. S. *Nature* **1958**, 181, 1199-1200.
24. Regnstrom, J.; Nilsson, J.; Tornvall, P.; Landou, C.; Hamsten, A. *Lancet* **1992**, *339*, 1183-1186.
25. Wallin, B.; Rosengren, B.; Shertzer H. G.; Camejo, G. *Anal. Biochem.* **1993**, *208*, 10-15.
26. Geleijnse, J. M.; Launer, L. J.; Hofman, A.; Pols, H. A.; Witterman, J. C. M. *Arch. Intern. Med.* **1999**, *159*, 2170-2174.
27. Hodgson, J. M.; Puddey, I. B.; Croft, K. D.; Burke, V.; Mori, T. A.; Caccetta, R. A.-A.; Beilin, L. J. *Am. J. Clin. Nutr.* **2000**, *71*, 1103-1107.
28. Richmond, W. *Clin. Chem.* **1973**, *19*, 1350-1356.
29. Wu, Y-J.; Hong, C.-Y.; Lin, S.-J.; Wu, P.; Shiao, M.-S. *Arterioscler. Thromb. Vasc. Biol.* **1998**, *18*, 481-486.
30. Demacker, P. N. M.; Hijimans, A. G. M.; Jansen, A. P. *Clin. Chem.* **1982**, *28*, 1765-1768.
31. Scheek, L. M.; Wiseman, S. A.; Tijburg, L. B. M.; Tol, A. V. *Atherosclerosis* **1995**, *117*, 139-144.

32. Abell, L. L.; Levy, B. B.; Kendall, F. E. I. *J. Biol. Chem.* **1952**, *195*, 357-366.
33. Bui, M. H. *J. Chromatogr.* **1994**, *645*, 129-133.
34. Kris-Etherton, P. M.; Dietschy, J. *Am. J. Clin. Nutr.* **1997**, *65*(suppl), 1590S-1596S.
35. Harris, W. S. *Am. J. Clin. Nutr.* **1997**, *65*(suppl), 1611S-1616S.
36. Ntanios, F. Y.; Jones, P. J. H. *Atherosclerosis* **1999**, *143*, 341-351.

Chapter 6

Protective Effects of Baicalein and Wogonin against Mutagen-Induced Genotoxicities

Yune-Fang Ueng[1], Chi-Chuo Shyu[2], Tsung-Yun Liu[2,3], Yoshimitsu Oda[4], Sang Shin Park[5], Yun-Lian Lin[1], and Chieh-Fu Chen[1,2]

[1]National Research Institute of Chinese Medicine, [2]Institute of Pharmacology, National Yang-Ming University, Taipei, Taiwan, Republic of China
[3]Department of Medical Research, Veterans General Hospital at Taipei, Taipei, Taiwan, Republic of China
[4]Osaka Prefectural Institute of Public Health, Osaka, Japan
[5]Ilchun Molecular Medicine Institute, Medical Research Center, Seoul National University, Seoul, Korea

The glucuronides of baicalein and wogonin are the main flavonoids of a Chinese herbal medicine *Scutellaria baicalensis*. Baicalein and wogonin reduced benzo(a)pyrene and aflatoxin B_1 (AFB_1) genotoxicities as monitored by *umu*C gene expression response in *Salmonella typhimurium* TA1535/pSK1002. *In vitro*, baicalein and wogonin inhibited benzo(a)pyrene hydroxylation (AHH) and AFB_1 oxidation (AFO) activities of mouse liver microsomes. One-week dietary treatment of baicalein and wogonin significantly decreased hepatic oxidations of benzo(a)pyrene and AFB_1. In benzo(a)pyrene-treated mice, hepatic DNA adduct formation was reduced by pre-treatment of wogonin. These *in vitro* and *in vivo* effects indicated that baicalein and wogonin might have beneficial effects against benzo(a)pyrene- and AFB_1-induced toxicities. However, the dietary treatments of baicalein and wogonin resulted in significant changes of cytochrome P450 1A2, 2E1, and 3A and UDP-glucuronosyltransferase activities in mouse liver. Thus, caution should be paid to the possible adverse effects resulted from interactions of baicalein/ wogonin and drugs mainly metabolized by these enzymes while the beneficial chemoprevention of flavonoids is noticed.

Introduction

Phase I and phase II drug-metabolizing enzymes play pivotal roles in the determination of biological fates of xenobiotics (*1,2*). Cytochrome P450 (CYP)-dependent monooxygenase is one of the major phase I enzymes catalyzing various oxidative and reductive metabolism. Oxidations catalyzed by microsomal monooxygenase system require a CYP, NADPH-CYP reductase, and phospholipids. Benzo(a)pyrene and aflatoxin (AF)B$_1$ are important environmental pollutants and their tumorigenic effects require the activation by CYPs (*3,4*). Benzo(a)pyrene, a polycyclic aromatic hydrocarbon mainly present in industrial and automobile emission, cigarette smoke, and charred food. AFB$_1$, a mycotoxin produced by *Asparagillus flavus*, is identified as a potent hepatocarcinogen in experimental animals and probably in humans (*5*). CYP-catalyzed oxidation occupies a key step in the formation of epoxide metabolites, benzo(a)pyrene 7,8-diol-9,10-epoxide and AFB$_1$ *exo*-epoxide, which have been known for their ability of forming DNA-adduct and subsequent tumorigenesis (*6,7*). Thus, suppression of the CYP-mediated activation may provide beneficial effect of reducing the risk of DNA-adduct formation.

UDP-glucuronosyltransferase (UGT) and glutathione S-transferase (GST) are the major phase II enzymes involved in the conjugation metabolism of xenobiotics. UGT and GST catalyze the formation of glucuronides and glutathione conjugates, respectively. In general, conjugation reactions convert xenobiotics to water soluble metabolites to enhance the excretion of xenobiotics. However, glucuronides of drugs can accumulate through enterohepatic circulation during long term therapy and may cause toxicity (*8*). CYP, UGT, and GST are responsive to the inductive and inhibitory effects of xenobiotics including natural products (*9*). Modulation of drug-metabolizing enzymes may change the pharmacological and toxicological effects of xenobiotics in humans and result in serious drug interactions.

Scutellariae Radix has been commonly used in traditional Chinese medicine for the treatment of inflammation and allergic disease. Baicalein and wogonin are the main active flavonoids of Scutellariae Radix and mainly present as glucuronides (*10,11*). After digestion, the glucuronides are readily hydrolyzed by intestinal bacteria (*12*). Several reports indicated the beneficial effects of extracts of Scutellariae Radix. The water extract of *Scutellariae Radix* suppressed the mutagenic activity of benzo(a)pyrene in *Salmonella typhimurium* TA98 and TA100 (*13*). The water extract of *Scutellariae baicalensis* also decreased the DNA-binding and metabolism of benzo(a)pyrene and AFB$_1$ activated by Aroclor

1254-induced rat hepatic S9 (*14*). Baicalein and wogonin reduced liver damage induced by acetaminophen, CCl₄ and β-galactosamine in rats (*10*). Baicalein inhibited 12-*O*-tetradecanoylphorbol-13-acetate-caused tumor promotion in benzo(a)pyrene-initiated CD-1 mouse skin (*15*). However, incubation of *S. typhimurium* TA100 with both baicalein and 2-amino-3-methylimidazo[4,5-f]quinoline (IQ) (or 2-(2-furyl)-3-(5-nitro-2-furyl) acrylamide (AF-2)) showed toxicity in Ames test (*16,17*). Elliger *et al.* (*18*) reported that wogonin showed mutagenicity in Ames test in the presence of NADPH and liver S9 from Aroclor1254-treated rats. This mutagenic effect was not diminished when microsomes were removed from the S9 by centrifugation. Thus, we have investigated the genotoxicities, protective effects against mutagens and interaction with drug-metabolizing enzymes of baicalein and wogonin.

Reduction of Genotoxicities and Oxidations of Benzo(a)pyrene and Aflatoxin B₁ by Baicalein and Wogonin *In Vitro*

Lee *et al.* (*15*) reported that topical treatment of mouse skin with baicalein dramatically decreased the skin tumor number in benzo(a)pyrene-initiated mouse skin. Lee *et al.* (*19*) reported that baicalein suppressed the mutagenicity of AFB₁ in *S. typhimurium* and reduced chromosome aberration induced by AFB₁ in CHL (Chinese hamster, lung) cells. These reports indicated the chemopreventive role of baicalein against carcinogenecity of benzo(a)pyrene and AFB₁. Our results showed that addition of baicalein and wogonin reduced the genotoxicities of benzo(a)pyrene and AFB₁ as monitored by the *umu*C expression response in *S. typhimurium* TA1535/pSK1002 (Table I) (*20*). Baicalein and wogonin at the concentration of 100 μM or less had no apparent genotoxicities in the presence of untreated mouse liver microsomes (Table I). There were also no genotoxicities detected in the presence of cytosol or a mixture of cytosol and microsomes. In this test system, benzo(a)pyrene (100 μM) showed mild genotoxicity and AFB₁ (5 μM) showed relatively strong genotoxicity. The expression unit of β-galactosidase induced by AFB₁ was 11 fold higher than the control value. In the presence of 50 μM baicalein and wogonin, benzo(a)pyrene genotoxicity was suppressed to a level close to the control value obtained from the incubation in the absence of benzo(a)pyrene (Table I). Baicalein at 10 μM significantly decreased the genotoxicity of AFB₁ and baicalein at 100 μM decreased this toxicity to a level approaching control value. Wogonin also decreased the genotoxicity of AFB₁, but showed less suppressive effects than baicalein.

Table I. Inhibition of Benzo(a)pyrene- and Aflatoxin B₁-induced
Genotoxicities by Baicalein and Wogonin

Mutagen	Flavonoid	umu response, units/mL
None	none	100.5 ± 9.3
	baicalein 100 μM	131.2 ± 6.7
	wogonin 100 μM	115.1 ± 1.6
Benzo(a)pyrene, 100 μM	none	231.1 ± 23.3
	baicalein 10 μM	175.3 ± 18.8
	baicalein 50 μM	146.3 ± 15.7 *
	baicalein 100 μM	148.3 ± 6.8 *
	wogonin 10 μM	172.9 ± 15.6
	wogonin 50 μM	154.4 ± 9.4 *
	wogonin 100 μM	157.2 ± 9.4 *
Aflatoxin B₁, 5 μM	none	1156 ± 182.6
	baicalein 10 μM	679.9 ± 86.4 *
	baicalein 50 μM	330.1 ± 23.0 *
	baicalein 100 μM	305.4 ± 22.5 *
	wogonin 10 μM	887.8 ± 165.5
	wogonin 50 μM	866.4 ± 61.4 *
	wogonin 100 μM	859.6 ± 9.7*

Genotoxicities were determined in *Salmonella typhimurium* TA1535/pSK1002 as described before (20). Induction of *umu* gene expression is presented as units of β-galactosidase activity standardized by bacterial growth. Results represent mean ± SEM of four separate experiments with duplicates. *Asterisks represent values significantly different from the values of mutagen only, $p < 0.05$.

Oxidations of benzo(a)pyrene and AFB₁ were determined using untreated mouse liver microsomes. At 100 μM benzo(a)pyrene, baicalein inhibited AHH activity with an IC₅₀ of 34 ± 1 μM. At 50 μM AFB₁, the IC₅₀ values of baicalein for the formation of AFQ₁ and AFB₁-epoxide were 23 ± 1 and 5 ± 1 μM, respectively (Table II) (20). Inhibition of AHH activity by 500 μM wogonin was 51%. In contrast, the inhibition by wogonin up to 500 μM was not sufficient for the IC₅₀ estimation. Inhibition of aflatoxin B₁ oxidation to form AFQ₁ and AFB₁-epoxide by 500 μM wogonin were 37% and 51%, respectively. These results indicated that both baicalein and wogonin inhibited microsomal benzo(a)pyrene hydroxylation (AHH) and aflatoxin B_1 oxidation (AFO) activities *in vitro*. Baicalein had stronger inhibitory effects than wogonin (Table II).

Table II. Inhibition of Benzo(a)pyrene Hydroxylation and Aflatoxin B_1 Oxidation Activities by Baicalein and Wogonin

Flavonoid	Benzo(a)pyrene hydroxylation	Aflatoxin B_1 oxidation	
		AFQ_1	AFB_1-epoxide
Baicalein	34 ± 1	23 ± 1	5 ± 1
Wogonin	351 ± 95	-[a]	-[a]

IC_{50}, μM

[a] The inhibition by wogonin up to 500 μM was not sufficient for the IC_{50} estimation.

Dietary Effects of Baicalein and Wogonin on Benzo(a)pyrene-DNA Adduct Formation and Oxidations of Benzo(a)pyrene and Aflatoxin B_1 in Mouse Liver

Mice were treated with a liquid diet containing 5 mM baicalein or wogonin for one week and were then treated with 200 mg/kg benzo(a)pyrene. Control group received a control diet and then was treated with benzo(a)pyrene. One-week pretreatment of mice with wogonin significantly decreased hepatic benzo(a)pyrene-DNA adduct level to 24% of the control as analyzed by ^{32}P-postlabeling analysis (control: 8.8 ± 0.9; baicalein: 12.2 ± 1.5; wogonin: 2.1 ± 0.3 nmol ^{2}N-guanyl adducts/10^7 nucleotides). However, baicalein-treatment had no effect on the benzo(a)pyrene-DNA adduct level. Consistent with the inhibitory effects *in vitro*, our *in vivo* study showed that baicalein- and wogonin-treatments significantly decreased AHH and AFO activities in mouse liver. Treatments of mice with baicalein and wogonin resulted in 29% and 43% decreases of hepatic AHH activities, respectively (Figure 1). Baicalein-treatment resulted in 39% and 32% decreases of AFQ_1 and AFB_1-epoxide formation, respectively. Wogonin-treatment resulted in 39% and 47% decreases of AFQ_1 and AFB_1-epoxide formation, respectively. Different from the stronger inhibition by baicalein *in vitro*, baicalein caused less inhibitory effect than wogonin on AHH activity *in vivo*. The actual reason for this discrepancy between *in vitro* and *in vivo* effects is not clear. Influence of pharmacokinetic parameters, pharmacodynamic effects, and other regulatory factors *in vivo* are possible causes of this discrepancy.

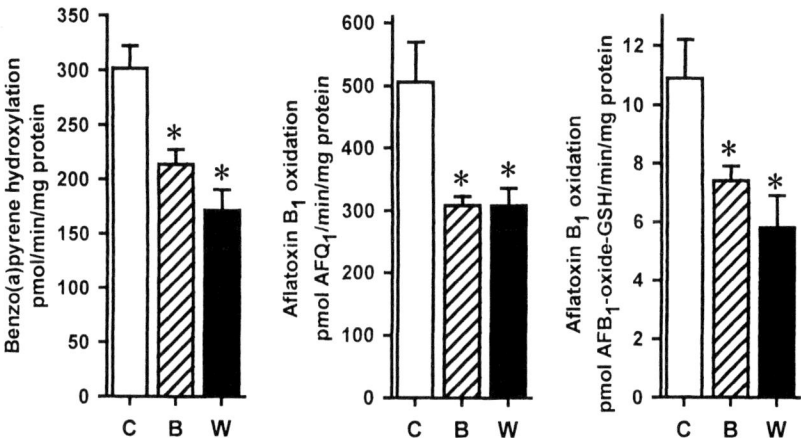

Figure 1. Effects of baicalein (B) and wogonin (W) on AHH and AFB_1 oxidation activities in mouse liver. Monooxygenase activities were determined after one-week treatment of liquid diets containing 5 mM flavonoids. Control (C) mice received a control diet. Formation of AFB_1-epoxide was determined as the glutathione-conjugate using rat glutathione S-transferase. Data represent the mean ±SEM of six mice. *Values significantly different from the control values, $p < 0.05$.

Low bioavailability is the general consideration for application of flavonoids. However, the concentrations of flavonoids *in vivo* was enough to evoke the pharmacological effect (21). In general, the aglycone of flavonoids was thought as their biological active form. After digestion, baicalein and wogonin were metabolized to form glucuronide and sulfate conjugates in rats and humans (22,23). Our results showed that there were no free baicalein and wogonin detected in sera from flavonoid-treated mice. After treatments of sera with glucuronidase/sulfatase, there were 45 ± 4 μM and 23 ± 7 μM of baicalein and wogonin in baicalein- and wogonin-treated groups, respectively. The serum concentrations of the conjugates of baicalein was one fold higher than the conjugates of wogonin. However, baicalein caused less inhibition of AHH activity than wogonin. Baicalein- and wogonin-treatments suppressed hepatic

AFO activity with similar potency. The potency of inhibition by dietary treatments of flavonoids was not correlated with the serum concentrations of metabolites.

Our *in vitro* and *in vivo* results together suggested that administrations of baicalein and wogonin reduced the carcinogenic risk caused by benzo(a)pyrene and AFB_1. Although the direct extrapolation from mouse study to human is difficult, our results together with previous reports indicate the contribution of baicalein and wogonin in the liver protective effects of *S. baicalensis*. One-week treatment of mice with a diet containing baicalein or wogonin had no effects on body and liver weights (24). We have also determined mouse serum alanine aminotransferase activity and there was no difference between control and flavonoid-treated groups. These results indicated that baicalein and wogonin were not hepatotoxic at the dosage for protection against benzo(a)pyrene and AFB_1 toxicities.

Dietary Effects of Baicalein and Wogonin on CYP, UGT, and GST in Mouse Liver

To have a better understanding of the effects of flavonoids on drug-metabolizing enzymes, mice were treated with liquid diets containing baicalein or wogonin. Treatments of mice with 5 mM or 7.5 mM baicalein or wogonin resulted in significant decreases of AHH activities (24). Thus, mice were fed with a diet containing 5 mM baicalein or wogonin in the following studies. Liver microsomal CYP and cytochrome b_5 contents and NADPH-CYP reductase activity were not affected by baicalein and wogonin (24). Baicalein-treatment resulted in 14% to 46% decreases of hepatic benzphetamine *N*-demethylation (BDM), *N*-nitrosodimethylamine *N*-demethylation (NDM), nifedipine oxidation (NFO), and erythromycin *N*-demethylation (EMDM) activities. Wogonin-treatment resulted in 21% to 35% decreases of hepatic BDM, NDM, NFO, and EMDM activities. However, baicalein and wogonin showed different modulation of CYP1A2-catalyzed activities of 7-methoxyresorufin *O*-demethylation (MROD) and caffeine 3-demethylation (CDM). Baicalein-treatment caused a 31% increase of MROD activity in mouse liver. In contrast, wogonin-treatment caused a 55% decrease of MROD activity. Consensus with the MROD determination, baicalein caused a slight increase of CDM activity, whereas wogonin-treatment caused a significant decrease of CDM activity. There were no significant changes in the hepatic EROD activity by baicalein- and wogonin-treatments. Yokoi *et al* (25) reported that addition of baicalein resulted in a decrease of UGT activity toward SN-38, the active metabolite of an anti-cancer drug CPT-11. The decrease of SN-38-glucuronide formation may prevent the enterohepatic circulation of SN-38 and protect against intestinal toxicity of CPT-

11 (*26*). Our results showed that dietary treatments of baicalein and wogonin also decreased UGT activities (*24*). In contrast, cytosolic GST was not affected by both flavonoids. Consistent with results of activity determination, immunoblot analysis of microsomal proteins showed that treatments of baicalein and wogonin increased and decreased the level of a CYP1A-immunoreactive protein, respectively (*24*). The mobility of this protein band was similar to that of CYP1A2 inducible by 3-MC. Baicalein- and wogonin-treatments decreased the levels of CYP2E1- and CYP3A-immunoreactive proteins. In contrast, microsomal CYP2B-immunoreactive protein was not affected by flavonoids.

Hepatic CYP 1A2 was induced and inhibited by baicalein and wogonin, respectively. However, catalytic activities of CYP2B (BDM), CYP2E1 (NDM), and CYP3A (NFO and EMDM) were generally decreased by both flavonoids. Since the modulation was not restricted to one CYP form, a general mechanism but not a gene specific modulation mechanism was likely contributing to this inhibition. Thus, regulation of CYP enzymes by flavonoids may involve a specific receptor-mediated pathway as described above for CYP1A and a general mechanism for other CYPs. The flavonoid antioxidant has been reported to affect cytokines and protein kinase (*27*). The influence of cellular redox potential, cytokines, and protein phosphorylation can be responsible for the modulation of CYPs by flavonoids. In addition to these pathways, direct interactions of CYP-catalyzed oxidations and a cross talk between regulatory factors may also be involved in the multiple effects of flavonoids. Further studies are required to elucidate the modulatory mechanism of flavonoids. Alteration of drug-metabolizing activities may cause serious drug interactions in a patient under herbal medicine and other drug treatments. Our results suggested that ingestion of baicalein and wogonin affected the activities of phase I and phase II drug-metabolizing enzymes (*24*). Thus, caution should be paid to possible drug interactions of patients concomitantly treated with natural products mainly containing baicalein or wogonin and other drugs metabolized by CYP and UGT enzymes.

Acknowledgements

This work was supported by the National Research Institute of Chinese Medicine and grant NSC 88-2314-B077-010 from the National Science Council, R.O.C.

References

1. Guengerich, F. P. *Toxicol. Lett.* **1994**, *70*, 133-138.
2. Nebert, D.W. *Crit. Rev. Toxicol.* **1989**, *20*, 153-174.
3. Bauer, E.; Guo, Z.; Ueng, Y. F.; Bell, L. C.; Zeldin, D.; Guengerich, F.P. *Chem. Res. Toxicol.* **1995**, *8*, 136-142.

4. Ueng, Y.F.; Shimada, T.; Yamazaki, H.; Guengerich, F.P. *Chem. Res. Toxicol.* **1995**, *8*, 218-224.
5. Qian, G.S.; Ross, R.K.; Yu, M.C.; Yuan, J.M.; Gao, Y.T.; Henderson, B.E.; Groopman, J.D. *Cancer Epidemiol. Biomarkers Prev.* **1994**, *3*, 3-10.
6. Conney, A. H.; Chang, R. L.; Jerina, D. M.; Wei, S. J. *Drug Metab. Dispos.* **1994**, *26*, 125-163.
7. Bechtel, D. H. *Regul. Toxicol. Pharmacol.* **1989**, *10*, 74-81.
8. Sperker,B.; Backman, J.T.; Kroemer, H.K. *Clin. Pharmacokinet.* **1997**, *33*, 18-31.
9. Nebert, D.W. *Biochem. Pharmacol.* **1994**, *47*, 25-37
10. Lin, C.C.; Shieh, D.E. *Phytotherapy Res.* **1996**, *10*, 651-654.
11. Michinori, K.; Matsuda, H.; Tani, T.; Arichi, S.; Kimura, Y.; Okuda, H. *Chem. Pharm. Bull.* **1985**, *33*, 2411-2415.
12. Manach, C.; Regerat, F.; Texier, O.; Agullo, G.; Demigne, C.; Remesy, C. *Nutr. Res.* **1996**, *16*, 517-544.
13. Sakai, Y.; Nagase, H.; Ose, Y.; Sato, T.; Kawai, M.; Mizuno, M. *Mutat. Res.* **1988**, *206*, 327-334.
14. Wong, B.Y.Y.; Lau, B.H.S.; Yamazaki, T.; Teel, R.W. *Cancer Lett.* **1993**, *68*, 75-82.
15. Lee, M.J.; Wang, C.J.; Tsai, Y.Y.; Hwang, J.M.; Lin, W.L.; Tseng, T.H.; Chu, C.Y. *Nutr. Cancer* **1999**, *34*, 185-191.
16. Edenharder, R.; von Petersdorff, I.; Rauscher, R. *Mutation Res.* **1993**, *287*, 261-274.
17. Ohtsuka, M., Fukuda, K., Yano, H., and Kojiro, M. *Jpn. J. Cancer Res.* **1995**, *86*, 1131-1135.
18. Elliger, C. A.; Henika, P. R.; MacGergor, J. T. *Mutation Res.* **1984**, *135*, 77-86.
19. Lee, B.H.; Lee, S.J.; Kang, T.H.; Kim, D.H.; Sohn, D.W.; Ko, G.I.; Kim, Y.C. *Planta Med.* **2000**, *66*, 70-71.
20. Ueng, Y.F.; Shyu, C.C.; Liu, T. Y.; Oda, Y.; Lin, Y. L.; Liao, J.F.; Chen, C.F. *Biochem.Pharmacol.* **2001**, *62*, 1653-1660.
21. Hollman, P. C. H.; Katan, M. B. *Food Chem. Toxicol.* **1999**, *37*, 937-942.
22. Li, C.; Homma, M.; Oka, K. *Biol. Pharm. Bull.* **1998**, *21*, 1251-1257.
23. Abe, K. i.; Inoue, O.; Yumioka, E. *Chem. Pharm. Bull.* **1990**, *38*, 208-211.
24. Ueng, Y.F.; Shyu, C.C.; Lin, Y.L.; Park, S.S.; Liao, J.F.; Chen, C.F. *Life Sci.* **2000**, *67*, 2189-200.
25. Yokoi, T.; Narita, M.; Nagai, E.; Hagiwara, H.; Aburada, M.; Kamataki, T. *Jpn. J. Cancer Res.* **1995**, *86*, 985-989.
26. Takasuna, K.; Kasai, Y.; Kitano, Y.; Mori, K.; Kobayashi, R.; Hagiwara, T.; Takihata, K.; Hirohashi, M.; Nomura, M.; Nagai, E.; Kamataki, T. *Jpn. J. Cancer Res.* **1995**, *86*, 978-984.
27. Gelboin, H. V. *Pharmacol. Rev.* **1993**, *45*, 413-453.

Chapter 7

Biological Activities of Flavonoids Isolated from Chinese Herb Huang Qui: Inhibition of NO and PGE$_2$ Production by Flavonoids

Yen-Chou Chen[1], Shing-Chuan Shen[2], and Foun-Lin Hsu[1]

[1]Graduate Institute of Pharmacognosy Science, School of Pharmacy, and [2]Department of Dermatology, School of Medicine, Taipei Medical University, Taipei, Taiwan, Republic of China

Huang Qui is one of the popular Chinese herbs, and has been used in treatment of several human diseases such as inflammation, allergy and artherosclosis for thousands of years. However the active components of Huang Qui are still undefined. Our recent studies demonstrated that flavonoids in Huang Qui including wogonin, quercetin, and oroxylin A showed the significant inhibition on lipopolysaccharide (LPS)-induced nitric oxide (NO) and prostaglandin E$_2$ (PGE$_2$) production, accompanied by inhibiting inducible nitric oxide synthase (iNOS) and cyclooxygenase 2 (COX-2) gene expression. The inhibitory mechanism of these compounds on LPS-induced responses was through inhibiting NF-kB activation. *In vivo* study showed that wogonin and quercetin were able to suppress LPS-induced NO production in the serum of Balb/c mice. In addition to NO inhibition, wogonin showed the apoptotic effect on human promyeloleukemia cells HL-60 and hepatocellular carcinoma cells SK-HEP-1 cells through activation of caspase 3-dependent cascade, and oroxylin A exhibited the significant relaxative effect in porcine cerebral arteries pre-constricted by U-46619 through activation of potassium channels. Results of our studies demonstrate that wogonin, quercetin, and oroxylin A are active components of Huang Qui and deserve several beneficial biological activities to be explored further.

© 2003 American Chemical Society

Medicinal plants have been used as traditional remedies for hundreds of years. *Scutellaria baicalensis* Georgi (Huang Qui) is one of the important medicinal herbs widely used for the treatment of various inflammatory diseases, hepatitis, tumors and diarrhea in East Asian countries such as China, Korea, Taiwan and Japan (*1*). The plant has been reported to contain a large number of flavonoids, frequently found as the glucosides and other constituents including phenethyl alcohols, sterols and essential oils and amino acids. Flavonoids have been identified as either simple or complex glycosides in many plants, and humans have been estimated to consume approximately 1 g flavonoids/day (*2*). Several previous studies (*3,4*) have demonstrated that flavonoids exhibit a wide variety of biological activities including antioxidant, free radical scavenging, anti-cancer and anti-inflammatory activities.

Although Huang Qui has been used extensively for a long time in the Chinese society, the active compounds are still undefined. Wogonin, oroxylin A and quercetin are components of Huang Qui, and seldom biological functions are identified. Quercetin is a prototypical polyphenolic plant flavonoid, and can be derived from rutin though hydrolization by glucosidase. Quercetin was a potent antioxidant and anti-inflammatory agent, prevented cisplatin-induced cytotoxicity in LLC-PK1 cells *in vitro*, and prevented tubular injury induced by acute renal ischemia *in vivo* (*5*). Rangan et al. reported that quercetin inhibited lipopolysaccharide-induced cytokines such as interleukin-1β, tumor necrosis factor-α productions though blocking nuclear factor-kappa B activation (*6*). In addition to beneficial effects, quercetin also has been implicated as a strong mutagen without microsomal activation, and the mutagenic activity of quercetin was increased significantly after microsomal activation (*7*). In contrast to quercetin, biological activities of wogonin and oroxylin A remained unclear.

Inhibition of Nitric Oxide and PGE$_2$ Production by Wogonin, Oroxylin A and Quercetin

The chemical structures of wogonin, quercetin, and oroxylin A are shown in Figure 1. In our study, effects of wogonin, oroxylin A, and quercetin on LPS-induced NO and PGE$_2$ production in RAW 264.7 macrophages were investigated. Nitrite accumulated in the culture medium was estimated by the Griess reaction as an index for NO synthesis from the cells. When the cells incubated with wogonin, quercetin, and oroxylin A alone, the amount of nitrite in medium was maintained at a background level similar to that in the unstimulated samples (data not shown). After treatment with LPS (100 ng/mL) for 12 hours, nitrite concentration in medium increases remarkable about 10 fold (~30 μM). When RAW 264.7 macrophages were treated with different concentration of wogonin, quercetin, and oroxylin A together with LPS (100 ng/mL) for 12 hours, significant concentration dependent inhibition of nitrite production was detected in the presence of wogonin, quercetin, and oroxylin A (Figure 2). Similarly,

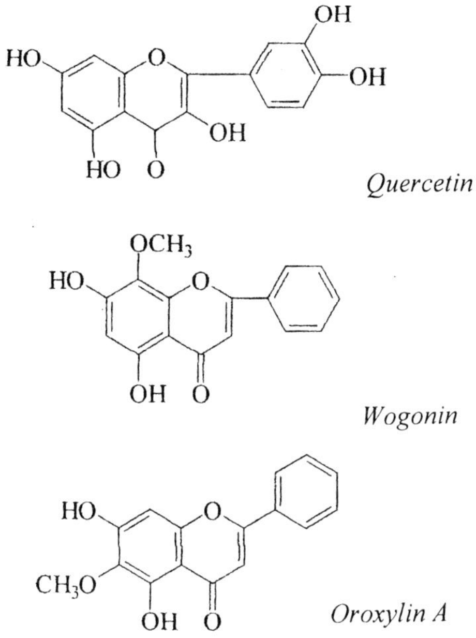

Figure 1. Chemical structure of quercetin, wogonin and oroxylin A.

wogonin, quercetin, and oroxylin A showed the inhibition on LPS-induced PGE$_2$ production (Figure 2). Examination of cytotoxicity of wogonin, quercetin, and oroxylin A in RAW 264.7 macrophages by MTT assay indicated that three compounds, even at the concentration of 40 μM, did not decrease cell viability in RAW 264.7 cells (data not shown). Therefore, inhibition of LPS-induced NO and PGE$_2$ production by wogonin, quercetin, and oroxylin A was not result of their cytotoxicity on cells.

Inhibition of iNOS and COX-2 Gene Expression, but not iNOS and COX-2 Activities, by Wogonin, Oroxylin A and Quercetin

Two possibilities of NO and PGE$_2$ inhibition by wogonin, quercetin and oroxylin A, one is inhibiting iNOS and COX-2 gene expression, the other is blocking iNOS and COX-2 activity. By Western blotting analysis, wogonin, quercetin and oroxylin A inhibited LPS-induced iNOS and COX-2 protein expression in cells (Figure 3). However, iNOS and COX-2 activities induced by LPS did not alter by these compounds by direct and indirect activity assays (8). It is suggested that inhibition of NO and PGE$_2$ production by wogonin, quercetin, and oroxylin A was mediated by inhibiting iNOS and COX-2 gene expression, but not their activities.

Induction of Apoptosis by Wogonin in Human Promyeloleukemia Cells and Hepatocellular Carcinoma Cells

Apoptosis is a programmed cell death, and cells from a variety of human malignancies have a decreased ability to undergo apoptosis in response to apoptotic stimuli (9-11). Therefore, developing various kinds of effective agents that can enhance the extent of apoptosis might be a promising strategy in the treatment of cancer. Apoptosis is characterized by cellular morphological change,

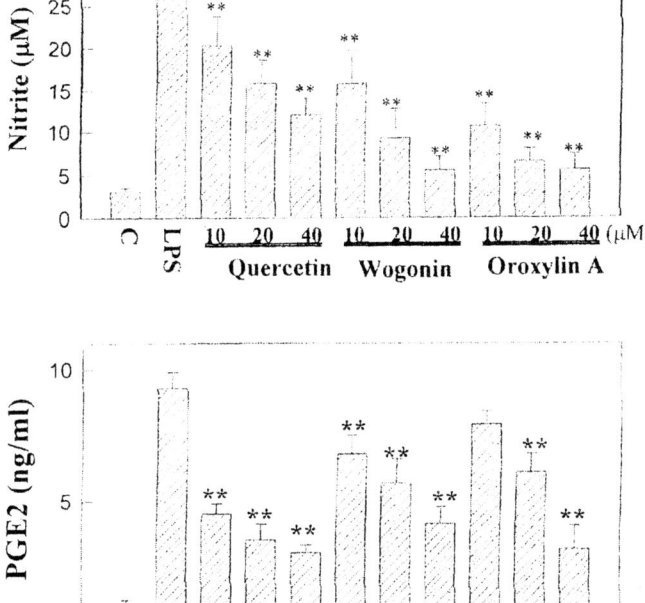

Figure 2. Inhibition of LPS-induced NO and PGE2 production by quercetin, wogonin and oroxylin A in RAW264.7 macrophages.

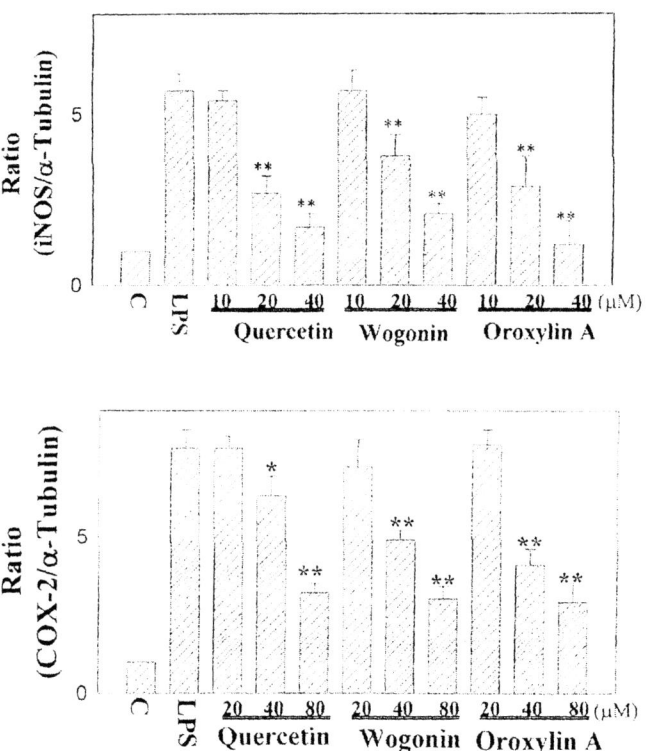

Figure 3. Inhibition of LPS-induced iNOS and COX-2 protein by quercetin, wogonin and oroxylin A in RAW264.7 macrophages.

chromatin condensation, and apoptotic bodies which are associated with DNA cleavage into ladders (*12,13*). Our recent study demonstrated that wogonin showed the effective apoptosis induction activity in human promyeloleukemia cells HL-60, and wogonin-induced apoptosis is caspase 3-dependent (*14*). P53 is a tumor suppressor gene and play an important role in the progression of apoptosis (*15*). In order to identify if p53 involved in wogonin-induced apoptosis, human hepatocellular carcinoma cells, which with wild–type p53 gene, were used in our study. Interestingly, wogonin induced apoptosis in SK-HEP-1 cells, accompanied by an increase in p53 protein and p53-controlled gene p21. Activation of caspase 3, but not caspase 1, was involved in wogonin-induced apoptosis in SK-HEP-1 cells (*16*). These data suggested that wogonin is an effective apoptosis inducer and deserves to be developed as an anti-cancer agent.

Induction of Vasorelaxation by Oroxylin A in Porcine Cerebral Arteries

In additional to NO and PGE_2 inhibition by oroxylin A, oroxylin A also showed the vasorelaxative effect in porcine cerebral arteries. As shown in Figure 4, addition of oroxylin A (0.1-30 μM) to the incubation medium resulted in concentration-dependent relaxation in intact arteries of circle of willis precontracted by U46619 (1 μM). Oroxylin A was able to induce maximal relaxation, and the medium effective concentration (EC50) was 7.1±2.4 μM. DMSO, a dissolution medium of oroxylin A, at the concentrations used (as much as 0.1 % vol/vol) had no effects on vasoreactivity precontracted by U46619 (data not shown). In endothelium-denuded arteries, oroxylin A show the same relaxation effects as that in intact arteries and EC50 was 5.87±3.1 μM. In order to study the cytotoxic effect of oroxylin A on arteries, two consecutive experiments were performed in same arteries with 60 min interval and 3 washes between two applications. The reproducible relaxations were obtained and they were not significant different. These results show that the relaxation elicited by oroxylin A are concentration-dependent and endothelium-independent, and oroxylin A did not have any cytotoxic or desensitive effect on arteries. Furthermore, vasorelaxative effect of oroxylin A was not appeared in KCl-preconstricted arteries, and blocked by potassium channel blockers tetraethylammonium (TEA), 4-aminopyridine (4-AP) and iberiotoxin (IBT). It is suggested that oroxylin A-induced vasorelaxation might be mediated by activation of potassium channels. (Figure 4)

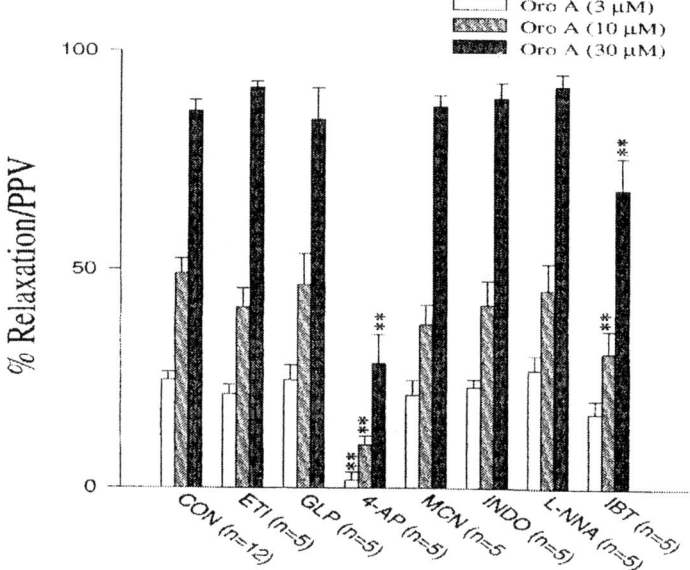

Figure 4. Attenuation of oroxylin A-induced Vasorelaxation by potassium channel blockers 4-AP and IBT. Eeicosatriyonic acid (ETI, 10 μM), 4-aminopyridine (4-AP, 10 mM), glipizide (GLP, 30 μM), miconazole (MCN, 5 μM), indomethacine (INDO, 30 μM), N-nitro-L-arginine (L-NNA, 60 μM), iberiotoxin (IBT, 100 ng/mL).

Conclusion

Results of our studies demonstrated that wogonin, quercetin and oroxylin A showed the potent inhibitory activities on LPS-induced responses including NO and PGE_2 production in RAW264.7, primary macrophages, and Balb/c mice. In addition to anti-inflammatory effects, wogonin showed the effective apoptosis inducing activity in tumor cells including human promyeloleukemia cells HL-60 and hepatocellular carcinoma cells SK-HEP-1 through a caspase 3-dependent pathway. And, oroxylin A exhibited vasorelaxative activity in cerebral arteries. These data provide scientific evidences to demonstrate that wogonin, quercetin and oroxylin A are functional compounds of Chinese herbs Huang Qui with a potential to be developed further.

Acknowledgement

This study was supported by the National Science Council of Taiwan (NSC 89-2314-B-038-035, NSC89-2320-B-038-054 and NSC89-2320-B-038-041).

References

1. Kubo, M.; Asano, T.; Shiomoto, H.; Matsuda, H. *Biol. Pharm. Bull.* **1994**, *17*, 1282-1286.
2. Manzanas, L.; Jesus del Nozal, M.; Marcos, M. A.; Cordero, Y.; Bernal, J. L.; Goldschmidt, P.; Pastor, J. C. *Exp. Eye Res.* **2002**, *74*, 23-28.
3. Chen, Y. C.; Shen, S. C.; Chen, L. G.; Lee, T. J. F.; Yang, L. L. *Biochem. Pharmacol.* **2001**, *61*, 1417-1427.
4. Shen, S. C.; Lee, W. R.; Lin, H. Y.; Huang H. C.; Ko, C. H.; Yang, L. L.; Chen, Y. C. *Eur. J. Pharmacol.* **2002**, *446*, 187-194.
5. Kuhlmann, M. K.; Burkhardt, G.; Horsch, E.; Wagner, M.; Kohler, H. *Free Rad. Res.* **1998**, *29*, 451-60.
6. Sato, M.; Miyazaki, T.; Kambe, F.; Maeda, K.; Seo, H. J. *Rheumatol.* **1997**, *24*, 1680-1684.
7. Bjeldanes, L. F. *Science* **1977**, *197*, 577-578.
8. Chen, Y. C.; Shen, S. C.; Lee, W. R.; Hou, W. C.; Yang, L. L.; Lee, T. J. F. *J. Cell. Biochem.* **2001**, *82*, 537-548.
9. Chen, Y. C.; Lin-Shiau, S. Y.; Lin, J. K. *J. Cell. Physiol.* **1998**, *177*, 324-333.
10. Chen, Y. C.; Tsai, S. H.; Shen, S. C.; Lin, J. K.; Lee, W. R. *Eur. J. Cell Biol.* **2001**, *80*, 201-256.
11. Ko, C. H.; Shen, S. C.; Lin, H. Y.; Hou, W. C.; Lee, W. R.; Yang, L. L.; Chen, Y. C. *J. Cell. Physiol.* (in press, 2002).
12. Yu, R.; Mandlekar, S.; Harvey, K. J.; Ucker, D. S.; Kong, T. A. N. *Cancer Res.* **2000**, *58*, 402-408.
13. Kawagoe, R.; Kawagoe, H.; Sano, K. *Leukemia Res.* **2002**, *26*, 495-502.
14. Lee, W. R.; Shen, S. C.; Lin, H. Y.; Hou, W. C.; Yang, L. L.; Chen, Y. C. *Biochem. Pharmacol.* **2002**, *63*, 225-236.
15. Mirza, A.; McGuirk, M.; Hockenberry, T. N.; Wu, Q.; Ashar, H.; Black, S.; Wen, S. F.; Wang, L.; Kirschmeier, P.; Bishop, W. R.; Nielsen, L. L.; Pickett, C. B.; Liu, S. *Oncogene* **2002**, *21*, 2613-2622.
16. Chen, Y. C.; Shen, S. C.; Lee, W. R.; Lin, H. Y.; Ko, C. H.; Shih, C. M. *Arch. Toxicol.* **2002**, *76*, 351-359.

Chapter 8

Induction of Apoptosis by Rosemary Polyphenols in HL-60 Cells

Shoei-Yn Lin-Shiau[1], Ai-Hsiang Lo[2], Chi-Tang Ho[3], and Jen-Kun Lin[2]

Institutes of [1]Toxicology and [2]Biochemistry and Molecular Biology, College of Medicine, National Taiwan University, Number 1, Section 1, Jen-Ai Road, Taipei, Taiwan, Republic of China
[3]Department of Food Science, Rutgers, The State University of New Jersey, 65 Dudley Road, New Brunswick, NJ 08901-8520

Rosemary (*Rosmarinus officinalis* Labiatae) and its extract are widely used in food processing and herbal medicine. Carnosic acid, carnosol, rosmarinic acid and ursolic acid are biologically active chemicals isolated from rosemary. Carnosol and ursolic acid have been shown to inhibit tumor initiation and promotion. In the present study, carnosic acid, carnosol, and ursolic acid were found to induce apoptotic (sub G1) peak and typical DNA fragmentation in a concentration- and time-dependent manner. Carnosic acid, carnosol and ursolic acid all activated caspase-3, -9 and induced the cleavage of down stream substrates such as poly-(ADP-ribose) polymerase (PARP) and the degradation of DNA fragmentation factor 45/inhibitor of caspase-activated DNase (DFF45/ICAD). Carnosic acid, carnosol and ursolic acid also induced mitochondrial membrane depolarization, intracellular reactive oxygen species (ROS) generation and the release of cytochrome c from mitochondria into cytosol. Taken together, these data suggest that carnosic acid, carnosol and ursolic acid induce apoptosis in HL-60 cells through the release of cytochrome c to the cytosol and activation of the caspases cascade. The induction of apoptosis by carnosic acid, carnosol and ursolic acid may be attributed to their cancer chemopreventive actions.

Introduction

It has been demonstrated that the commonly used spice and flavoring agent rosemary, derived from the leaves of the plant *Rosmarinus officinalis* Labiatae, has important antioxidant properties (*1*). The antioxidant activity of an extract from rosemary leaves is comparable to known antioxidants (*1,2*), such as butylated hydroxyanisole (BHA) and butylated hydroxytoluene (BHT). Among the antioxidant compounds in rosemary leaves, about 90% of the antioxidant activity can be attributed to carnosol and carnosic acid (*2*). Topical application of either rosemary extract, carnosol or ursolic acid to mouse skin inhibited the covalent binding of benzo[a]pyrene (B[a]P) to epidermal DNA, tumor initiation by 7,12-dimethyl benz[a]anthracene (DMBA), TPA-induced tumor promotion, ornithine decarboxylase activity and inflammation (*1*). Ursolic acid was widely studied for its growth inhibitory (*3*) and anti-metastasis effects (*4*). Rosmarinic acid was known to possess antimicrobial and complement inhibition properties (*5,6*). Additional studies have revealed that rosemary extracts, carnosic acid and carnosol strongly inhibited phase I enzyme, CYP450 activities and induced the expression of phase II enzyme, glutathione S-transferase (*7*). These results give insights into the different mechanisms involved in the chemopreventive action of carnosic acid, carnosol, rosmarinic acid and ursolic acid.

Many substances that possess antioxidant activity inhibit tumor initiation and promotion in mouse skin. Antioxidant chemicals found in herbals and edible plants, such as quercetin (*8*), curcumin (*9*), EGCG (*10*), and apigenin (*11*) possess antioxidant and/or reactive oxygen scavenging activity and inhibit or prevent the biochemical events associated with tumor formation. In the light of the potential impact of these compounds on human health, it is important to elucidate the mechanisms involved.

Apoptosis, or programmed cell death, characterized by cell shrinkage, membrane blebbing, nuclear break down and DNA fragmentation, is essential for embryonic development, tissue homeostasis and regulation of the immune system in eukaryote organisms (*12-14*). Malfunctions of apoptosis have been implicated in human disease including ischemic stroke, neurodegenerative disorders and cancer (*15,16*).

The molecular machinery that drives the apoptotic program consists of a family of cysteine protease, the caspases that cleave their substrates after specific aspartic acid residues (*17-19*). Active caspases can typically amplify the apoptotic response by their ability to cleavage their own precursor forms as well as those of other caspases. It is widely believed that the initiator caspases lie at the apex of separate apoptotic cascades that converge on activation of down stream effecter caspases. Diverse apoptotic signals for caspase activation converge at the mitochondrial level (20). Disruption of the mitochondrial membrane by apoptotic stimuli results in the release of cytochrome c into the cytoplasm. The increased cytosolic cytochrome c during apoptosis triggers the formation of a complex containing oligomerized Apaf-1 (apoptotic-protease-activating factor 1) and procaspase-9 (31-33). Active caspase-9 can subsequently auto-process and process downstream effecter procaspases such as procaspase-3 (17,24). The processing of these caspases participates in the execution phase of apoptosis and is followed by the cleavage of apoptotic death substrates, such as lamins, poly (ADP-ribose) polymerase (PARP) (25,26), and the inhibitor of caspase-activated DNase (ICAD) . Previous report has been focused on the role of apoptosis in mediating the lethal effects of diverse antineoplastic agents in leukemia cells (27).

The aim of this study is to elucidate whether carnosic acid, carnosol, rosmarinic acid and ursolic acid have any effects on apoptotic induction, since ursolic acid was previously reported to possess this effect previously (28). We used HL-60 cells to investigate the molecular mechanisms involved. We examined the effects of these rosemary phytochemicals on DNA fragmentation, activation of caspases, altering the mitochondrial function, ROS generation and releasing of cytochrome c from mitochondria. In the present study, we have demonstrated carnosic acid, carnosol, and ursolic acid induced apoptosis in HL-60 cells and activated caspase-3 and caspase-9 via provoking the release of cytochrome c.

Materials and Methods

Chemicals and Enzymes

Carnosol, carnosic acid, rosmarinic acid and ursolic acid were isolated from rosemary (1). Their structures are shown in Figure 1. RNase A, proteinase K and propidium iodide were purchased from Sigma (St. Louis, MO). Substrates for caspase-1, Ac-YVAD-AMC; casppase-3, Ac-DEVD-AMC; caspase-8, Ac-IETD-AMC; and caspase-9, Ac-LEHD-AMC were obtained from AnaSpec Inc. (San Jose, CA). DiOC6 (3) and DCFH-DA were obtained from Molecular Probes (Eugene, OR).

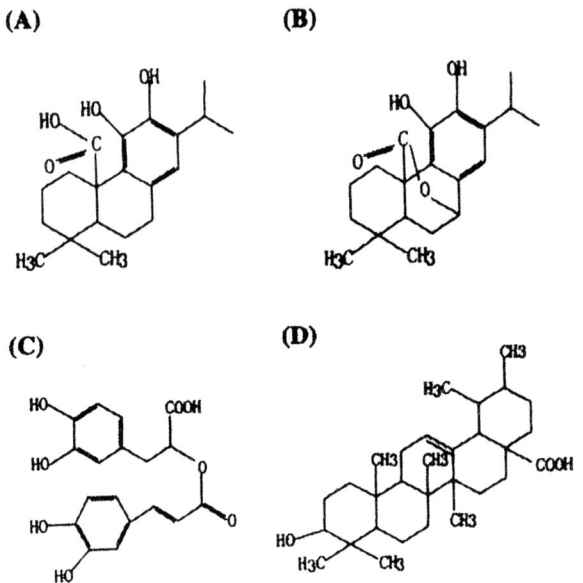

Figure 1. Structures of rosemary phytochemicals. (A) Carnosic acid; (B) Carnosol; (C) Rosmarinic acid; (D) Ursolic acid.

Cell Culture

Human promyelocytic leukemia HL-60 cells (ATCC CRL-1964) were grown in RPMI 1640 with 15% fetal bovine serum (GIBCO BRL, Grand Island, New York, U.S.A.) and supplemented with 2 mM glutamine (GIBCO). The cells were grown in a humidified incubator at 37°C, 5% CO_2 atmosphere up to density of 1.6×10^6 cells/mL. Cells were then plated at a density of 2×10^5 cells/mL immediately prior to assay.

Cell Viability Assay

Cells were plated at a density of 2×10^5 cells/mL and treated with different concentrations of tested compounds for indicated periods. The final concentration of DMSO, as vehicle, was less than 0.04%. Cell viability was assayed with an ATP-Lite M non-radioactive cell proliferation assay kit (Packard Bioscience, Meriden, CT) using ATP as a marker for cell viability. The reaction of ATP with added luciferase and D-luciferin emits light. The luminescence intensity was measured by Topcount Microplate Scintillation and Luminescence Counter (Packard 9912v).

Flow Cytometry Analysis of DNA Content

Cells (2×10^5/mL) were treated with different tested compounds for indicated periods. At the end of treatment, cells were harvested, washed with PBS twice and fixed in ice-cold 70% ethanol at 4 °C for 6 h. The cell pellets were washed and resuspended in PBS with 0.5 µg/mL RNase A at 37 °C for 30 min. Finally, propidium iodide solution (50 µg/mL) was added and the cells were stored at 4 °C in the dark for 1 h. Fluorescence emitted from the PI-DNA complex (564-606 nm) was examined using argon laser excitation at 488 nm by a FACScan flow cytometer (Becton-Dickinson). At least ten thousand cells were analyzed in each sample. The numbers of cells in apoptotic (sub-G1) phase were determined.

DNA Fragmentation Assay

Cells were harvested, washed with PBS twice and then lysed with buffer containing 0.5% sarkosyl, 50 mM Tris pH 8.0, 10 mM EDTA and 0.5 µg/mL proteinase K at 37 °C for 6 h, and then treated with 0.5 µg/mL RNase A for 24 h at 50 °C. The DNA was extracted by phenol/chloroform/isoamyl (25/24/1) and analyzed by 2% agarose gel electrophoresis. The laddering pattern of DNA fragmentation was visualized and photographed under UV light after staining the gel with ethidium bromide.

Measurement of Caspase Activity and Activation of PARP or Degradation of DFF45/ICAD

HL-60 cells were lysed with buffer containing 10% glycerol, 1% Triton X-100, 1 mM sodium orthovanadate, 1 mM EGTA, 5 mM EDTA, 10 mM NaF, 1 mM sodium pyrophosphate, 20 mM Tris-HCl pH 7.9, 100 µM β-glycerophosphate, 137 mM NaCl, 1 mM phenylmethylsulfonyl fluoride (PMSF), 10 µg/mL aprotinin and 10 µg/mL leupeptin. The supernatant thus obtained was used for immunoblotting of PARP and DFF45/ICAD and for caspase activity assay.

Caspase activity in the cell lysate was determined by a fluorogenic assay. 50 µg of total protein was pre-incubated in buffer containing 25 mM HEPES pH7.5, 5 mM $MgCl_2$, 5 mM EDTA, 0.1 M NaCl and 5 mM DTT at 30 °C for 30 min. Then 25 µM of various caspase substrates, such as Ac-YVAD-AMC for caspase-

1, Ac-DEVD-AMC for caspase-3, Ac-IETD-AMC for caspase-8 and Ac-LEHD-AMC for caspase-9 were added and incubated at 30 °C for an additional 1 h. The fluorescence of released 7-amino-4-methylcoumarin (AMC) was measured by excitation at 360 nm and emission at 460 nm using a fluorescence spectrophotometer (Hitachi F4500).

Identification of Cytochrome c Release

Cells were harvested and resuspended in ice-cold buffer A containing 250 mM sucrose, 20 mM HEPES pH7.5, 10 mM KCl, 1.5 mM $MgCl_2$, 1 mM EDTA, 1 mM EGTA, 1 mM DTT and 0.1 mM PMSF. The cells were homogenized with a needle and the homogenates were centrifuged at 750 g for 10 min at 4 °C to pellet the nuclei and unlysed cells. Then the supernatants were centrifuged at 10,000 g for 15 min at 4 °C. The resulting pellet was resuspended in buffer A and represented the mitochondrial fraction used for cytochrome c immunoblotting. The supernatant was centrifuged again for 30 min at 12,000 g. The final supernatant represented the cytosolic fraction and was used for identification of cytochrome c released from mitochondria.

Western Blotting

The total cell lysate (containing 50 µg of protein) was subjected to SDS-PAGE (8% for PARP, 10% for DFF45) and electro-transferred to PVDF membrane (Millipore, Bedford, MA). The membrane was blocked with blocking solution containing 0.01% Tween 20, 1% bovine serum albumin and 0.2% sodium azide in PBS buffer. The membrane was then immunoblotted with anti-poly- (ADP-ribose) polymerase polyclonal antibody (UBI, Lake Placid, NY), or anti-DFF45/ICAD antibody (MBL, NaKa-ku, Nagoya, Japan). After incubation with horseradish peroxidase- or alkaline phosphatase-conjugated anti-rabbit/goat/mouse IgG antibody, the immunoreactive bands were visualized with enhanced chemiluminescence reagents (ECL, Amersham), or with coloregenic substrates: nitro blue tetrazolium (NBT) and 5-bromo-4-chloro-3-indolyl-phosphate (BCIP), as suggested by the manufacturer (Sigma Chemical Co.). 20 µg of the mitochondrial and cytosolic fractions from subcellular fractionation were used for immunoblot analysis of cytochrome c release (15% SDS-PAGE) using anti-cytochrome c antibody (Research Diagnostic, Flanders, NJ).

Analysis of Mitochondria Transmembrane Potential, and Generation of Reactive Oxygen Species (ROS)

At the end of treatment, HL-60 cells were stained with 40 nM 3,3'-dihexyloxacarbocyanine (DiOC6 (*3*), Molecular Probes, Eugene, OR) for 15 min at 37 °C. Mitochondrial transmembrane potential was measured by flow cytometry. At least 10,000 cells were collected in each sample. Generation of ROS was monitored with 20 µM 2',7'-dichlorofluorescein diacetate (DCFH-DA) (Molecular Probes, Eugene, OR) staining for 1 h at 37 °C and assayed by flow cytometry.

Results

Cytotoxicity of Rosemary Phytopolyphenols on HL-60 Cells

Figure 2 illustrates the effects of carnosic acid, carnosol, rosmarinic acid and ursolic acid on cell viability in human promeylocytic leukemia HL-60 cells. The doubling time of HL-60 cells was about 24 h. Cells were treated with different concentrations of tested compounds for 24 h. As shown in Figure 2, carnosol had a LD_{50} of 17.5 µM and appeared to have a concentration-dependent inhibition of cell viability between 10 µM and 20 µM, but this increased concentration of carnosol did not possess more cytotoxicity. Carnosic acid and ursolic acid significantly affected cell viability after 20 µM treatment. In contrast, rosmarinic acid only slightly affected cell viability up to 100 µM treatment (data not shown). Ursolic acid reduced cell viability 6 h after treatment. Carnosic acid and carnosol showed this effect after 12 h (data not shown). We chose the optimal concentration range of each compound to conduct the following experiments.

Induction of Apoptosis by Rosemary Phytochemicals

Apoptotic cell death is characterized by morphology changes, such as chromatin condensation, membrane blebbing, internuleosomal degradation of DNA and apoptotic body formation. We used a fluorescent DNA-binding dye, acridine orange, to characterized the cell death induced by carnosic acid, carnosol, or ursolic acid (data not shown). Chromatin condensation and morphological change were observed under a laser-inverted microscope. The DNA content of treated cells was quantified by flow cytometry to confirm the state of apoptosis

(Figure 3, Table 1). Cells were treated with carnosic acid (30 μM), carnosol (20 μM), or ursolic acid (40 μM) for different times, fixed and stained with propidium iodide. The sub-G1 (sub-2N) cell population increased with time after treatment (Table 1), without the arrest of growth in any phase of the cell cycle (Figure 3). Concentration-dependent induction of sub-G1 peak was also shown (Table 1). DNA fragmentation, which is typical of apoptosis, was demonstrated by treating cells with different concentrations of carnosic acid, carnosol, or ursolic acid (Figure 4A) or for different time periods (Figure 4B). Clear DNA ladders were visible at 12-18 h and showed a concentration-dependent manner. Ursolic acid possessed the strongest potency on induction of DNA fragmentation.

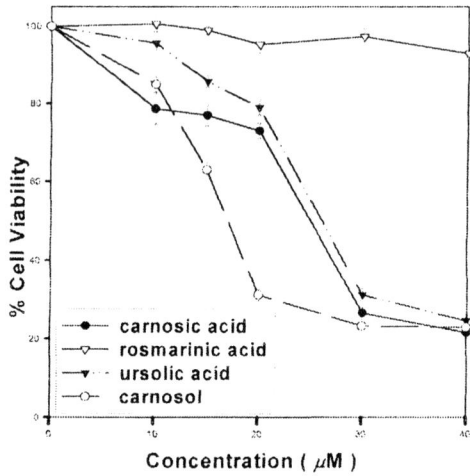

Figure 2. Effects of test compounds on HL-60 cells viability. Cells were treated with different concentrations of tested compounds for 24 h. Cell viability was assayed as described in Materials and Methods. Data represent means ±SE of triplicate tests.

Effects of Rosemary Phytopolyphenols on Various Caspase Activities in HL-60 Cells

Caspases are effecters in mediating various apoptotic stimuli. Caspases are activated in a sequential cascade of cleavages from their inactive forms. Once activated, caspases can cleave their substrates at specific sites. Here, we examined the effects of carnosic acid, carnosol and ursolic acid on various caspase activities in HL-60 cells to determine which caspases are involved in

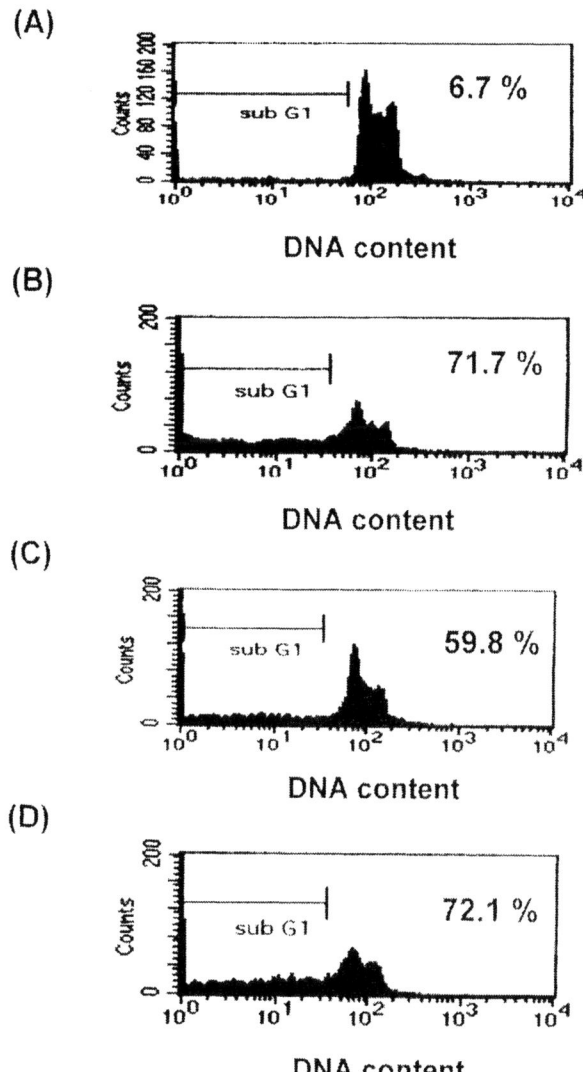

Figure 3. Flow cytometry analysis of apoptotic cells. HL-60 cells were treated with vehicle alone (A), 30 µM carnosic acid (B), 20 µM carnosol (C), or 40 µM ursolic acid (D) for 18 h. After 70% ethanol fixation and propodium iodide (PI) staining, DNA content was analyzed by FACs laser flow cytometer.

apoptosis. We used four fluorogenic peptide substrates, Ac-YVAD-AMC for caspase-1, Ac-DEVD-AMC for casppase-3, Ac-IETD-AMC for caspase-8 and Ac-LEHD-AMC for caspase-9. The induction of caspase-3 activity by carnosic acid, carnosol, or ursolic acid was significant compared to the control group (Figure 5A, B), and consistent with the appearance of the DNA ladder (Figure 4). Longer treatment of ursolic acid and carnosol yielded some diminished caspase-3 activity, possibly due to severe apoptotic cell death followed by acute necrosis. To rule out non-specific protease activity, we added specific caspase-3 inhibitor (DEVD-CHO) and found the inhibitor completely abolished the induction of caspase-3 activity (Figure 5B). Negligible caspase-1 and caspase-8 activities were observed (data not shown). Although the activity of caspase-9 (Figure 5C) was lower than that of caspase-3, it was clearly activated by carnosic acid, carnosol, or ursolic acid. It is known that caspase-3 function at late stages of protease cascade and activation of caspase-9 is linked with apoptosis through mitochondrial dysfunction.

Table I. Flow Cytometry Analysis of subG1 Percentage of Apoptotic cells[*]

Control		6.9 ± 1.2%		6.7 ± 1.4%
carnosic acid	6 hr	34.7 ± 9.3%	10 μM	38.1 ± 2.7%
	12 hr	57.7± 4.0%	20μM	69.1 ± 4.8%
	18 hr	70.5 ± 2.4%	30μM	71.7 ± 0.9%
carnosol	6 hr	29.4 ± 5.0%	10μM	21.8 ± 3.9%
	12 hr	48.7 ± 4.2%	15μM	49.1 ± 4.5%
	18 hr	54.5 ± 6.2%	20μM	59.8 ± 1.8%
ursolic acid	6 hr	26.1 ± 3.8%	20μM	19.6 ± 2.1%
	12 hr	67.6 ± 1.0%	30μM	57.7 ± 5.5%
	18 hr	74.3 ± 8.5%	40μM	72.1 ± 2.2%

* HL-60 cells were treated with indicated concentrations of carnosic acid, carnosol or ursolic acid for 18 h (right column) and treated with carnosic acid (30 μM), carnosol (20 μM) or ursolic acid (40 μM) for different time periods (left column). After 70% ethanol fixation and propidium iodide (PI) staning, DNA content was analyzed by FACs laser flow cytometer. Percentages represent apoptotic cells among each group. Data represents mean ± SE of triplicate tests.

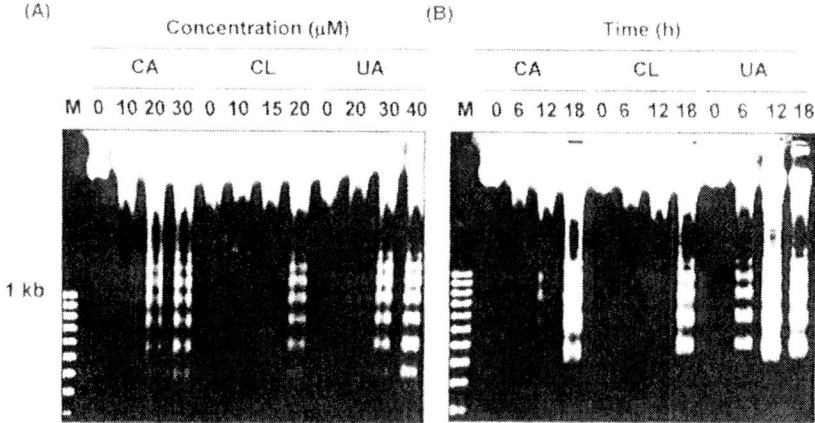

Figure 4. Induction of DNA fragmentation in HL-60 cells. (A) Cells were treated with increasing concentrations of tested compounds for 18 h. Abbreviations: carnosic acid (CA), carnosol (CL), ursolic acid (UA). (B) Cells were treated with 30 µM carnosic acid, 20 µM carnosol, or 40 µM ursolic acid for different time periods. DNA fragments were separated by electrophoresis in 2% agarose gel M, marker.

Rosemary Phytopolyphenols Cause Cleavage of Poly- (ADP-ribose) Polymerase and DFF45/ICAD

Activation of caspases leads to the cleavage of some down stream proteins, such as poly- (ADP-ribose) polymerase (PARP), lamin and DFF45 / ICAD. The cleavage of PARP is the hallmark apoptosis. PARP (116-kDa) is cleaved to produce an 85-kDa fragment. During carnosic acid-, carnosol-, or ursolic acid-induced apoptosis, the full-sized 116-kDa molecule disappeared in a time-dependent manner (Figure 6A), and was consistent with DNA fragmentation (Figure 4) and caspase-3 activity (Figure 5). DFF (DNA fragmentation factor) has been identified as a heterodimeric protein composed of DFF-45 (ICAD, inhibitor of caspase-activated deoxyribonuclease) and CAD (caspase-activated deoxyribonuclease). Caspase-3 activated by apoptotic signals acts to degrade and dissociate DFF-45 from CAD, allowing CAD to enter the nucleus and degrade DNA. Here we show the cleavage of DFF-45 (Figure 6B) following treatment of carnosic acid, carnosol, or ursolic acid with a time-course paralleled to the activation of caspase-3 (Figure 5) and the pattern of the DNA ladder (Figure 4).

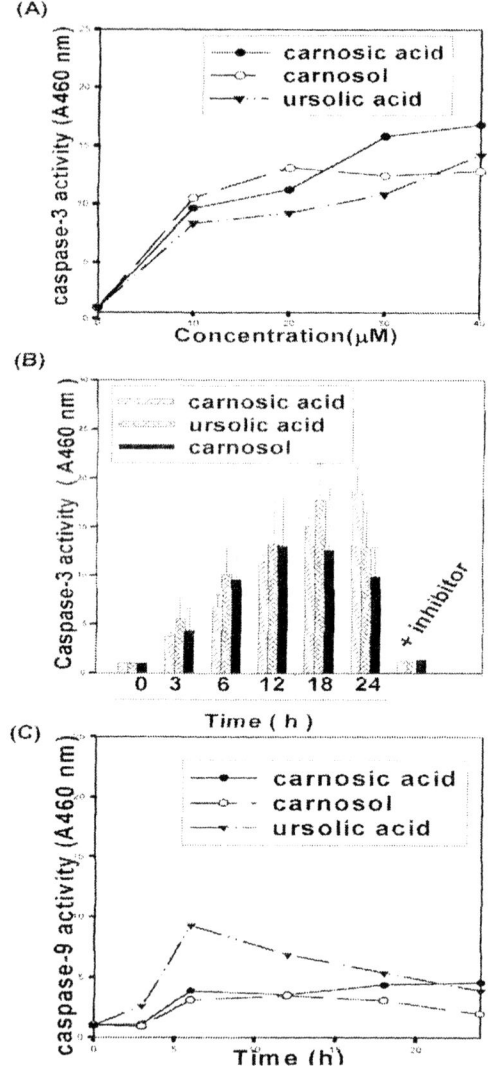

Figure 5. Induction of caspase-3, -9 activity in HL-60 cells. 50 μg total cell lysate was incubated with fluorogenic substances. Emission spectra were measured by fluorescence spectrophotometer and normalized to control. (A) Concentration-dependent activation of caspase-3 by tested compounds at 24 h. (B) Kinetics of caspase-3 activation. Cells were treated with 30 μM carnosic acid, 20 μM carnosol or 40 μM ursolic acid for different time periods. Data represent means ± SE of triplicate tests. (C) Kinetics of caspase-9 activity.

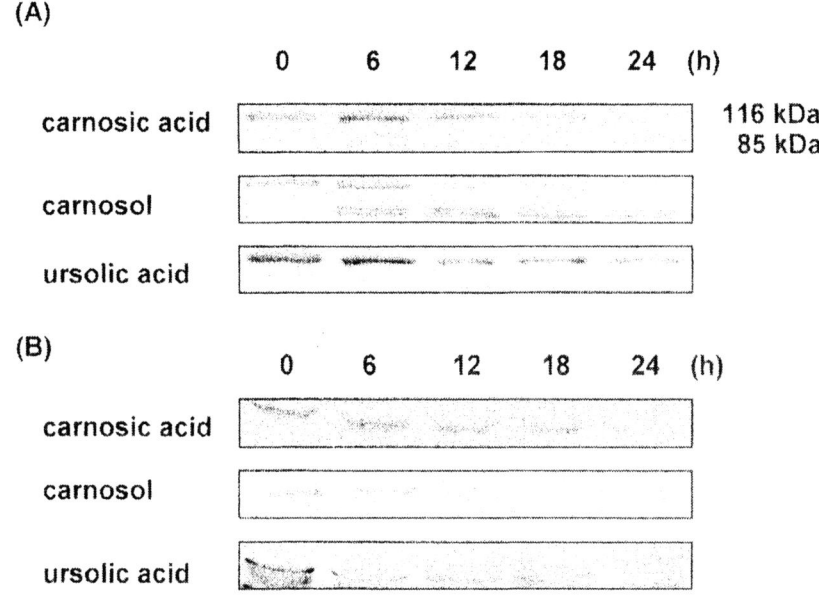

Figure 6. (A) Time course of the poly-(ADP-ribose) polymerase cleavage. HL-60 cells were treated with 30 µM carnosic acid, 20 µM carnosol, or 40 µM ursolic acid for indicated time and total cell lysates were subjected to Western blot analysis as described in Material and Methods. Similar results were obtained in three separate experiments. (B) Degradation of DNA fragmentation factor-45/inhibitor of caspase-activated DNase (DFF45/ICAD).

Rosemary Phytopolyphenols Induced Loss of Mitochondrial Transmembrane Potential (Ψm), Intracellular ROS Generation and Cytochrome c Release

Mitochondria play a central role in response to apoptotic stimuli. There is increasing evidence that altered mitochondrial function is linked to apoptosis and a decreasing mitochondrial transmembrane potential (Ψm) is associated with mitochondria dysfunction. The MPT (mitochondria permeability transition) is a permeability increase of the mitochondria membrane coupled with depolarization of the membrane and disruption of mitochondrial membrane integrity. We measured ΔΨm using a fluorescent probe DiOC6(3) which specially accumulated in polarized membranes and was monitored by flow

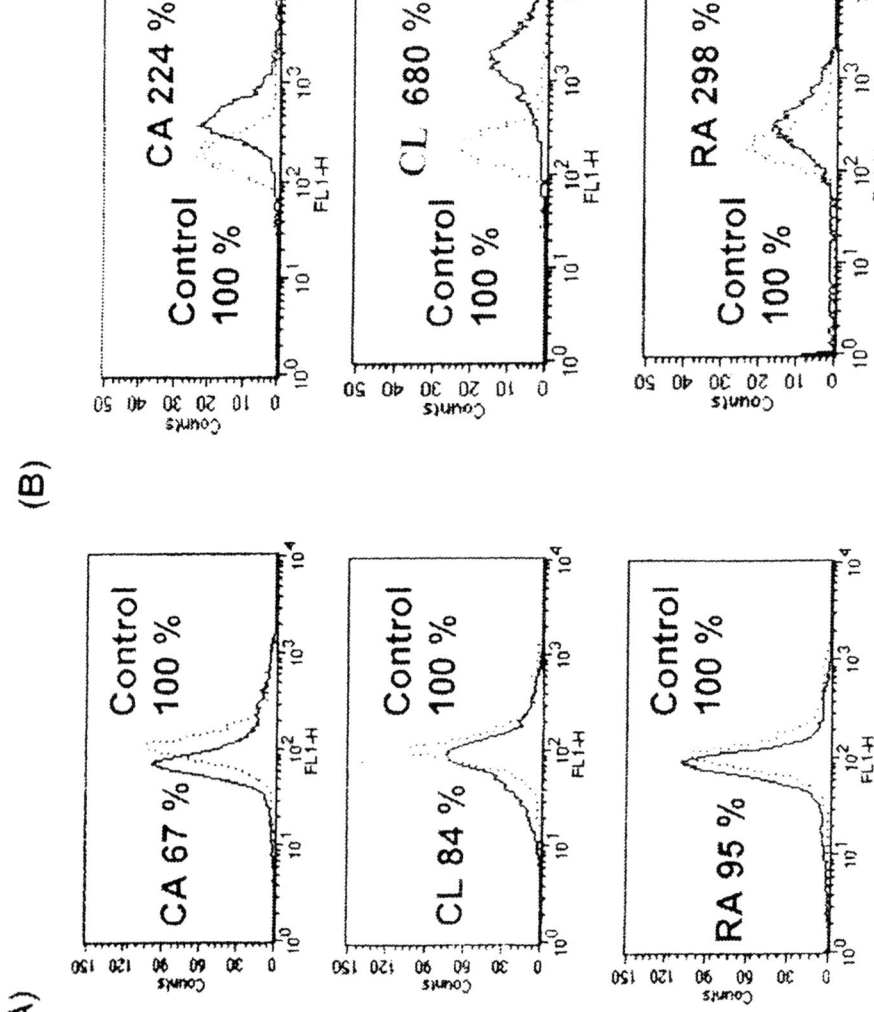

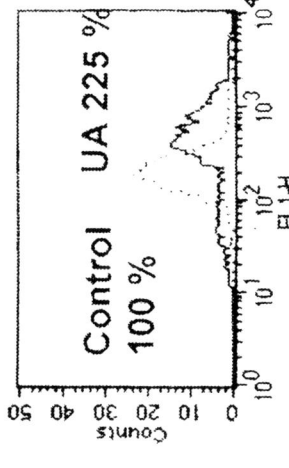
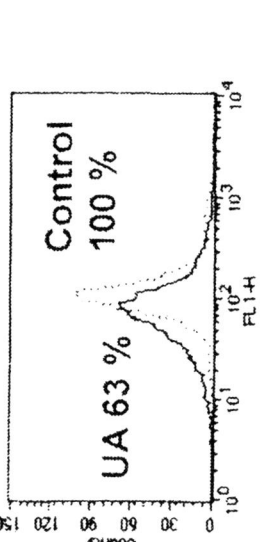

Figure 7. Induction of mitochondrial membrane potential loss and ROS generation in HL-60 cells. (A) Cells were treated with 30 μM carnosic acid (CA), 20 μM carnosol (CL), 100 μM rosmarinic acid (RA), or 40 μM ursolic acid (UA) for 1 hr and incubated with DiOC6(3) for the last 15 min, then analyzed by flow cytometry. The percentage was the mean value of DiOC6(3) intensity normalized to control. (B) Analysis if reactive oxygen species (ROS) generation after 2 h treatment. Cells were incubated with 20 μM DCFH-DA for the last 1 h, then analyzed by flow cytometry. The percentage was the mean value of DCF fluorescence intensity normalized to control.

cytometry. As shown in Figure 7(A), HL-60 cells exposed to carnosic acid, carnosol, or ursolic acid for only 1 h showed a sharp decline in DiOC6(3) fluorescence mean value. Treatment of rosmarinic acid did not interfere mitochondrial transmembrane potential (Ψm). Loss of mitochondrial transmembrane potential (Ψm) is usually accompanied by reactive oxygen species (ROS) generation. DCFH-DA permeates the cell membrane passively. Once cleaved by intracelullar esterase, DCFH can be oxidized and assayed by flow cytometry. As shown in Figure 7(B), significant ROS was generated after 2 h treatment of carnosic acid, carnosol or ursolic acid. Treatment with rosmarinic acid also increased intracellular ROS. The release of cytochrome c from the mitochondrial intermembrane space plays an important role in the induction of apoptosis and is related to MPT-induced dysfunction of mitochondria. Cytochrome c released to cytosol binds to an Apaf-1-procaspase-9 complex and induces caspase cascade activation. To demonstrate the involvement of cytochrome c release from mitochondria into the cytosol during treatment of carnosic acid, carnosol, or ursolic acid, mitochondrial fraction and cytosolic fraction were subjected to Western blot analysis using anti-cytochrome c antibody. An increase in the amount of cytochrome c in the cytosol paralleled to a decrease in amount of cytochrome c in the mitochondria fraction after treatment (Figure 8). These results indicate that mitochondrial dysfunction caused cytochrome c release and the cytosolic cytochrome c might form an Apaf complex in the cytosol and contribute to the activation of caspase-9 and caspase-3.

Discussion

The inhibition of skin carcinogenesis by rosemary and its constituents, carnosol and ursolic acid, has been demonstrated by Huang et al. (*1*). Furthermore, the suppression of rat mammary tumorigenesis induced by DMBA has been demonstrated by Singletary et al. (*29*). Meanwhile, the formation of mammary DMBA-DNA adducts *in vivo* was dose-dependently inhibited by carnosol and ursolic acid (*29*). Rosemary components have the potential to decrease activation and increase detoxification of benzo(a)pyrene, identifying them as promising chemopreventive agents (*30*).

The present studies show that carnosic acid, carnosol, and ursolic acid induced apoptosis in HL-60 cells in a concentration- and time-dependent manner. Rosmarinic acid did not interfere HL-60 cell viability. We also showed that carnosic acid, carnosol, or ursolic acid activated caspase-3 and -9, and caused cleavage and degradation of down stream death substrates PARP and DFF45/ICAD. These results indicate that the typical death protease cascade mediates carnosic acid-, carnosol-, or ursolic acid-induced apoptosis in HL-60 cells.

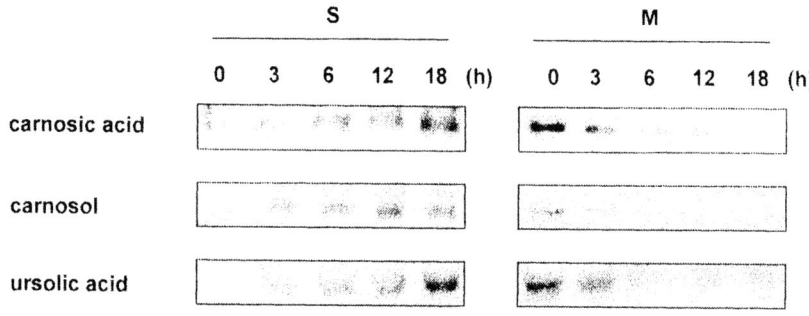

Figure 8. Release of cytochrome c from mitochondria in HL-60 cells. The cells were treated with 30 µM carnosic acid, 20 µM carnosol, or 40 µM ursolic acid for the periods indicated, and cytochrome c in cytosolic (S) and mitochondrial (M) fractions was detected by Western blot with anti-cytochrome c antibody. Similar results were obtained in three separated experiments.

Accumulated evidence from recent research indicates that mitochondria-derived factors, such as cytochrome c, have an important role in the apoptosis of some cells. Previous reports show that cytochrome c and caspase-9 participate in Apaf1 apoptosome, a complex important for caspase-3 activation. Cytochrome c was released from mitochondria of HL-60 cells during treatment with carnosic acid, carnosol or ursolic acid. Previous evidence showed that mitochondria permeability transition (MPT) coupled with depolarization of the membrane potential induces the release of cytochrome c *(31-33)*. It also has been reported that cytochrome c released from mitochondria can precede dissipation of the voltage gradient (the mitochondrial transmembrane potential; ψ_m) across the membrane, suggesting that the escape of cytochrome c from mitochondria occurs prior to permeability transition pore opening (loss of mitochondrial transmembrane potential) *(34,35)*. Here we observed the depolarization of the mitochondrial membrane potential in HL-60 cells by treatment with carnosic acid, carnosol, or ursolic acid for 1h. Cytochrome c was released after 3h treatment of carnosic acid, carnosol, or ursolic acid.

Bcl2 and Bax are mitochondrial proteins possessing anti-apoptotic and pro-apoptotic properties, respectively. Overexpression of Bcl2 prevents apoptosis and cytochrome c release induced by various stimuli *(36,37)*. However, in our study, we did not observe changes of Bcl2/Bax ratio in carnosic acid-, carnosol-, or ursolic acid-treated HL-60 cells (data not shown). Hsp70 is stated to negatively regulate Apaf1 apoptosome (*38-40*). Again, we did not observe any change in Hsp70 expression (data not shown).

Many papers have reported the involvement of reactive oxygen species in the apoptosis of different cells induced by different ligands. Our unpublished data and other investigations showed strong antioxidant effects of rosemary phytochemicals. But in this study, we found that carnosic acid, carnosol, rosmarinic acid and ursolic acid increased ROS in HL-60 cells. As reported, antioxidative phenolic phytochemicals can actually induce DNA single-strand breaks, as determined by comet assay (*41*). The pro-oxidant effects of phenolic phytochemicals appears to be due to the presence of metal ions, which can facilitate the formation of ROS (*42*). It seems that carnosic acid, carnosol, rosmarinic acid, and ursolic acid can have either antioxidant or pro-oxidant properties depending on the experimental conditions, like previously reported EGCG (*43*), quercetin (*44*) and apigenin (*11*). More specifically, in the presence of pro-oxidants or substances that generate ROS, carnosic acid, carnosol, rosmarinic acid and ursolic acid are probably scavengers. In striking contrast, in the absence of other added pro-oxidants but in the presence of trace amounts of metal ions in cultured cells, carnosic acid, carnosol, rosmarinic acid, and ursolic acid may promote the formation of ROS. As such, carnosic acid, carnosol, rosmarinic acid, and ursolic acid may be able to either cause DNA damage or interfere mitochondrial re-dox cycling and electron transport to initiate or augment apoptosis. Interestingly, rosmarinic acid did not cause apoptosis in HL-60 cells while it functioned as a pro-oxidant. This may be related to the fact that no significant mitochondrial transmembrane potential change ($\Delta\Psi m$) was observed in rosmarinic acid treated cells (Figure 4A).

Increased intracellular calcium ions ($[Ca^{2+}]_i$) have been demonstrated to act as an important mediator of apoptosis in a variety of cells (*45*). Baek et al. (*28*) reported that ursolic acid induced apoptosis in HL-60 cells and increased intracellular calcium ions ($[Ca^{2+}]_i$). Although the precise mechanism by which $[Ca^{2+}]_i$ mediates apoptosis is not known, the Ca^{2+}/Mg^{2+}-dependent endonuclease, which cleaves double-strand DNA at nucleosome linker regions, remains an attractive target of Ca^{2+} (*16*). Some calcium-dependent/independent mechanisms may be involved in the apoptosis-inducing activity of carnosic acid, carnosol, and ursolic acid and remain a subject for further study.

Taken together, these results suggest that carnosic acid, carnosol, and ursolic acid exhibit anti-proliferate effects by inducing apoptosis in HL-60 cells through cytochrome c release and caspase activation. As demonstrated by Singletary et al. the anti-carcinogenic effects of rosemary phytopolyphenols might be through suppressing the formation of carcinogen-DNA adducts (*29*). Our present findings on apoptosis induction may at least provide an alternative mechanism of their anti-tumor effects and suggest reason for further studies on their possible application in human leukemia therapy.

Acknowledgements

This study was supported by the National Science Council NSC 89-2320-B-002-245 and NSC 89-2320-B-002-223; by the National Health Research Institute NHRI-EX 90-8913BL; by the National Research Institute of Chinese Medicine, NRICM-90102 and by the Ministry of Education, ME 89-B-FA01-1-4.

References

1. Huang, M.T.; Ho, C.-T.; Wang, Z.Y.; Ferraro, T.; Lou, Y.R.; Stauber, K.; Ma, W.; Georgiadis, C.; Laskin, J.D.; Conney, A.H. *Cancer Res.* **1994**, *54*, 701-708.
2. Aruoma, O.I.; Halliwell, B.; Aeschbach, R.; Loligers, J. *Xenobiotica* **1992**, *22*, 257-268.
3. Es-saady, D.; Simon, A.; Ollier, M.; Maurizis, J.C.; Chulia, A.J.; Delage, C. *Cancer Lett.* **1996**, *106*, 193-197.
4. Cha, H.J.; Park, M.T.; Chung, H.Y.; Kim, N.D.; Sato, H.; Seiki, M.; Kim, K.W. *Oncogene* **1998**, *16*, 771-778.
5. Bult, H.; Herman, A.G.; Rampart, M. *Br.J. Pharmacol.* **1985**, *84*, 317-327.
6. Rampart, M.; Beetens, J.R.; Bult, H.; Herman, A.G.; Parnham, M.J.; Winkelmann, J. *Biochem. Pharmacol.* **1986**, *35*, 1397-1400.
7. Mace, K.; Offord, E. A.; Harris, C.C.; Pfeifer, A.M. *Arch. Toxicol. Suppl.* **1998**, *20*, 227-236.
8. Kato, R.; Nakadate, T.; Yamamoto, S.; Sugimura, T. *Carcinogenesis* **1983**, *4*, 1301-1305.
9. Huang, M.T.; Smart, R.C.; Wong, C.Q.; Conney, A.H. *Cancer Res.* **1988**, *48*, 5941-5946.
10. Huang, M.T.; Ho, C.-T.; Wang, Z.Y.; Ferraro, T.; Finnegan-Olive, T.; Lou, Y.R.; Mitchell, J.M.; Laskin, J.D.; Newmark, H.; Yang, C.S. *Carcinogenesis* **1992**, *13*, 947-954.
11. Wang, I.K.; Lin-Shiau, S.Y.; Lin, J. K. *Eur. J. Cancer* **1999**, *35*, 1517-1525.
12. Steller, H. *Science* **1995**, *267*, 1445-1449.
13. Vaux, D.L.; Haecker, G.; Strasser, A. *Cell* **1994**, *76*, 777-779.
14. White, E. *Genes Dev.* **1996**, *10*, 1-15.
15. Jacobson, M. D.; Weil, M.; Raff, M.C. *Cell* **1997**, *88*, 347-354.

16. Thompson, C.B. *Science* **1995**, *267*, 1456-1462.
17. Alnemri, E.S. *J. Cell Biochem.* **1997**, *64*, 33-42.
18. Cohen, G.M. *Biochem. J.* **1997**, *326 (Pt 1)*, 1-16.
19. Cryns, V.; Yuan, J. *Genes Dev.* **1998**, *12*, 1551-1570.
20. Green, D.R.; Reed, J.C. *Science* **1998**, *281*, 1309-1312.
21. Li, P.; Nijhawan, D.; Budihardjo, I.; Srinivasula, S.M.; Ahmad, M.; Alnemri, E.S.; Wang, X. *Cell* **1997**, *91*, 479-489.
22. Liu, X.; Kim, C.N.; Yang, J.; Jemmerson, R.; Wang, X. *Cell* **1996**, *86*, 147-157.
23. Zou, H.; Henzel, W.J.; Liu, X.; Lutschg, A.; Wang, X. *Cell* **1997**, *90*, 405-413.
24. Nicholson, D.W.; Ali, A.; Thornberry, N.A.; Vaillancourt, J.P.; Ding, C.K.; Gallant, M.; Gareau, Y.; Griffin, P.R.; Labelle, M.; Lazebnik, Y.A. *Nature* **1995**, *376*, 37-43.
25. Tewari, M.; Quan, L.T.; O'Rourke, K.; Desnoyers, S.; Zeng, Z.; Beidler, D.R.; Poirier, G.G.; Salvesen, G.S.; Dixit, V.M. *Cell* **1995**, *81*, 801-809.
26. Vaux, D.L.; Strasser, A. *Proc. Natl. Acad. Sci. U.S.A.* **1996**, *93*, 2239-2244.
27. Kaufmann, S.H. *Cancer Res.* **1989**, *49*, 5870-5878.
28. Baek, J.H.; Lee, Y.S.; Kang, C.M.; Kim, J.A.; Kwon, K.S.; Son, H.C.; Kim, K. W. *Int. J. Cancer* **1997**, *73*, 725-728.
29. Singletary, K.; MacDonald, C.I,; Wallig, M. *Cancer Lett.* **1996**, *104*, 41-48.
30. Offord, E. A.; Mace, K.; Ruffieux, C.; Malnoe, A.; Pfeifer, A. M. A. *Carcinogenesis* **1995**, *16*, 2057-2062.
31. Li, H.; Zhu, H.; Xu, C.J.; Yuan, J. *Cell* **1998**, *94*, 491-501.
32. Luo, X.; Budihardjo, I.; Zou, H.; Slaughter, C.; Wang, X. *Cell* **1998**, *94*, 481-490.
33. Tatton, W. G.; Olanow, C.W. *Biochim. Biophys. Acta* **1999**, *1410*, 195-213.
34. Kluck, R.M.; Bossy-Wetzel, E.; Green, D.R.; Newmeyer, D.D. *Science* **1997**, *275*, 1132-1136.
35. Yang, J.; Liu, X.; Bhalla, K.; Kim, C.N.; Ibrado, A.M.; Cai, J.; Peng, T.I.; Jones, D.P.; Wang, X. *Science* **1997**, *275*, 1129-1132.
36. Inoue, Y.; Sato, Y.; Nishimura, M.; Seguchi, M.; Zaitsu, Y.; Yamada, K.; Oka, Y. *Anticancer Res.* **1999**, *19*, 3989-3992.
37. Kuo, M.L.; Huang, T.S.; Lin, J.K. *Biochim. Biophys. Acta* **1996**, *1317*, 95-100.
38. Beere, H.M.; Wolf, B.B.; Cain, K.; Mosser, D.D.; Mahboubi, A.; Kuwana, T.; Tailor, P.; Morimoto, R.I.; Cohen, G.M.; Green, D.R. *Nat. Cell Biol.* **2000**, *2*, 469-475.
39. Li, C.Y.; Lee, J.S.; Ko, Y.G.; Kim, J.I.; Seo, J.S. *J. Biol. Chem.* **2000**, *275*, 25665-25671.

40. Saleh, A.; Srinivasula, S.M.; Balkir, L.; Robbins, P.D.; Alnemri, E.S. *Nat. Cell Biol.* **2000**, *2*, 476-483.
41. Duthie, S.J.; Johnson, W.; Dobson, V.L. *Mutat. Res.* **1997**, *390*, 141-151.
42. Halliwell, B.; Gutteridge, J.M. *Methods Enzymol.* **1990**, *186*, 1-85.
43. Long, L.H.; Clement, M.V.; Halliwell, B. *Biochem. Biophys. Res. Commun.* **2000**, *273*, 50-53.
44. Richter, M.; Ebermann, R.; Marian, B. *Nutr. Cancer* **1999**, *34*, 88-99.
45. Dowd, D.R. *Adv. Second Messenger Phosphoprotein Re*s. **1995**, *30*: 255-280.

Chapter 9
Anticaries Effect of Wasabi Components

Hideki Masuda, Toshio Inoue, and Yoko Kobayashi

Material R&D Laboratories, Ogawa & Company, Ltd., 15–7, Chidori, Urayasushi, Chiba 279–0032, Japan

Isothiocyanates, the main volatile components in wasabi, are well-known to have a characteristic pungent odor. In addition, they have many biological functions, such as an antimicrobial effect. However, the effect of isothiocyanates on the mutans streptococci, which are the cause bacteria for dental caries, has not been studied in detail. In this study, the anticaries effect of isothiocyanates was confirmed by *in vitro* and *in vivo* testing. The anticaries mechanism of isothiocyanates was also reported.

Mutans streptococci is well-known to be the most important factor causing dental caries (*1, 2*). Water-insoluble glucan, which constitutes a dental plaque, is synthesized from sucrose by the action of glucosyltransferases (GTases) produced from mutans streptococci. Mutans streptococci, fixed in the glucan on the surface of tooth, produces organic acids via the glycolytic pathway. The enamel of teeth is dissolved by the organic acids to give dental caries. Therefore, the anticaries strategy is as follows: 1) the use of substitute sweeting agents, 2) the sterilization of mutans streptococci, 3) the inhibition of GTases activity, and 4) the lysis of glucan. The anticaries effect of the active principle of natural substances, such as a poly-phenol, a diterpene, and organic acids, has already been reported (*3-8*). The anticaries activity of polyphenol in green tea and oolong tea was studied in detail (*9-14*). Isothiocyanates, the representative volatile components of wasabi, are known to have a significantly pungent odor and many biological fuctions (*15-18*). However, the effect of isothiocyanates on dental caries has not been systematically studied (*19*). Therefore, we have focused on the anticaries effect of many isothiocyanates by *in vitro* and *in vivo* testing. In addition, in order to clarify the anticaries mechanism of isothiocyanates, we studied both the inhibition effect of GTases activity and the antibacterial effect using an *in vitro* test.

Experimental

Materials

Four ω-alkenyl isothiocyanates (except for allyl isothiocyanate) were prepared by isomerization of the corresponding ω-alkenyl thiocyanates (*20*). Five ω-alkenyl isothiocyanates were converted to the corresponding ω-methylthioalkyl isothiocyanates (*21*). Five ω-methylthioalkyl isothiocyanates were oxidized to the corresponding ω-methylsulfinylalkyl isothiocyanates using m-chloroperbenzoic acid. Allyl isothiocyanate was purchased from commercial sources. Chinese oolong tea extract (13 g) was obtained from dry Suisen leaf in Fukken-sho (51 g) by extraction with 50% ethanol for 24 h at room temperature, followed by filtration, evaporation and freeze-drying. Taiwanese oolong tea extract (17 g) was obtained from dry Tootyo leaf in Taiwan (50 g) by extraction with 50% ethanol for 24 h at room temperature, followed by filtration, evaporation, and freeze-drying.

Inhibitory Effect on Sucrose Dependent Adherence by Growing Cells of Mutans Streptococci

Streptococcus mutans (*S. mutans*) IFO 13955 was obtained from the Institute for Fermentation, Osaka, Japan. *S. mutans* IFO 13955, was cultured on BHI (Brain Heart Infusion) broth, and each isothiocyanate was added in 80% aqueous N,N-dimethylformamide (0.05 mL) or 80% aqueous ethanol solution (0.05 mL), and then added to the test tube containing 4.94 mL of BHI broth and 1% sucrose. The test tube was allowed to stand at an angle of 30° for 24-48 h at 37°C. The BHI broth was removed and the residue was dried. The weight of the residue was then determined. The sucrose dependent adherence of *S. mutans* IFO 13955 was represented by the relative percentage compared with the adherence in the absence of isothiocyanate. All tests were run in triplicate and averaged.

Antibacterial Effect

Streptococcus sobrinus (*S. sobrinus*) 6715 and *Streptococcus mutans* (*S. mutans*) MT8148 were obtained from the Institute of Physical and Chemical Research (RIKEN), Saitama, Japan. *S. sobrinus* 6715, *S. mutans* MT8148, and *S. mutans* IFO 13955 were precultured in BHI broth overnight to prepare the seeded solution. The precultured mutans streptococci was adjusted to 10^6 cfu/ml with BHI broth by optical density (660 nm). Each isothiocyanate, in an 80% aqueous methanol solution (50 μL), was added to the test tube containing five mL of each mutans streptococci BHI broth (10^6 cfu/mL). The test tube was incubated for 24 h at 37°C and the minimum inhibitory concentration (MIC) value was measured. All tests were run in triplicate and averaged.

Preparation of Gtases

S. sobrinus 6715 was cultured in 500 ml of BHI broth overnight and the culture supernatant (centrifugation: 6000×g for 20 min at 4 °C) was then salted out with saturated ammonium sulfate after adjustment at pH 7 with 1N sodium hydroxide solution. After ultracentrifugation (12000×g for 30 min at 4°C), the precipitate was dialyzed against 50 mM potassium phospate buffer (pH 6.5) to yield a water-insoluble GTases (*22*).

Effect of the Isothiocyanate on Glucan Synthesis by GTases

The crude GTases solution (0.1 mL), which was diluted 20 times by 0.1 M potassium phosphate buffer, was added to potassium phosphate buffer (0.1 M, 1.5 mL, pH 6.5), 10% sucrose in 0.2% sodium azide aqueous solution (0.3 mL), the isothio-cyanate in methanol solution (0.03 mL), and water (1.07 mL), followed by incubation for 18 h at 37 °C. After centrifugation (15000 rpm, 10 min), the precipitate was washed and followed by another centrifugation. Water (5 mL) was added to the precipitate and then sonication was performed. A five % phenol solution (0.5 mL) and concentrated sulfuric acid (2.5 mL) were added to the suspension (0.5 mL) obtained by sonication. After standing for 30 min, a transmittance of 470 nm was measured using a Bausch & Lomb Spectronic 20 Spectrophoto-meter. All tests were run in triplicate and averaged.

Animals and Treatment

Eighteen-day-old male Wistar rats were purchased from Shimizu Laboratory Supplies (Kyoto, Japan). The animals were kept in stainless steel cages and housed in an air-conditioned room maintained at 22-26°C and humidity of 40-70%. All animals were weaned for 18 days after birth. After weaning, the rats were fed on a normal diet containing tetracycline (4 mg/g) and drinking water containing penicillin G (4000 unit/L) ad libitum for two days. Three days later, all animals were randomly distributed into three groups. The control: the rats were infected with *S. sobrinus* 6715 and fed on diet 2000 throughout the experiment period. The isothiocyanate treatment groups: the rats infected with *S. sobrinus* 6715 were fed on diet 2000 containing 50 and 100 ppm of 6-methylthiohexyl isothiocyanate. Fifty-five days later, the rats were killed and their jaws were removed.

Assay for GOT and GPT

The serum GOT and GPT activities were measured using the colorimetric test reported by Reitman and Frankel with S.TA-test Wako kits from Wako Pure Chemical Industries, Ltd. (Osaka, Japan).

Data and Statistical Analysis

All data are presented as means±S.E. A statistical analysis was performed using the Kruskal-Wallis and Steel test. A probability value of less than 0.05 was considered significant.

Results and Discussion

As shown in Figures 1-4, sucrose dependent adherence by growing cells of mutans streptococci was dependent on the concentration of isothiocyanate (ally- (**1**), 3-butenyl- (**2**), 4-pentenyl- (**3**), 5-hexenyl- (**4**), 6-heptenyl- (**5**), 3-methylthiopropyl- (**6**), 4-methylthiobutyl- (**7**), 5-methylthiopentyl- (**8**), 6-methylthiohexyl- (**9**), 7-methylthioheptyl- (**10**), benzyl- (**11**), 2-phenethyl- (**12**), 3-methylsulfinylpropyl- (**13**), 4-methylsulfinylbutyl- (**14**), 5-methylsulfinylpentyl- (**15**), 6-methylsulfinylhexyl- (**16**), 7-methylsulfinylheptyl- (**17**)). In general, the longer the side chain, the lower the sucrose dependent adherence by the growing

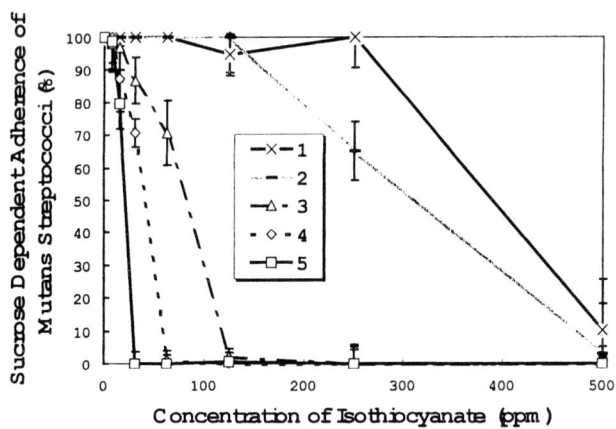

Figure 1. Inhibitory effect of the ω-alkenyl isothiocyanates on sucrose dependent adherence by growing cells of S. mutans IFO 13955.
(Reproduced with permission from reference 19, Copyright 1999 Kluwer Academics.)

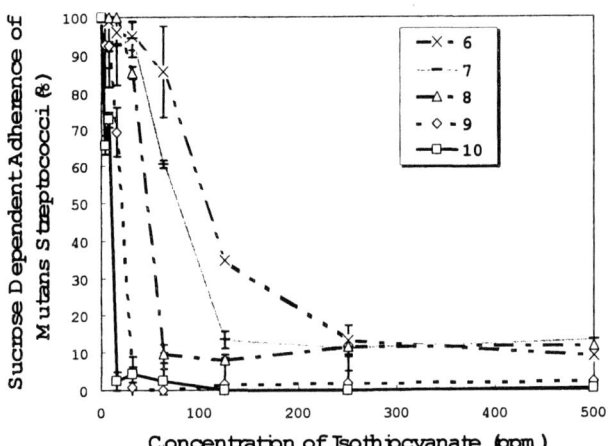

Figure 2. Inhibitory effect of the ω-methylthioalkyl isothiocyanates on sucrose dependent adherence by growing cells of S. mutans IFO 13955.
(Reproduced with permission from reference 19, Copyright 1999 Kluwer Academics.)

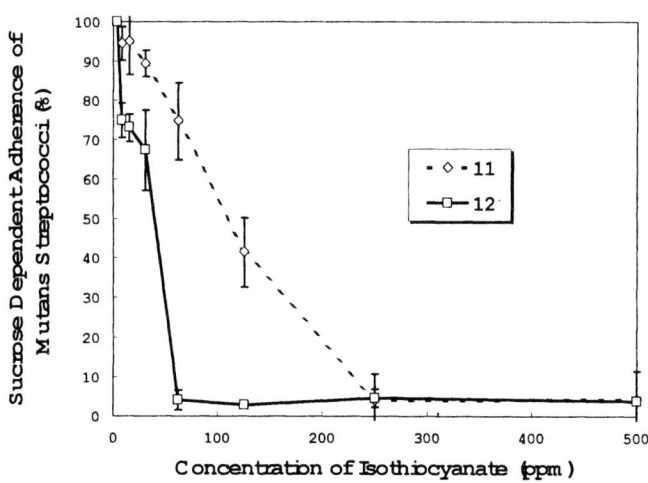

Figure 3. Inhibitory effect of the aryl isothiocyanates on sucrose dependent adherence by growing cells of S. mutans IFO 13955.
(Reproduced with permission from reference 19, Copyright 1999 Kluwer Academics.)

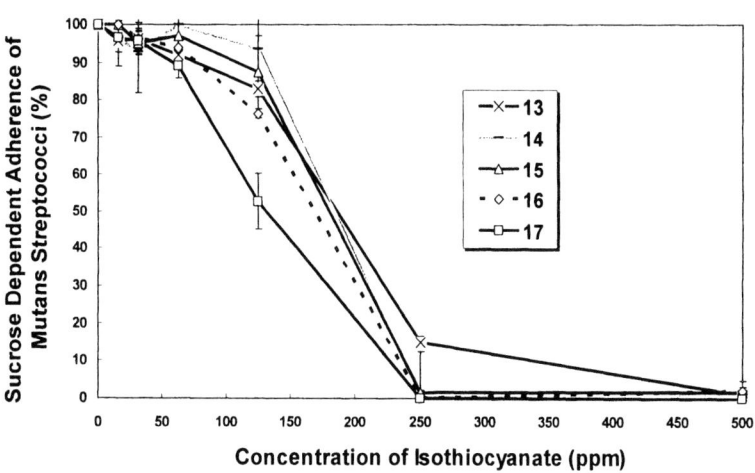

Figure 4. Inhibitory effect of the ω-methylsulfinylalkyl isothiocyanates on sucrose dependent adherence by growing cells of S. mutans IFO 13955.

cells of mutans streptococci. Taking into account the 100% inhibition concentration of isothiocyanate 9, 25 ppm in Figure 2, this higher concentration of isothiocyanate 9, 50 and 100 ppm, was used to confirm anticaries in the *in vivo* test using some rats. As shown in Figure 5, the body weight of the rats after being fed the diet containing 50 ppm of isothiocyanate 9 was nearly equal to that of the control. Therefore, the amount of isothiocyanate 9 used in the diet was determined to be 50 ppm. The caries score of the rats, after being fed the diet containing 50 ppm of isothiocyanate 9, was found to be lower than that of the control (Figure 6). Keeping the result that the glutamic-oxaloacetic transaminase (GOT) and the glutamic-pyruvic transaminase (GPT) values were nearly equal to that of the corresponding control in mind (Figure 7), the anticaries effect of the isothiocyanate by *in vivo* testing shown in Figure 6 was considered to be significant. In order to confirm the anticaries mechanism of the isothiocyanate, the inhibitory effect on both the glucan synthesis by the GTases and the antibacterial effect was studied. Figure 8 shows that isothiocyanate 9 was found to have no inhibition against glucan formation at a 62.5 ppm concentration, which was a higher value than a 50 ppm concentration *in vivo*. In addition, there was almost no inhibition against glucan formation at a 125 ppm concentration of isothiocyanate 9. Similarly, isothiocyanate 10 at 62.5 and 125 ppm concentration gave almost no inhibition against glucan synthesis (Figure 8). In order to confirm the antibacterial mechanism of the isothiocyanate on the mutans streptococci, the MIC value was measured. The MIC values of isothio-cyanate 9 on three species of mutans streptococci, *i.e.*, *S. sobrinus* 6715, *S. mutans* MT8148, and *S. mutans* IFO 13955, were found to be less than a 62.5 ppm concentration resulting in no inhibition against glucan synthesis (Figure 9). In addition, isothiocyanate 10 showed lower MIC values (Figure 9) than a 62.5 ppm concentration (Figure 8), which gave almost no inhibition against glucan synthesis. From the result of the almost non-inhibition effect against GTases (Figure 8) and the high antibacterial effect (Figure 9), the anticaries mechanism of the isothiocyanate seems not to be responsible for the inhibition effect against the GTases of mutans streptococci, but the antibacterial effect against mutans streptococci. On the other hand, oolong tea polyphenol is known to have an anticaries effect (*13, 14*). Contrary to isothiocyanate, the inhibition effect of oolong tea extract against the GTases was remarkable at concentarations of 62.5 and 125 ppm (Figure 10). However, the MIC values of the oolong tea extract against three species of mutans streptococci, *i.e.*, *S. sobrinus* 6715, *S. mutans* MT8148, and *S. mutans* IFO 13955, were more than 1000 ppm (Figure 11). Therefore, in contrast to the anticaries mechanism of isothiocyanate, the cause of the anticaries effect of oolong tea extract was not an antibacterial effect, but a GTases inhibition effect (*13*).

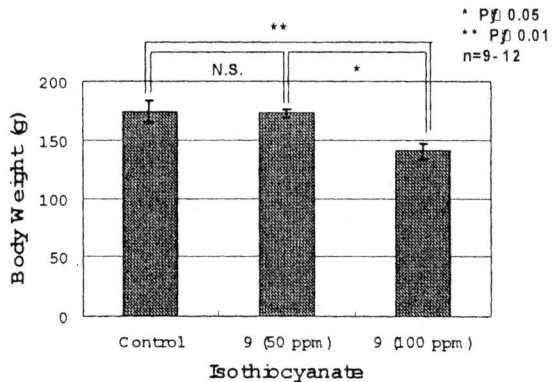

*Figure 5. Body weight of rats after being fed diet with or without isothiocyanate **9**. P: Observed significance level of the test. n: Number of rats. N.S.: No significance.*

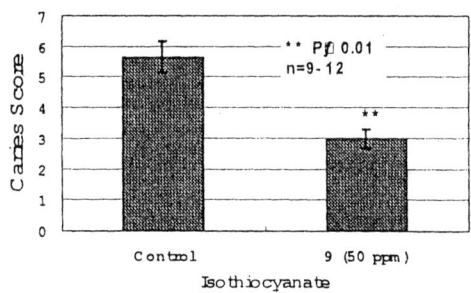

*Figure 6. Caries score of rats after being fed diet with or without isothiocyanate **9**. P: Observed significance level of the test. n: Number of rats.*

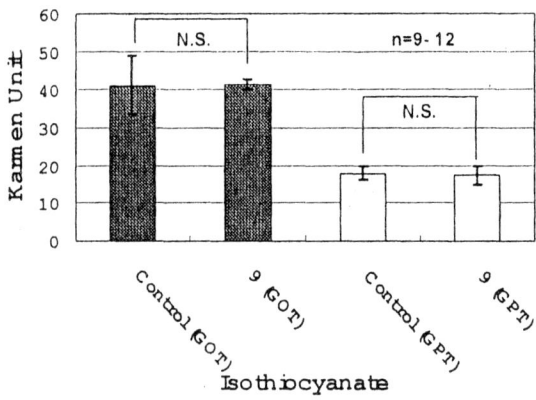

Figure 7. Karmen unit (Serum) of rats after being fed diet with or without isothiocyanate 9. n: Number of rats. N.S.: No significance.

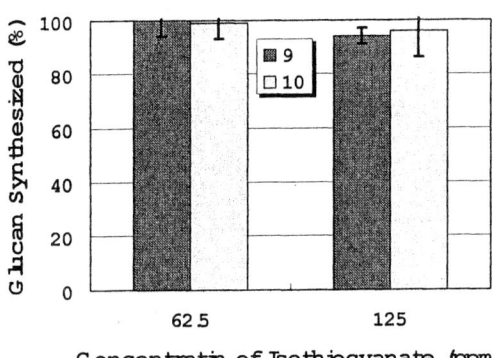

Figure 8. Inhibitory effect of isothiocyanates 9 and 10 on insoluble glucan synthesis by GTases from S. sobrinus 6715.

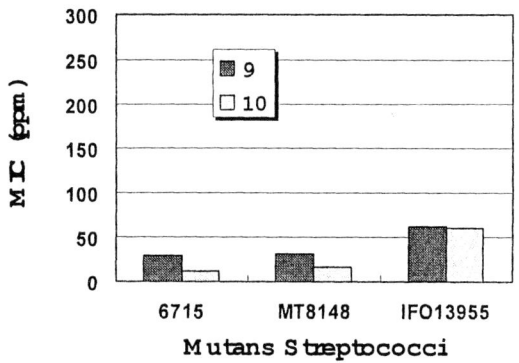

*Figure 9. MIC values of isothiocyanates **9** and **10** against S. sobrinus 6715, S. mutans MT8148, and S. mutans IFO 13955.*

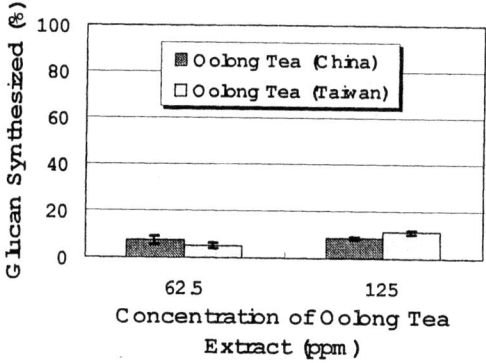

Figure 10. Inhibitory effect of the oolong tea extract on insoluble glucan synthesis by GTases from S. sobrinus 6715.

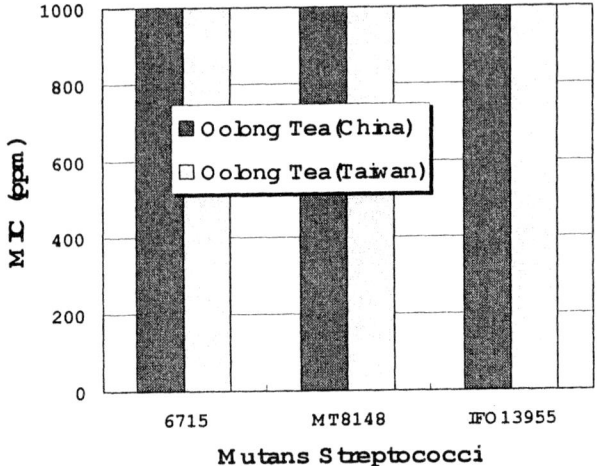

Figure 11. MIC values of oolong tea extract against S. sobrinus 6715, S. mutans MT8148, and S. mutans IFO 13955.

Conclusion

Isothiocyanate has been found to have an inhibitory effect on the sucrose dependent adherence by growing cells of mutans streptococci. In addition, the anticaries effect of isothiocyanate was confirmed by *in vivo* tests using rats. The anticaries mechanism of isothiocyanate seems to be an antibacterial effect rather than the inhibition of enzymatic glucan synthesis. The anticaries mechanism by isothiocyanate is considered to be opposite that of oolong tea extract.

References

1. Hamada, S.; Slade, H. D. *Microbiological Reviews*, **1980**, *44*, 331-384.
2. Loesche W. J. *Microbiological Reviews*, **1986**, *50*, 353-380.
3. Kakiuchi, N.; Hattori, M.; Nishizawa, M.; Yamagishi, T.; Okuda, T.; Namba, T. *Chem. Pharm. Bull.*, **1986**, *34*, 720-725.
4. Hattori, M.; Hada, S.; Watahiki, A.; Ihara, H.; Shu, Y.-Z.; Kakiuchi, N.; Mizuno, T.; Namba, T. *Chem. Pharm. Bull.*, **1986**, *34*, 3885-3893.
5. Saeki, Y.; Ito, Y.; Shibata, M.; Sato, Y.; Okuda, K.; Takazoe, I. *Bull. Tokyo dent. Coll.*, **1989**, *30*, 129-135.
6. Osawa, K.; Yasuda, H.; Maruyama, T.; Morita, H.; Takeya, K.; Itokawa, H. *Chem. Pharm. Bull.*, **1992**, *40*, 2970-2974.

7. Osawa, K.; Yasuda, H.; Maruyama, T.; Morita, H.; Takeya, K.; Itokawa, H.; Okuda, K. *Chem. Pharm. Bull.*, **1994**, *42*, 922-925.
8. Muroi, H.; Kubo, I. *J. Agric. Food Chem.*, **1993**, *41*, 1780-1783.
9. Sakanaka, S.; Kim, M.; Taniguchi, M.; Yamamoto, T. *Agric. Biol. Chem.* **1989**, *53*, 2307-2311.
10. Sakanaka, S.; Sato, T.; Kim, M.; Yamamoto, T. *Agric. Biol. Chem.* **1990**, *54*, 2925-2929.
11. Sakanaka, S.; Shimura, N.; Aizawa, M.; Kim, M.; Yamamoto, T. *Biosci. Biotech. Biochem.* **1992**, *56*, 592-594.
12. Yamamoto, T.; Juneja, L. R.; Chu, D.-C.; Kim, M. In *Chemistry and application of green tea*; Sakanaka, S., Ed.; CRC Press, LLC: USA, 1997; pp. 87-101.
13. Ooshima, T.; Minami, T.; Aono, W.; Izumitani, A.; Sobue, S.; Fujiwara, T.; Kawabata, S.; Hamada, S. *Caries. Res.* **1993**, *27*, 124-129.
14. Ooshima, T.; Minami, T.; Aono, W.; Tamura, Y.; Hamada, S. *Caries. Res.* **1994**, *28*, 146-149.
15. Masuda, H. Harada, Y.; Tanaka, K.; Nakajima, M.; Tateba, H. In *Biotechnology for Improved Foods and Flavors*; Takeoka, G. R.; Teranishi, R.; Williams, P. J.; Kobayashi, K. Eds.; ACS Symposium Series 637; American Chemical Society: Washington, DC, 1996; pp. 67-78.
16. Kishimoto, N; Tano, T.; Harada, Y.; Masuda, H. *J. of the Japan Association of Food Preservation Scientists* **1999**, *25*, 7-13.
17. Masuda, H.; Harada, Y.; Kishimoto, N.; Tano, T. In *Aroma Active Compounds in Foods*; Takeoka, G. R.; Guntert, M.; Engel, K.-H. Ed.; ACS Symposium Series 794; American Chemical Society: Washington, DC, 2001; pp. 229-250.
18. Masuda, H. In *Functional Food Ingredients: Trends and Prospects*; Shahidi, F.; Shibamoto, T.; Osawa, T. Eds.; Proceedings of 2000 International Chemical Congress of Pacific Basin Societies. Honolulu, Hawaii, Dec. 14-19, 2000, submitted.
19. Masuda, H.; Harada, Y.; Inoue, T,; Kishimoto, N.; Tano, T. In *Flavor Chemistry of Ethnic Foods*; Shahidi, F.; Ho C.-T. Eds.; Kluwer Academic/Plenum Publishers, N.Y., USA, 1999; pp. 85-104.
20. Masuda, H.; Tsuda, T.; Tateba, H.; Mihara, S., Japan Patent 90,221,255, 1990.
21. Harada, Y.; Masuda, H.; Kameda, W., Japan Patent 95,215,931, 1995.
22. Tanabe, M.; Kanda, T.; Yanagida, A.; Shimoda, S., Japan Patent 96,259,453, 1996.

Chapter 10

Bioactivity of Lycopene-Rich Carotenoid Concentrate Extracted from Tomatoes

John Shi

Food Research Center, Agriculture and Agri-Food Canada, 93 Stone Road West, Guelph, Ontario N1G 5C9, Canada

Licopene-rich carotenoid concentrate as a functional food is extracted from tomatoes by food processing. Lycopene would undergo degradation via isomerization and oxidation during extraction under different processing conditions, which impacts its bioactivity and reduces the functionality for health benefits. The study has been carrying on to determine the antioxidant properties of lycopene-rich carotenoid concentrate on lipid peroxidation which was generated from the oxidation of α-linolenic acid. The effects of thermal treatment, O_2 and light on the stability of lycopene bioactivity were determined. Results have shown that lycopene bioactivity potency depends on the extent of isomerization, and the composition of lycopene-rich carotenoid concentrate. *cis*-Isomers have high bioactivity but are less stable. Lycopene-rich carotenoid concentrate has higher bioactivity than puree lycopene. Synergistic effect of lycopene-rich carotenoid concentrate is based on the content of lycopene, and other carotenoids in the extract concentrate. A true assessment of helath benefits of lycopene-rich carotenoid concentrate depends on the lycopene content of all *trans*-isomer, *cis*-isomers, other carotenoids, and their composition.

Recently there has been growing interest in the ability of lycopene to act as a cancer-preventative agent. Lycopene functions as an antioxidant and exhibits a high physical quenching rate constant for singlet oxygen *in vitro*. The quenching constant of lycopene has been reported to be more than double that of β-carotene and 10 times more than α-tocopherol. Increasing clinical evidence supports the role of lycopene as an important micronutrient, since it may provide protection against prostate cancer, lung cancer and a broad range of other epithelial cancers (*1*). Levy et al. (*2*) studied the inhibitory effect of lycopene comparing it with that of α-and β-carotene on the growth of several human cancer cells. The number of servings of lycopene-rich food, such as tomatoes, tomato sauce, and pizza, significantly correlated with a low risk for prostate cancer (*3*). Determination of the bioactivity related to the degree of lycopene oxidation and isomerization would provide better insight into the potential health-promoting benefits of the lycopene-based products.

Chemical Property of Lycopene

Lycopene is known to exist in a variety of isomeric forms, including the all-*trans*, mono-*cis*, and poly-*cis* forms. With very few exceptions, the natural configuration of lycopene in plants is all-*trans*. Thermodynamically, this corresponds to the most stable configuration (*4,5*). All-*trans*- isomer lycopene ($C_{40}H_{56}$) is an acyclic, open chain polyene hydrocarbon with 13 double bonds, of which 11 are conjugated in a linear array (Figure 1). But 7 bonds can isomerize from the *trans*-form to the mono or poly-*cis* form under the influence of heat, light, and certain chemical reactions. *Cis*-isomers have distinct physicochemical characteristics (and hence bioactivity and bioavailability) compared to their all-*trans* counterparts. In general, the *cis*-isomers are more soluble in oil and hydrocarbon solvents than all-*trans* isomers. They are less prone to crystallization because of their kinked structures. The appearance of a distinct absorption maximum in the UV region ("*cis*-peak") is useful for distinguishing between the different isomers. The 5-*cis*, 9-*cis*, and 15-*cis*- isomers of lycopene have been assayed in various foods and in human tissues using NMR spectroscopy (*6*).

With its acyclic structure, large array of conjugated double bonds, and important hydrophobicity, lycopene exhibits a range of unique and distinct biological properties. Of these properties, its antioxidant properties continue to arouse substantial interest. The system of conjugated double bonds allows lycopene molecules to efficiently quench the energy of notably deleterious forms of oxygen (singlet oxygen) and to scavenge a large spectrum of free radicals. Of all naturally occurring carotenoids, lycopene is the most efficient quenchers of singlet oxygen (*7,8*).

The antioxidant activity of lycopene is related to their ability to quench singlet oxygen (O_2^-) and to trap peroxyl radicals (ROO·) (9). The quenching activity of the different carotenoids depends essentially on the number of conjugated double bonds. It is modulated by end groups or the nature of the substitutes in the carotenoids that contain cyclic end groups (10). Lycopene's superior ability to quench singlet oxygen vs. that of γ- and β-carotene is related to the opening of the β-ionone ring. Lycopene also allows for the formation of peroxyl radicals capable of acting as prooxidants and of undergoing autooxidation themselves. Oxidative degradation can occur at either end of the C_{40}-carbon skeleton. Ultimately, the final non-carotenoid fragments of lycopene oxidative degradation form as the result of direct oxidative cleavage of carbon-carbon double bonds.

Bioactivity potency of lycopene is dependent on the extent of degradation due to isomerization and oxidation. The main causes of lycopene degradation during food processing are oxidation and isomerization. Determination of the extent of lycopene isomerization would provide better insights into the potential health benefits of processed food products. In processed foods, oxidation is a complex process and depends upon many factors, such as processing conditions, moisture, temperature, and the presence of pro- or antioxidants and of lipids. The characterization and quantification of isomers would be desirable to more accurately assess the bioactivity than just the total lycopene content with no knowledge of its isomeric composition.

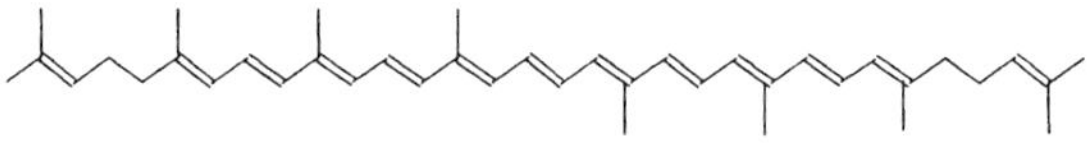

Figure 1. Molecular structure of lycopene.

Lycopene in Tomatoes

Lycopene is the principal pigment found in tomatoes and is responsible for the characteristic deep-red color of ripe tomato fruits and tomato products. Lycopene is the most abundant carotenoid in tomatoes, and represent about 83% of the total pigments present (11). Tomatoes and tomato products are the major source of lycopene, and are considered to be an important contributor of carotenoids to the human diet. Deep-red fresh tomato fruits are considered to contain high concentrations of lycopene. Processing conditions such as high temperature, long processing time, light, and oxygen have been shown to have effects on lycopene degradation. Degradation of lycopene not only affects the attractive color of the final products, but also their nutritive value. Losses of lycopene during the tomato processing are of commercial significance. Although processing of tomatoes by cooking, freezing, or canning does not usually cause significant changes in the total lycopene content in the conventional processing of tomatoes, much of the bioactivity of lycopene can be lost due to the conversion of the all-*trans*- isomers to *cis*- isomers which are susceptible to oxidation. These changes are mainly due to heat stress imposed by the relatively harsh thermal process required to achieve the shelf-life stability of processed tomato products. Coupled with exposure to oxygen and light, heat treatment which disintegrate tomato tissue can result in destruction of lycopene (*12,13*).

Effect of Heat on Lycopene Degradation

The changes of content of *trans*- and *cis*-isomers of lycopene concentrate during heat treatments are shown in Figure 2. Increasing the temperature from 100 to 180°C increased the degradation of *trans*-isomer and *cis*-isomer of lycopene. It was observed that the *cis*-isomers increased with thermal treatment at 100°C, but dropped significantly with treatment at 180°C. An increase in temperature from 100°C to 180°C caused a 76% decrease in total lycopene content (Figure 3). After 90 min treatment at 180°C resulted in lycopene degradation and the greater concentration of total lycopene loss, compared to *cis*-isomer formation. Total lycopene concentration decreased over treatment time, but *cis*-isomers mostly appeared within the first hour of heating. After 1 hours of heating, the rate of *cis*-isomer accumulation decreased.

The main cause of lycopene degradation is isomerization and oxidation. It is widely presumed that lycopene in general undergoes isomerization with thermal processing. This isomerization resulted in conversion of the all-*trans* isomers to the *cis*-isomers. The *cis*-isomers were formed in samples but oxidized with temperature and time during heat treatment. A large loss of lycopene during processing would indicate a longer and more drastic thermal procedure. The results suggest that oxidation of lycopene was the main mechanism of lycopene loss when heated above 100°C. The changes in lycopene content and the formation of *cis*-isomers may result in a reduction in bioactivity potency (*4,5,14,15*).

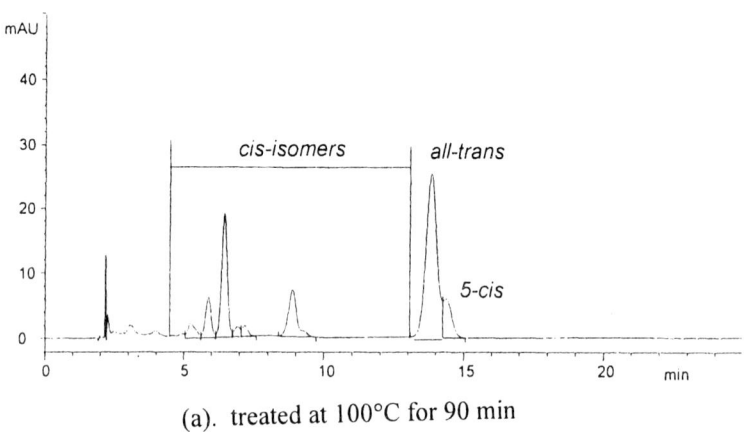

(a). treated at 100°C for 90 min

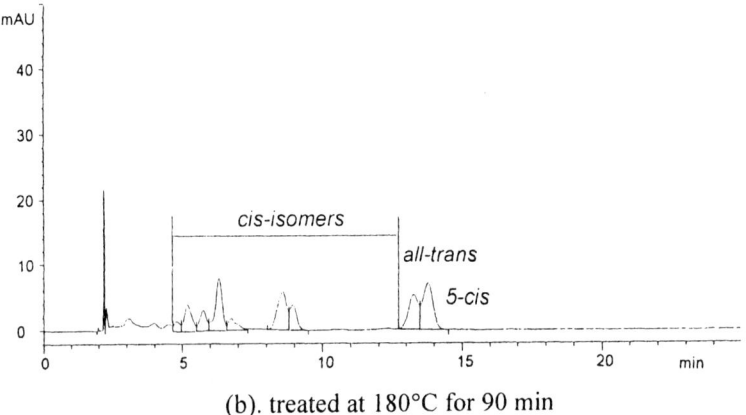

(b). treated at 180°C for 90 min

Figure 2. Effect of heat treatment on lycopene oxidation and isomerization (Reproduced with permission from reference 16. Copyright 2002 Korea Society of Food Science and Nutrition.)

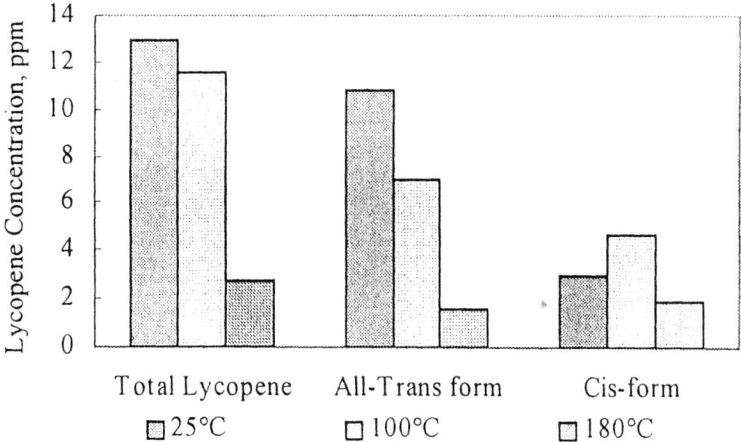

Figure 3. Lycopene oxidation and isomerization after being heated 90 min (Reproduced with permission from reference 16. Copyright 2002 Korea Society of Food Science and Nutrition.)

Effect of Light-Irradiation on Lycopene Degradation

The effects of light irradiation on the content of total lycopene, *trans*-isomers and *cis*-isomers in lycopene sample under different irradiation intensities of 2010, 900, 650 and 140 :mol m^{-2} s^{-1} for 24 hrs are shown in Figure 4. The loss in total lycopene, *trans*-isomers and *cis*-isomers increased significantly as the intensity of the light irradiation increased. The small amount of *cis*-isomers formation during light irradiation indicate either less *cis*-isomers were formed, or that the oxidation reactions were predominating. The increase in *cis*-isomers began to occur at the onset of exposure to light. Total lycopene loss as well as *cis*-isomer loss increased with light irradiation time. This suggests that the light irradiation causes losses of total lycopene content. The amount of *cis*-isomer formed during light irradiation appeared to decrease quickly as the intensity of light increased, suggesting that *cis*-isomer oxidation was the main reaction pathway. A possible explanation for this phenomenon is that the *trans*-isomer first isomerizes into *cis*-isomers, which then preferentially follow the oxidative pathway. The rate of *cis*-isomer oxidation was much greater under both heat and light treatment than rate of *cis*-isomer formation. It is possible that any *cis*-isomer formed was quickly degraded into oxidative by-products, which indicate that the rate of *cis*-isomer oxidation during light irradiation was much greater than formation of *cis*-isomers. Comparing to treatments at 2010 :mol m^{-2} s^{-1} and at 140 :mol m^{-2} s^{-1} for 24 hrs, results showed 89% of total lycopene was

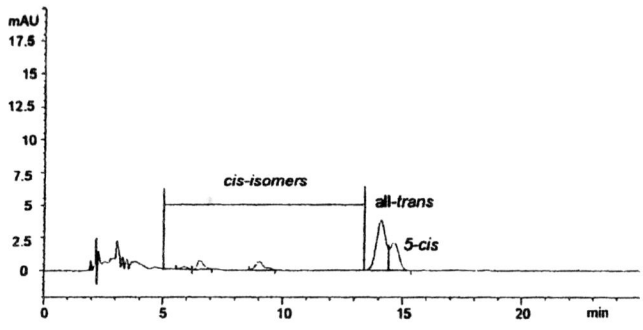

(a). at 2010 μmol m^{-2} s^{-1} intensity.

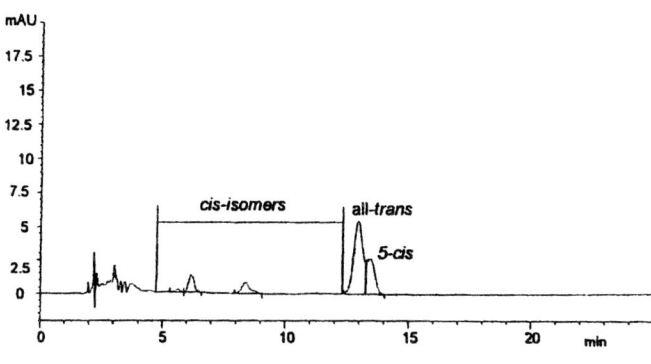

(b). at 900 μmol m^{-2} s^{-1} intensity.

Figure 4. Changes of content of all-trans-and cis-isomers of lycopene during sun-light irradiation

(Reproduced with permission from reference 16. Copyright 2002 Korea Society of Food Science and Nutrition.)

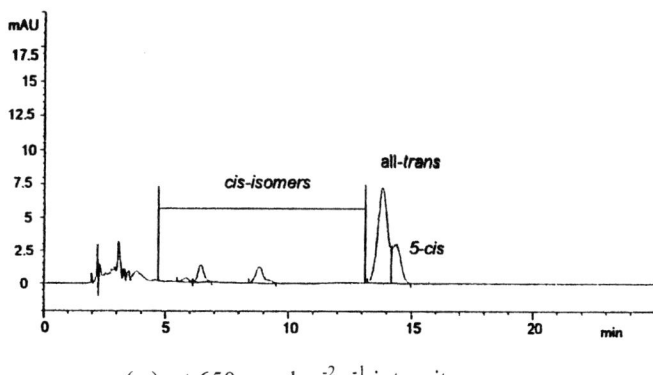

(c). at 650 μmol m^{-2} s^{-1} intensity

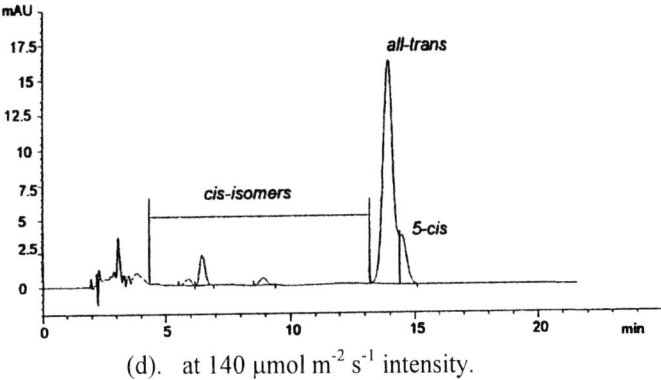

(d). at 140 μmol m^{-2} s^{-1} intensity.

Figure 4. *Continued.*

lost. The content of total lycopene, *trans*-isomers and *cis*-isomers in lycopene extract decreased under all light irradiation cases, which indicate light irradiation more quickly induce lycopene oxidation.

Health-Promoting Effect and Lycopene Degradation

The changes in lycopene content and the distribution of *trans-cis* isomerization will result in a reduction in biological potency, when lycopene concentrate are subjected to processing (*4,5,14,15*). Lycopene would undergo degradation via isomerization and oxidation under different processing conditions, which impacts its bioactivity and reduce the functionality for health benefits. It is the key process how to maintain high bioactive property during food processing and storage. The isomerization and oxidation of lycopene greatly depends on the treatments used since each treatment produces a different form of energy (such as heat, light, respectively). It was found that the light irradiation caused more losses in total lycopene than the heating treatment. In all treatments, the rate of *trans*-isomer loss was greater than the *cis*-isomer formation, which suggests that degradation through oxidation of lycopene was the predominant mechanism, as compared to isomerization of *trans*-isomer to *cis*-isomer. The loss could be coupled by the conversion of *trans*- to *cis*-isomers, and followed by further direct degradation to smaller molecules of oxidized by-products which were not isomers of lycopene (*17*). *Cis*-isomers are unstable, while *trans*-isomers are in a stable ground state. Because lycopene is a highly unsaturated molecule, comprising many conjugated double bonds, it is very susceptible to oxidation. The pathway of lycopene degradation is described in Figure 5.

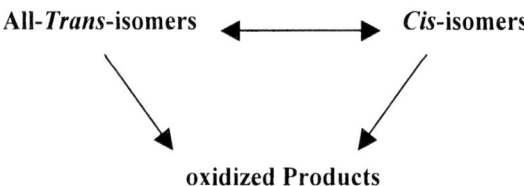

Figure 5. Pathway of lycopene degradation.

Bioactivity potency depends on the content of total lycopene and the extent of isomerization (*4,5,14,18*). Thus isomerization would lead to change the bioactivity of lycopene. Characterization and quantification of lycopene isomers would provide a better understanding of the potential bioactive properties and health benefits of the lycopene-based products. Controlling isomerization and oxidation of lycopene during lycopene-based food preparation can be of benefit in improving product quality. Changes in lycopene content and in the distribution of *trans- cis-* isomers result in modifications of its biological properties. Coupled with exposure to oxygen and light, heat treatments that disintegrate tomato tissue can result in substantial destruction of lycopene.

A true assessment of the relationship between nutritional quality and health benefits of dietary lycopene depends not only on the total lycopene content, but also on the distribution of lycopene isomers. Better characterization and quantification of lycopene isomers would provide better insight into the potential nutritional quality and health benefits. The control of lycopene oxidation and isomerization during production and storage can be of benefit improving the retention biological activity and health-promoting effect.

Acknowledgements

The authors gratefully acknowledge the contribution of the Guelph Food Research Center, Agriculture and Agri-Food Canada (AAFC Journal Series No. S128).

References

1. Micozzi, M.S.; Beecher, G.R.; Taylor, P.R.; Khachik, F. *J. Natl. Cancer Inst.* **1990**, *82*, 282-288.
2. Levy, J.; Bosin, E.; Feldman, B.; Giat, Y.; Minster, A.; Danilenko, M.; Sharoni, Y. *Nutr. Cancer* **1995**, *24*, 257-266.
3. Giovannucci, E.; Ascherio, A.; Rimm, E.B.; Stampfer, M.J.; Colditz, G.A.; Willett, W.C. J. Natl. Cancer Inst. **1995**, *87*, 1767-1777.
4. Emenhiser, C.; Sander, L.C.; Schwartz, S.J. *J. Chromatogr. A.* **1995**, *707*, 205-216.
5. Wilberg, V.C.; Rodriguez-Amaya, B.D. *Lebensm Wiss. und Technol.* **1995**, *28*, 474-480.
6. Zumbrunn, A.; Uebelhart, P.; Eugster, C.H. *Helv. Chim. Acta.* **1985**, *68*, 1540-1542.
7. Di Mascio, P.; Kaiser, S.; Sies, H. *Arch. Biochem. Biophys.* **1989**, *274*, 532-538.
8. Conn, P.F.; Schalch, W.; Truscott, T.G. *J. Photochem. Photobiol. B.* **1991**, *11*, 41-47.
9. Foote, C.S.; Denny, R.W. *J. Am. Chem. Soc.* **1986**, *90*, 6233-6235.

10. Stahl, W.; Sundquist, A.R.; Hamusch, M.; Schwarz, W.; Sies, H. *Clin. Chem.* **1993**, *39*, 810-814.
11. Gould, W.A. *Tomato Production, Processing and Technology*, 3rd ed., 1992, CTI Publ., Baltimore.
12. Schierle, J.; Bretzel, W.; Buhler, I.; Faccin, N.; Hess, D.; Steiner, K.; Schuep, W. *Food Chem.* **1996**, *59*, 459-465.
13. Nguyen, M.; Schwartz. S. *Proc. Soc. Exp. Biol. Med.* **1998**, *218*, 101-105.
14. Khachik, F.; Beecher, G.R.; Goli, M.B.; Luby, W.R.; Smith, J.C. *Anal. Chem.* **1992**, *64*, 2111-2122.
15. Stahl, W.; Sies, H. *Arch. Biochem. Biophys.* **1996**, *336*, 1-9.
16. Shi, J.; Wu, Y.; Bryan, M.; Le Maguer, M. *Nutriceuticals and Food.* **2002**, *7*, 179-183.
17. Boskovic, M.A. *J. Food Sci.* **1979**, *44*, 84-86.
18. Stahl, W.; Sies, H. *J. Nutr.* **1992**, *122*, 2161-2166.

Antioxidants

Chapter 11

Antioxidants in Herbs of Okinawa Islands

Nobuji Nakatani

Graduate School of Human Life Science, Osaka City University, Osaka, Japan

In the course of our searching for antioxidants from natural sources, we have focused on the plants which are traditionally used as food, tea and medicine in Okinawa, Japan. Several spices, such as *Peucedanum japonicum* Thunb (Botanbofu in Japanese), *Musa balbisiana* Colla (Ryukyubasho), *Alpinia flabellata* Ridley (Iriomotekumatakeran), *Alpinia speciosa* K. Schum (Getto), *Smilax nervo-marginate* Hayata (Sasabasankirai), *Smilax china* L var. *kuru* Sakaguchi ex Yamamoto (Okinawasarutoriibara)., *Livistona chinensis* R. Br var. *subglobosa* Becc (Biro), and *Artemisia campestris* L (Ryukyuyomgi). The present study describes the chemistry and antioxidant activity of these eight Okinawan herbs.

There are increasing much attention and interest in natural antioxidant, one of the most important nutraceuticals, not only for food preservation but for the prevention of numerous diseases such as atherosclerosis, diabetes and cancer, induced by oxidative stress (*1*). Our group has been engaging in isolation and structure elucidation of efficient antioxidants from natural resources, particularly from tropical spices and herbs (*2-5*). Recently we have focused on the subtropical plants, which grow in Yaeyama Islands, Okinawa, the southernmost part of Japan. Okinawa is well known as the island famous for world's highest longevity. The plants used in this study are *Peucedanum japonicum* Thunb

(Botanbofu in Japanese), *Musa balbisiana* Colla (Ryukyubasho), *Alpinia flabellata* Ridley (Iriomotekumatakeran), *Alpinia speciosa* K. Schum (Getto), *Smilax nervo-marginate* Hayata (Sasabasankirai), *Smilax china* L var. *kuru* Sakaguchi ex Yamamoto (Okinawasarutoriibara)., *Livistona chinensis* R. Br var. *subglobosa* Becc (Biro), and *Artemisia campestris* L (Ryukyuyomgi). The leaves of these eight plants are traditionally used for vegetables and teas, and for flavoring rice cakes and meats by wrapping with.

The present study describes antioxidant activity of the eight Okinawan herbs, their constituents and antioxidant evaluation measured by several methods.

Experimental

Plant Materials

The leaves of eight herbs were collected on Iriomote Island, one of The Yaeyama Islands, Okinawa.

Extraction Procedure for Antioxidant Screening

Dried and pulverized leaves were extracted with ten times volume (w/v) of acetone-water (8:2). After evaporation of acetone from the extract, ethyl acetate was added to the aqueous residue, then partitioned to give the ethyl acetate-soluble fraction and the water-soluble fraction.

General Procedure for Isolation of Antioxidants from Okinawan Herbs

Dried and pulverized leaves was successively extracted with hexane, dichloromethane, and 70% aqueous acetone at room temperature (Figure 1). Organic solvent of the 70% aqueous acetone extract was evaporated *in vacuo*. The residual aqueous solution was successively partitioned with ethyl acetate, n-butanol to give the ethyl acetate fraction, the n-butanol fraction and the water-soluble fraction. Each fraction was examined for antioxidant assay. The active fractions were purified with several kinds of chromatographic technique to isolate active constituents.

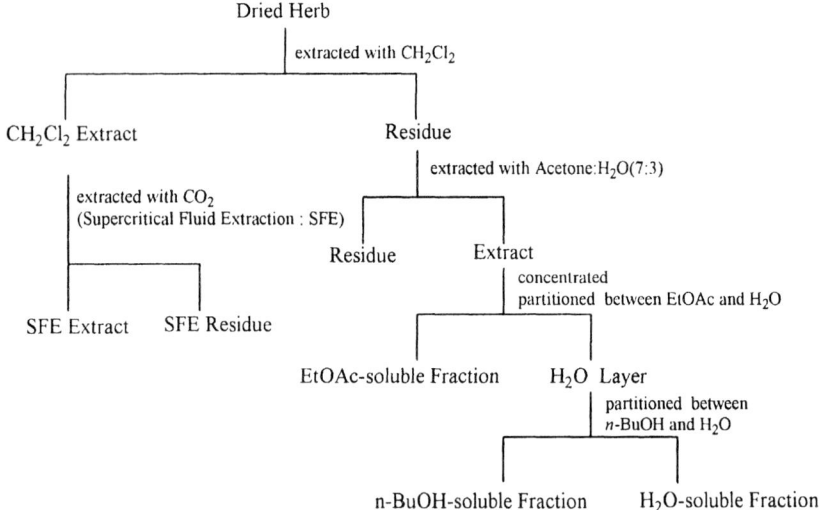

Figure 1. Extraction and fractionation of Herbs.

Results and Discussion

Antioxidant Screening of Okinawan Herbs

The antioxidant activities of the ethyl acetate-soluble and the water-soluble fractions were measured by the ferric thiocyanate method (6). The activity of both soluble fractions of eight species decreased in the order BHT> *P. japonicum* > *S. nervo-marginate* > *L. chinensis* > *Art. campestris* > *S. china* > *M. balbisiana* > *Alp. speciosa* > *Alp. flabellata* > α-tocopherol. Similar results were obtained when the antioxidant of these specimens were measured by the TBA method. The antioxidant activity of both fractions of all herbs was stronger than that of α-tocopherol (natural antioxidant) at the same concentration of 0.02%. The activity of the ethyl acetate-soluble and the water-soluble fractions of *Art. campestris* was comparable to that of butylated hydroxy toluene (BHT, synthetic antioxidant).

DPPH Radical Scavenging Activity of Okinawan Herbs

The DPPH radical scavenging activity (6) of each fraction was evaluated by direct ESR measurements. The relative peaks height of the DPPH radical to the standard signal intensity of the manganese oxide appears dose-dependent. The concentration at which the scavenging activity of the tested fractions showed 50% of control (IC$_{50}$ value) was determined for the ethyl acetate-soluble, the water-soluble fractions, α-tocopherol, and L-ascorbic acid. The activity of the ethyl acetate-soluble fraction of each species decreased in the order α-tocopherol ≥ L-ascorbic acid > *Alp. flabellata* ≥ *S. china* ≥ *Art. campestris* ≥ *P. japonicum* > *M. balbisiana* > *Alp. speciosa* > *L. chinensis* based on the statistical analysis. *S. nervo-marginate* did not show any significant differences against five species (*P. japonicum, M. balbisiana, Alp. flabellata, S. china,* and *Art. campestris*). The activity of the water-soluble fractions decreased in the order α-tocopherol ≥ L-ascorbic acid > *P. japonicum* ≥ *Alp. flabellata* > *S. china* ≥ *M. balbisiana* ≥ *Art. campestris* ≥ *Alp. speciosa* ≥ *L. chinensis* > *S. nervo- marginate*. In addition, the radical scavenging activity of the water-soluble fractions also showed a stronger activity than those of the ethyl acetate-soluble fractions.

Evaluation of the Scavenging Activity on Superoxide Anion Radical (O$_2^-$)

Superoxide anion radical (O$_2^-$) was generated by hypoxanthine (HPX) / xanthine oxidase (XOD) (6). The O$_2^-$ scavenging activity of the water-soluble fractions was measured by the ESR spin-trapping technique. The activity decreased in the order L-ascorbic acid > *P. japonicum* ≥ *S. china* ≥ *Alp. flabellate* ≥ *S. nervo-marginate* > *Art. campestris* > *M. balbisiana* ≥ *L. chinensis*. The activity of *Alp. speciosa* was stronger than that of *L. chinensis*. There was no significant difference between the activity of *Alp. speciosa* and that of *M. balbisiana* or *Art. campestris*.

Evaluation of the Inhibitory Activity Against Xanthine Oxidase (XOD) (17)

As for the evaluation of the scavenging activity against O$_2^-$ in the HPX/XOD system, it should be taken in consideration that each tested fraction might inhibit XOD itself which generates O$_2^-$ (6). All the fractions did not show any significant inhibitory activity against XOD except for *S. china*, which inhibited 30% of the enzyme activity. This finding indicates that all plants except for *S. china* directly scavenged O$_2^-$.

In conclusion based on the total results on screening tests, it is clear that the antioxidant and the radical scavenging activities of the water-soluble fraction of

P. japonicum showed the highest activity among all fractions of tested herbs. Further study on the isolation and structure determination of antioxidants were done on *Peucedanum japonicum*, *Artemisia campestris* and *Alpinia speciosa*.

Compounds Isolated from n-Butanol Soluble Fraction of *Peucedanum japonicum*

P. japonicum belongs to Umbelliferae. The leaves and roots of this plant have been used as a wholesome vegetable and a folk medicine for the treatment of cough, diuretic, tonic, and nerve sedative in the Yaeyama Islands and southeast Asian countries. It is reported that some coumarins isolated from this plant were found to possess pharmacological activities (7-9). The n-butanol fraction of *P. japonicum* was purified by silica gel column chromatography to afford several phenolic compounds. Based on NMR and MS spectroscopic analyses, the structure of isolated compounds were determined as quercetin 3-*O*-glucoside (**1**, isoquercitrin), quercetin 3-O-rhamnosyl-(1,6)-glucoside (**2**, rutin), 3-O-caffeoylquinic acid (**3**), 4-*O*-caffeoylquinic acid (**4**), 5-*O*-caffeoylquinic acid (**5**, chlorogenic acid) and four prenylated coumarins such as praeroside II (**6**), praeroside III (**7**) and aspterin (**8**) and esclin (**9**). Figure 2 shows their structures and IC$_{50}$ values [in the parentheses] of scavenging activity against DPPH radical. Isoquercitrin (**1**), rutin (**2**) and caffeoylquinic acids (**3-5**) showed stronger radical scavenging activity than those of α-tochopherol and ascorbic acid. There were no differences among the activity of three caffeloylquinic acid isomers. Four coumarins did not show any activity compared with control.

Compounds Isolated from Ethyl Acetate Soluble Fraction of *Artemisia campestris*

Art. Campestris, called tall wormwood, belonging to Compositae, has been traditionally used as a folk medicine for tonsils, gallstone in Yaeyama Islands and for cold, cough and tuberculosis by American Indians (*10*), and reported that the water or ethanol extracts of this plant showed antioxidant and hepatoprotective actions (*11*).

The ethyl acetate fraction of *Art. campestris* was purified to give 3,4-dicafeoylquinic acid (**10**), 3,5-dicafeoylquinic acid (**11**) and 4,5-dicaffeoylquinic acid (**12**), methyl 3,5-dicaffeoylquinate (**13**) together with monocaffeoylquinic acid isomers and flavonoids. Figure 3 shows the DPPH radical scavenging activity of caffeoylquinic acid derivatives. Monocaffeoylquinic acids showed the activity comparable to α-tocopherol and ascorbic acid. On the other hand, dicaffeoylquinic acid group and caffeic acid showed remarkably higher activity than standard antioxidants.

Flavonoids Isolated from Ethyl Acetate Soluble Fraction and Water Soluble Fraction of *Alpinia speciosa*

The leaves of *Alpinia speciosa* and *Alp. Flabellate*, belonging to Zingiberaceae, have been also used as a wrapping and flavoring material for foods in the Yaeyama area. Some compounds isolated from *Alp. speciosa* are known to show strong antifungal activity (*12*). We previously reported the isolation and structural determination of five novel polymethoxylated phenolics and two unique flavonol-phenylbutadiene adducts from the leaves of *Alp.*

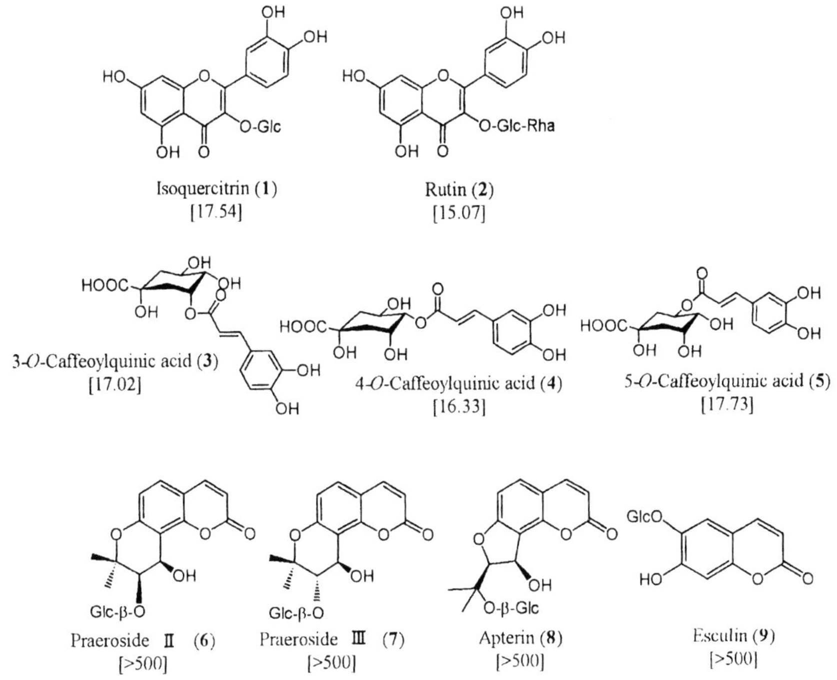

Isoquercitrin (1) [17.54]

Rutin (2) [15.07]

3-O-Caffeoylquinic acid (3) [17.02]

4-O-Caffeoylquinic acid (4) [16.33]

5-O-Caffeoylquinic acid (5) [17.73]

Praeroside II (6) [>500]

Praeroside III (7) [>500]

Apterin (8) [>500]

Esculin (9) [>500]

Figure 2. DPPH radical scavenging activity of the compounds isolated from Peucedanum japonicum. *[IC$_{50}$ values(μM) of isolated compounds tested against DPPH radical].*

flabellata (*13,14*). These compounds showed an antibacterial activity and a regulating effect on ovalbumin transport in human intestinal Caco-2 cells (*15*). From the ethyl acetate fraction of *Alp. speciosa*, kaempherol 3-*O*-glucoside (**14**), kaempherol 3-*O*-glucuronide (**15**), and quercetin 3-*O*-glucoside (**16**) were obtained along with *p*-coumaric acid and ferulic acid. From the water soluble fraction, kaempherol 3-*O*-glucuronide (**17**), quercetin 3-*O*-glucuronide (**18**) and quercetin 3-*O*-rhamnosyl-(1,6)-galactoside (**19**) were determined (Figure 4 for structures **14-19**).

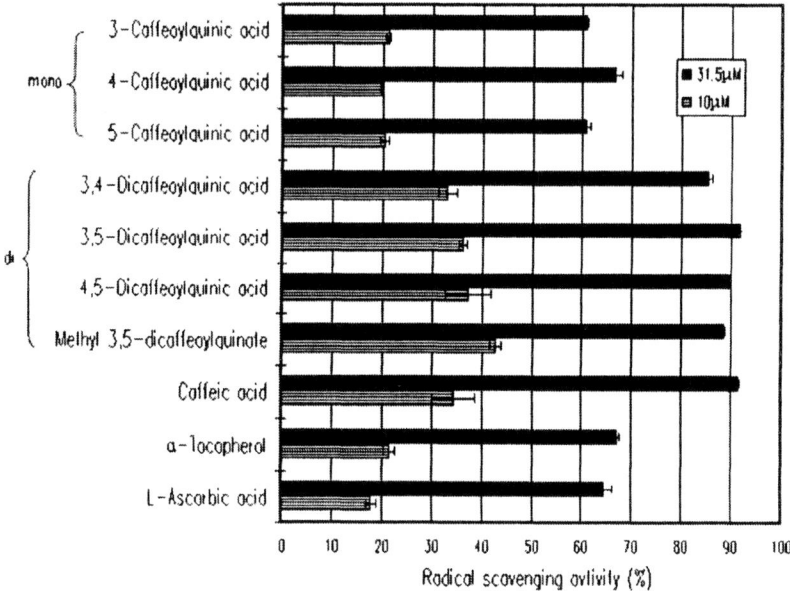

Figure 3. DPPH radical scavenging activities of caffeoylquinic acids isolated from Okinawan herbs. (DPPH conc. 100μM).

Figure 5 shows the DPPH radical scavenging activity of flavonol 3-*O*-glycosides and aglycones (kaempherol and quercetin). The activity of kaempherol-glycosides was very low except kaempherol, aglycone, which exhibited the activity as high as α-tochopherol and ascorbic acid. Quercetin and its glycosides shows the activity higher than standard antioxidants. It is suggested that a partial structure of *ortho*-dihydroxyphenyl moiety in B-ring of flavonoid is important to exhibit antioxidant effectiveness.

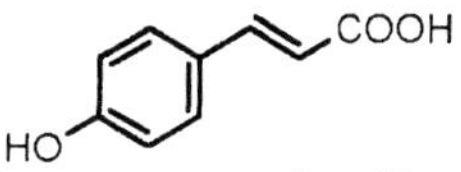

173

p-coumaric acid

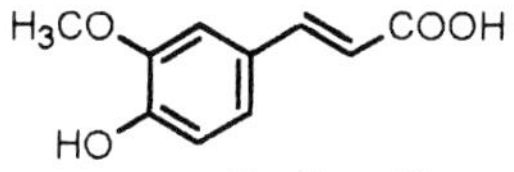

ferulic acid

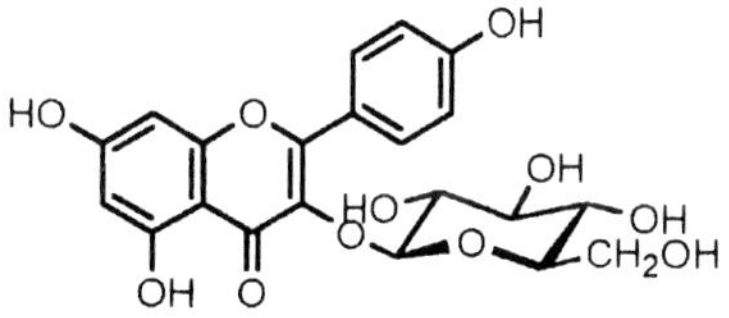

kaempferol 3-*O*-β-D-glucopyranoside (**14**)

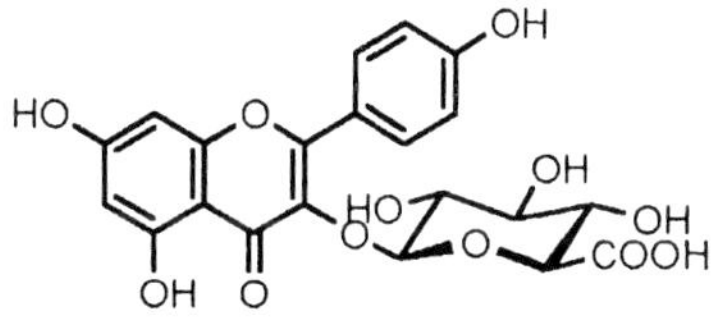

kaempferol 3-*O*-β-D-glucuronopyranoside (**15**)

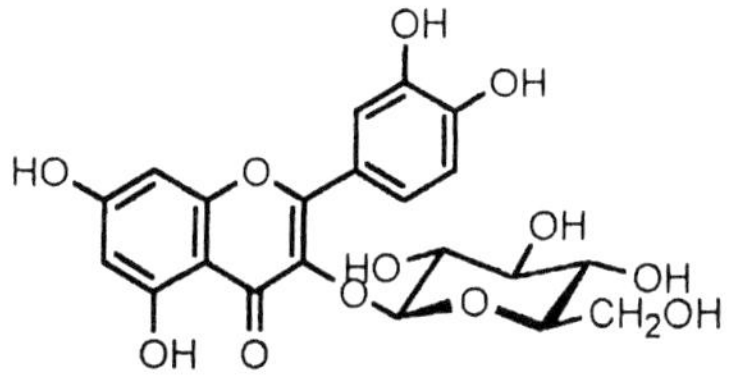

quercetin 3-*O*-β-D-glucopyranoside (**16**)

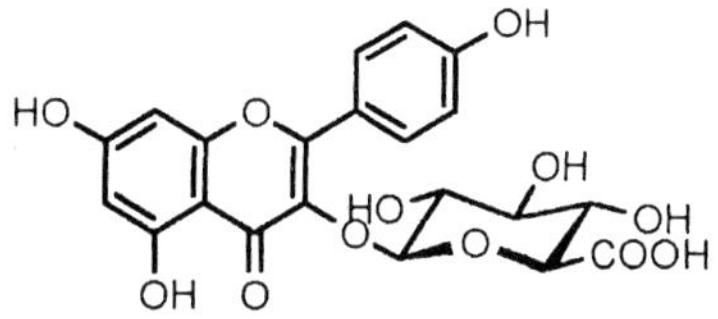

kaempferol 3-*O*-β-D-glucuronopyranoside (**17**)

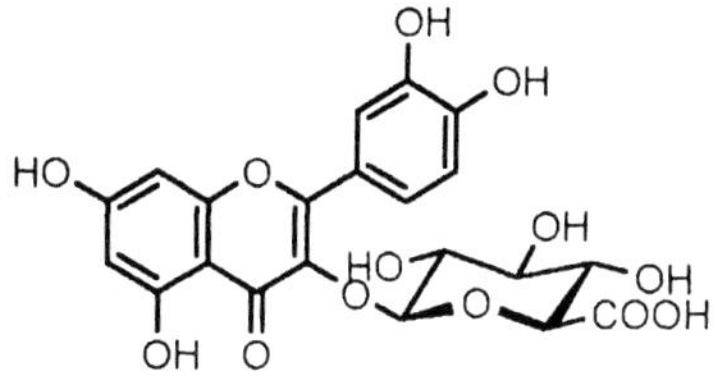

quercetin 3-*O*-β-D-glucuronopyranoside (**18**)

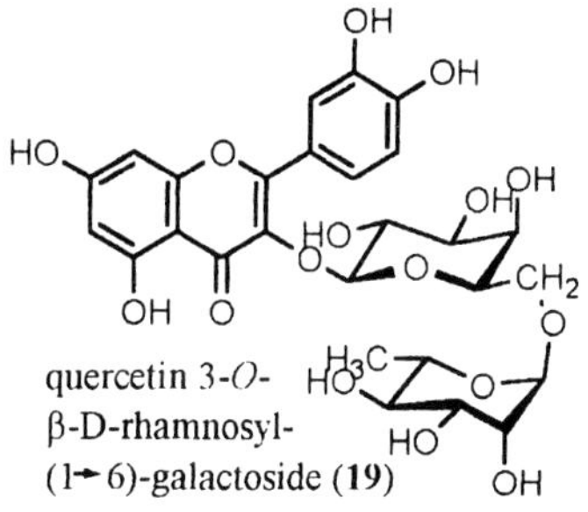

quercetin 3-*O*-β-D-rhamnosyl-(1→6)-galactoside (**19**)

Figure 4. Compounds isolated from Alpinia speciosa.

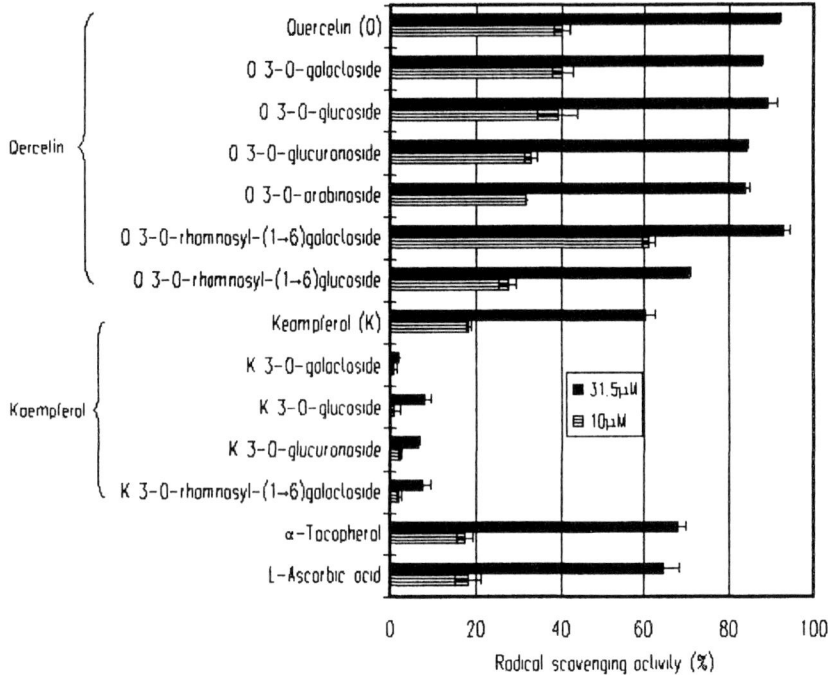

Figure 5. DPPH radical scavenging activities of flavonoids isolated from Okinawan herbs.(DPPH conc. 100μM).

Conclusion

Natural antioxidants are widely found as secondary metabolites in the vegetable kingdom of tropical and subtropical zones on the globe. In Okinawa, many different types of antioxidative polyphenols are included in edible and fork medicinal plants, which have been traditionally used for a long time. These antioxidants play an important and promising role in the preventive effect on diseases caused by oxidative stress. It is strongly suggested that chemical structure of antioxidants and activity should be studied more systematically, which is still continuously going on in our group.

Acknowledgements

This study was supported by the Program for Promotion of Basic Research Activities for Innovative Biosciences (BRAIN) and by Grants-in-Aid for Scientific Research from the Ministry of Education, Culture, Sports, Science and Technology of Japan (MEXT).

Reference

1. Shahidi, F. In *Natural Oxidants: chemistry, health effects, and applications*; Shahidi, E., Ed.; AOCS Press, Champaign, Illinois, 1997; pp 1-11.
2. Nakatani, N. In *Spices, Herbs and Edible Fungi;* G Charalambous, G., Ed.; Elsevier, Amsterdam; 1994, pp 251-271.
3. Nakatani, N.; Kikuzaki, H. In *Quality Management of Nutraceuticals*; Ho, C.-T.; Zheng, Q. Y., Eds.; American Chemical Society, Washington, DC; 2002, pp 230-240.
4. Kikuzaki, H. In *Herbs, Botanicals & Teas*; Mazza, G.; Olmah, B. D., Eds.; Technomic Publishing Co., Lancaster; 2000, pp 75-105.
5. Kikuzaki, H. *J. Food Sci.* **1993**, *58*, 1407-1410.
6. Hisamoto, H.; Kikuzaki, H.; Yonemori, S.; Nakatani, N. *ITE Letters on Batteries, New Technologies & Medicine* 2002, *3*, 63-68.
7. Jong,T. T.; Hwang, H-C.; Jean, M-Y. *J. Nat. Prod.* **1992**, *55*, 1396-1402.
8. Chen, I-S.; Chang, C-T.; Sheen, W-S.; Teng, C-M.; Tsai, I-L.; Duh, C-Y.; Ko, F-N. *Phytochem.* **1996**, *41*, 525-530.
9. Hsiao, G.; Ko, F-N.; Jong, T-T.; Teng, C-M. *Biol. Pharm. Bull.* **1998**, *21*, 688-692.
10. Foster, S.; Dulee, J.A. In *A Field Guide to Medical Plants: Eastern and Central North America*. Peterson, E.T., Ed.; Houghton Mifflin Company, New York, 1990, pp 222.
11. Aniya, Y.; Shimabukuro, M.; Shimoji, M.; Kohatsu, M.; Gyamfi, M.A.; Miyagi, C.; Kunii, D.; Takayama, F.; Egashira, T. *Biol. Pharm. Bull.* **2000**, *23*, 309-312.
12. Tawata, S.; Taira, S.; Kobamoto, N.; Ishihara, M.; Toyama, S. *Biosci. Biotechnol. Biochem.* **1996**, *60*, 1643-1645.
13. Kikuzaki, H.; Tesaki, S.; Yonemori, S.; Nakatani, N. *Phytochem.* **2001**, *56*, 109-114.
14. Tesaki, S.; Kikuzaki, H.; Yonemori, S.; Nakatani, N. *J. Nat. Prod.* **2001**, *64*, 515-517.
15. Tesaki, S.; Kikuzaki, H.; Tanabe, S.; Watanabe, M.; Nakatani, N. *ITE Letters on Batteries, New Technologies & Medicine* **2001**, *2*, 106-110.

Chapter 12

Antioxidants from Some Tropical Spices

Hiroe Kikuzaki

Division of Food and Health Sciences, Graduate School of Human Life Science, Osaka City University, Osaka, Japan

Antioxidants of some tropical spices originated from Southeast Asia were characterized. Five 2,3-dimethylbutane lignans and three tetrahydrofuran lignans were isolated from Papua mace, the aril of *Myristica argentea* (Myristicaceae). Several diarylheptanoids were found in the fruits of *Amomum subulatum* and the rhizome of *Zingiber officinale*, both of which belong to Zingiberaceae. From the leaves of *Murraya koenigii* (Rutaceae), ten carbazoles were obtained. Antioxidant properties of these isolated compounds evaluated by some different assay systems were overviewed.

Antioxidant activity is one of the characteristic properties of spices. Spices belonging to Labiatae, Myrtaceae and Zingiberaceae are generally well-known to possess antioxidant activity (*1*). Our previous studies on the antioxidants from spices and herbs resulted in finding totally more than a hundred active components in Labiatae herbs such as rosemary, sage, oregano and thyme (*2*), in some tropical ginger rhizomes (*3*) and in allspice belonging to the family Myrtaceae (*4,5*).

There are many native spices and herbs in Southeast Asia as well as popular spices such as pepper and cinnamon, which have been used not only for flavoring food but also in traditional medicine for the treatment of ailments by the local inhabitants. In most case, they are still unexplored phytochemically and biologically. Our purpose of this study is to get new information on antioxidants from such tropical spices and herbs.

Screening of the Antioxidant Activity of Tropical Spices

Thirty-three tropical spices and herbs originated in Southeast Asia were collected, two of which were belonging to the family Lauraceae, two to Myristicaceae, two to Myrtaceae, one to Ranunculaceae, one to Rutaceae, and 25 to Zingiberaceae. In preliminary screening of their antioxidant activity, extraction of each plant using some solvents with a different polarity afforded the non-polar and polar extracts.

Their antioxidant activity was evaluated on the basis of the inhibitory effects on the oxidation of a bulk methyl linoleate under aeration and heating using the oil stability index (OSI) method (6). Among the tested spices, fourteen species in the Table I showed a strong activity in their non-polar and/or polar extracts. In this study, we selected four species (*Myristica argentea*, *Amomum subulatum*, *Zingiber officinale* and *Murraya koenigii*) which we could obtain enough for the isolation of active compounds.

Antioxidant Assay

For the assessment of antioxidant properties of the isolated compounds from these four spices, we used two other systems as well as the OSI method. The electron-donating ability of chemical substances results in their antioxidant activity toward lipid oxidation. Therefore the radical scavenging activity against 1,1-diphenyl-2-picrylhydrazyl (DPPH) was measured as one of the antioxidant assay (7). The inhibitory effect on autoxidation of linoleic acid in an ethanol-buffer system was adopted as another method. This assay system is one of the simple conditions of lipid oxidation in multiple phases like food for evaluation of effects of antioxidants (8).

Antioxidants from Some Tropical Spices

Myristica argentea Warb.

Myristica argentea is belonging to the family Myristicaceae and originated in Irian Jaya and now cultivated in Papua New Guinea. Average yield of fruits is about 2000 fruits per year. The fruit has an oblong-cylindrical seed (up to 4 cm) called Papua nutmeg and a red aril covering the seed called Papua mace. Papua

Table I. Antioxidative Spices and Herbs in Southeast Asia

Taxonomic name	English name	Parts used	Origin [a]	Usage
Family Myristicaceae				
Myristica fragrans	Nutmeg	Seed	I	Spice, Stimulant, Carminative, Aphrodisiac
Myristica argentea	Papua nutmeg	Seed	I	Spice, Hypnotic, Antifebrile, Suppression of cough, Purgative
	Papua mace	Aril	I	Spice
Family Myrtaceae				
Syzygium aromaticum	Clove	Bud	I	Spice, Bactericide, Nematicide
Syzygium polyanthum	Salam leaf	Leaves	I	Spice, Diarrhea remedy
Family Rutaceae				
Murraya koenigii	Curry leaf	Leaves	M	Spice, Digestive, Medicine for vomiting
Family Zingiberaceae				
Alpinia malaccensis		Rhizome	M	Poultice
Alpinia nutans	Shell-flower	Rhizome	M	
Alpinia vitteliana		Rhizome	M	
Amomum blumenum		Rhizome	M	
Amomum subulatum	Greater cardamom	Fruit	India	Spice, Antifungal
Amomum tsao-ko	Tsao-ko	Fruit	C	Spice, Stomachache remedy
Curcuma xanthorrhiza		Rhizome	I	Spice, Medicine for anorexia, eczema, Anticonvulsant
Elettaria cardamomum	Cardamom	Fruit	India	Spice, Stomachache remedy
Zingiber officinale	Ginger	Rhizome	C	Spice, Medicine for anorexia, cough, diarrhea, vomiting

[a] I: Indonesia, M: Malaysia, C: China

nutmeg and mace are used as a spice like common nutmeg and mace (*M. fragrance*). In addition, Papua nutmeg is used as a traditional medicine in Indonesia to induce hypnosis, to suppress fever and coughs and to treat diarrhea (*9*).

Isolation of Lignans from Papua Mace

Dried Papua mace was extracted successively with dichloromethane and 70% aqueous acetone. A potion of the dichloromethane extract was further extracted with supercritical fluid carbon dioxide. The 70% aqueous acetone extract was separated into the ethyl acetate-soluble and the water-soluble fractions. The supercritical fluid CO_2 extract and the ethyl acetate-soluble fraction showed superior DPPH radical-scavenging activity to the dichloromethane extract and the water-soluble fraction. The supercritical fluid CO_2 extract was purified using several chromatographic techniques to give seven compounds (compounds **1~7**). The ethyl acetate-soluble fraction was purified in the same manner to afford compound **8**. Several spectroscopic analyses achieved to finding that these nine compounds were lignans, in which five were 2,3-dimethylbutane type lignans and the other three were tetrahydrofuran type lignans (Figure 1).

Radical-scavenging and Antioxidant Activities of the Isolated Lignans

Based on the substitution patterns of the aromatic rings of lignans **1~8**, they could be structurally classified into three groups (Groups A~C) as shown in Figure 1. Every lignan had one 4-hydroxy-3-methoxyphenyl group in the molecule. Another aromatic ring was a methylenedioxyphenyl group in Group A, a 4-hydroxy-3-methoxyphenyl group in Group B and a 3,4-dihydroxyphenyl group in Group C.

Figure 1 shows the DPPH radical-scavenging activity of the isolated lignans at the concentration of 40 μM. In this test, the concentration of DPPH ethanolic solution was 100 μM. DPPH radical-scavenging activity of each lignan was expressed in %=[(*A* in the absence of compound – *A* in the presence of compound at the concentration of 40 μM) / A in the absence of compound] x 100. Here, *A*=Absorbance at 520 nm. At the same concentration, α-tocopherol, a natural antioxidant, scavenged 84% DPPH radical and butyl hydroxytoluene (BHT), a synthetic antioxidant, scavenged 35% of DPPH radical. The activity decreased in the order, Group A>α-tocopherol>Group B>Group C, BHT. The scavenging ability of these lignans with two 3,4-disubstituted phenyl moieties against DPPH radical seemed to be dependent on the number of hydroxyl groups

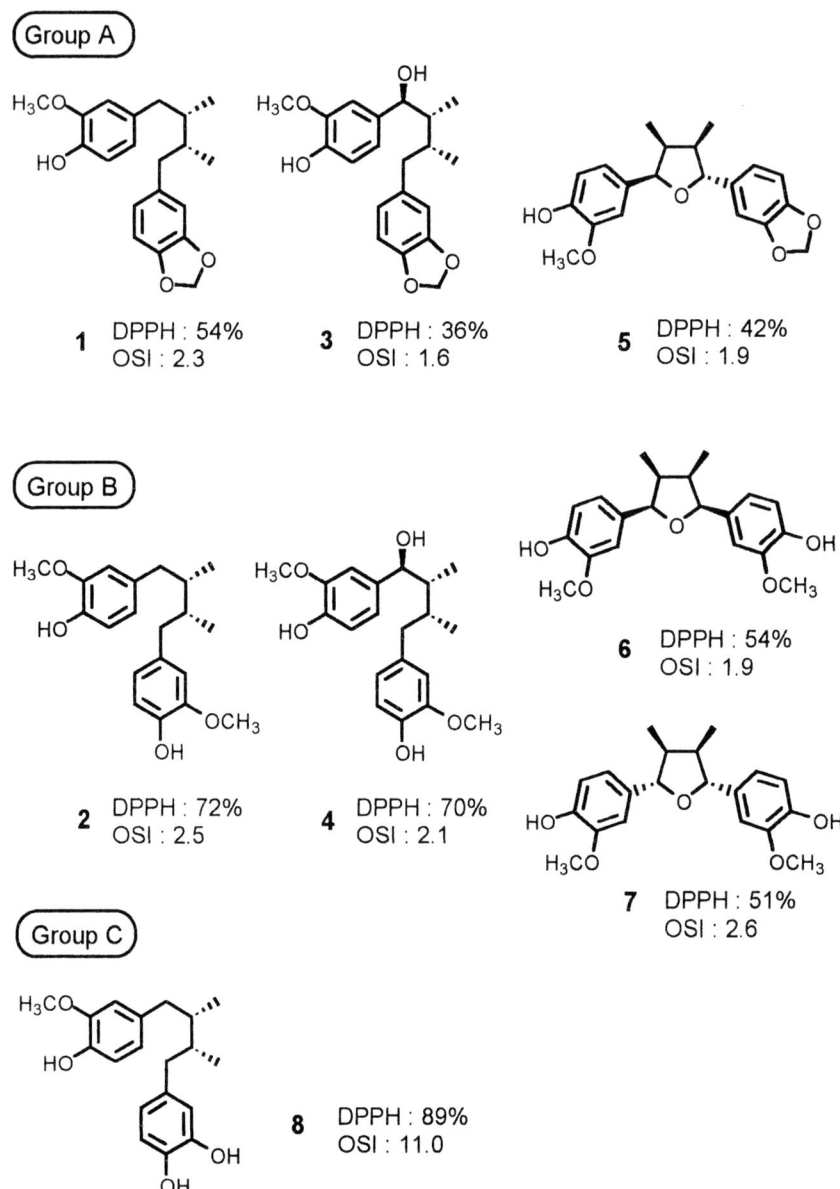

Figure 1 Structure of lignans isolated from Papua mace and their DPPH radical-scavenging and antioxidant activities.

on the benzene rings. In addition, compound **1** was more active than **3** and **5**. Similarly **2** showed a stronger activity than **4**, **6** and **7**. These findings might indicate that lignans without oxygen function on the alkyl chain have higher radical-scavenging effects than lignans with an oxygen function on the alkyl chain.

Each OSI value in Figure 1 is defined as the induction time of lipid oxidation of a substrate oil with each lignan at the concentration of 0.2μmol/g substrate oil over the induction time of a substrate oil with no additive (control) at 90 °C. α-Tocopherol and BHT showed 5.6 and 3.4 of OSI under the same condition, respectively. High OSI values mean a high level of antioxidant activity. Compounds **1~7** showed about two times longer induction time than control. On the other hand, compound **8** having a 3,4-dihydroxyphenyl moiety in the molecule remarkably extended the induction period. Such high radical-scavenging ability of **8** was consistent with its high antioxidant activity against the bulk methyl linoleate, suggesting that the antioxidant efficiencies of such lignans result in their electron-donating ability.

In the case of the ethanol-buffer system, the antioxidant activity was evaluated as the ratio of the length of time at which linoleic acid with each lignan (the final concentration of 25 μM) was 50% oxidized to that of control. High ratios indicated the high activity. The ratios were more than four for lignans of Group B, about three for lignans of Group A and about two for compound **8** (Group C). In this assay, 4-hydroxy-3-methoxyphenyl group was more effective than 3,4-dihydroxyphenyl group. These results differed from those obtained by the radical-scavenging assay and the OSI method.

Amomum subulatum (10)

Amomum subulatum is one kind of cardamoms belonging to Zingiberaceae. It is a native of the eastern Himalayas and now cultivated in Nepal and India. Because the fruits of *A. subulatum* are straw-colored and at least six times the size of the common greenish cardamoms (*Elettaria cardamomum*), they are commonly called either greater cardamom or black cardamom. They give a characteristic intense flavor to food and are also used as a traditional medicine.

Isolation of the Components from Greater Cardamom

Extraction and fractionation of the components of greater cardamoms were carried out using a similar way to those of papua mace. DPPH radical-scavenging activity at the concentration of 100 μg/mL of each extract and fraction were as follows; the dichloromethane extract: 15%, the ethyl acetate-soluble fraction: 93%, and the water-soluble fraction: 24%. Further purification of the ethyl acetate-soluble fraction which showed the highest ability resulted in the isolation of four compounds, protocatechualdehyde (9), protocatechuic acid (10) and two diarylheptanoids (11 and 12) (Figure 2). The complete structure determination of 11 and 12 were achieved on the basis of ^{1}H, ^{13}C and 2D NMR, UV, IR and MS measurements. Compound 12 was a new rare type of diarylheptanoid. All these compounds had one or two 3,4-dihydroxyphenyl groups in the molecules.

Radical-scavenging Activity of the Isolated Compounds

The DPPH radical-scavenging activity of all isolated compounds increased with concentration in the range of 5 to 80 µM. Figure 2 shows the scavenging effects at a concentration of 20µM. Compounds 10 and 12 scavenged about 50% of DPPH radical, comparable to α-tocopherol. Furthermore 9 and 11 scavenged more than 80% of the radical. The difference in the activity between 9 and 10 might be due to their electron-withdrawing capacity (*11*). The strongest activity of 11 might be ascribable to delocalization of the unpaired electron by its conjugated side chain (*12*).

Zingiber officinale

The rhizome of *Zingiber officinale* (the Family Zingiberaceae), common ginger is one of the most popular spices. It is native to Southeast Asia and nowadays used all over the world. In addition, it is well-known that ginger has many biological activities such as antimicrobial, antihepatotoxic, anti-ulcer, antitumor, cardiotonic activities, inhibition of prostaglandin-synthetase, inhibition of platelet aggregation, and so on (*13*). We have studied on the chemical components of ginger with the purpose of finding antioxidants. Our previous works resulted in the isolation and structure determination of more than twenty new gingerol related compounds and diarylheptanoids (*14-19*). Further investigation led to the isolation of some diarylheptanoids, which had the same heptane chain, a hept-4-en-3-one, and a variety of aromatic rings (Figure 3). The DPPH radical-scavenging activity of them was dependent on their aromatic ring structure as observed in lignans from Papua mace. The activity decreased in the order, 14, 15, 13>16>17, which indicated that the number of hydroxy groups on the benzene ring and *ortho* substitution with the electron donor methoxy group contribute to the radical scavenging ability of these diarylheptanoids.

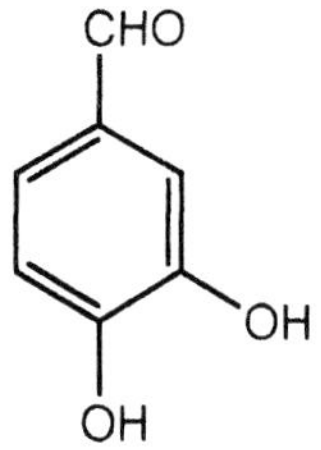

9 DPPH : 82%

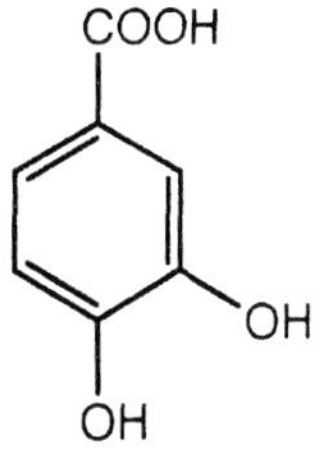

10 DPPH : 52%

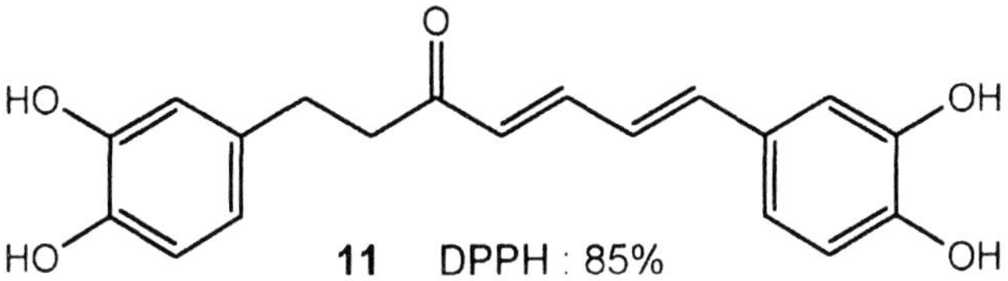

11 DPPH : 85%

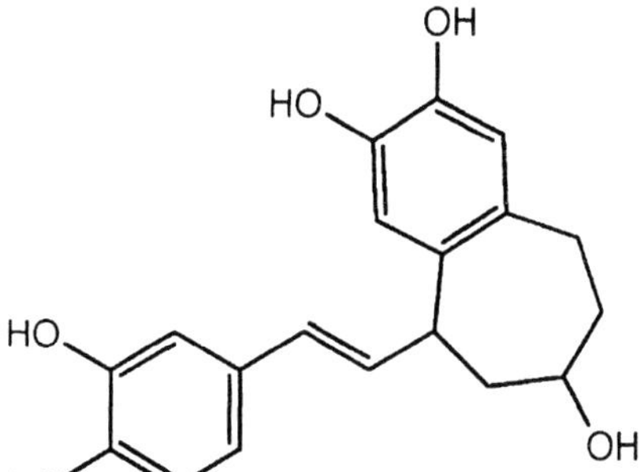

12 DPPH : 50%

Figure 2 Structure of phenolics isolated from greater cardamom and their DPPH radical-scavenging activity. (DPPH radical-scavenging activity (%) at the sample concentration of 20μM; DPPH conc. 100 μM; α-tocopherol: 39%)

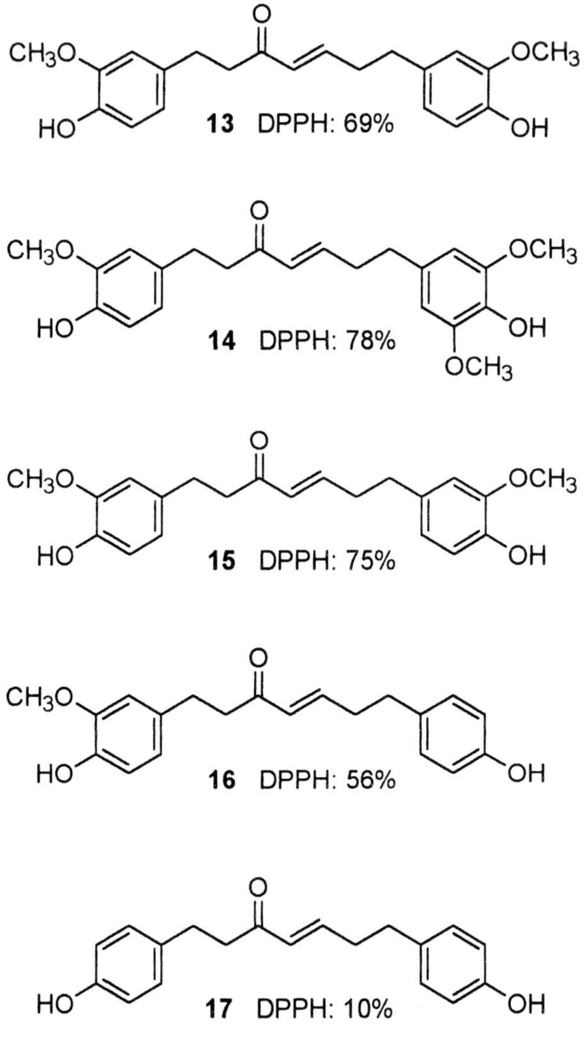

Figure 3 Structure of diarylheptanoids isolated from ginger and their DPPH radical-scavenging activity. (DPPH radical-scavenging activity (%) at the sample concentration of 40μM; DPPH conc. 100 μM. α-tocopherol: 83%, BHT: 35%)

Murraya koenigii

Murraya koenigii belonging to the family Rutaceae is originated from India, which grows wild as far as northern Thailand. It is cultivated in Malaysia garden as well as India. The leaves are called curry leaf and a necessary ingredient of southern Indian cooking. The leaves are also available for folk medicine both internally and externally as well as the roots and bark. The leaves increase digestive secretions and relieve nausea, indigestion and vomiting. Previous phytochemical studies on the plant *M. koenigii* have resulted in the isolation of carbazoles (*20*).

Isolation of Carbazoles from Curry Leaves

Table II showed the radical-scavenging and the OSI values of the extract and fractions obtained from curry leaves. The dichloromethane extract and the ethyl acetate-soluble fraction showed both high radical scavenging and antioxidant activities. Especially, these curry leaf extracts had remarkable antioxidant effectiveness against the bulk methyl linoleate by the OSI method, comparable to α-tocopherol and BHT (*21*). The dichloromethane extract was purified by several chromatographic techniques to give ten carbazoles (**18~27**) (Figure 4).

Table II showed the radical-scavenging and the OSI values of the extract and fractions obtained from curry leaves. The dichloromethane extract and the ethyl acetate-soluble fraction showed both high radical scavenging and antioxidant activities. Especially, these curry leaf extracts had remarkable antioxidant effectiveness against the bulk methyl linoleate by the OSI method, comparable to α-tocopherol and BHT (*21*). The dichloromethane extract was purified by several chromatographic techniques to give ten carbazoles (**18~27**) (Figure 4).

Radical-scavenging and Antioxidant Activities of the Isolated Carbazoles

The DPPH radical-scavenging activity of the isolated carbazoles was expressed as IC$_{50}$ values in Figure 4. Compounds **18**, **19**, **21**, **24** and **27** showed the positive radical-scavenging activity, while remaining five compounds (**20**, **22**, **23**, **25** and **26**) were not effective. All active carbazoles had a free hydroxy group on the carbazole skeleton. It was noteworthy that compounds **18**, **21**, **24** and **27** exhibited about two times as high OSI values as α-tocopherol at 110 °C. There have been few compounds from spices that possessed such a high activity at a high temperature of 110 °C.

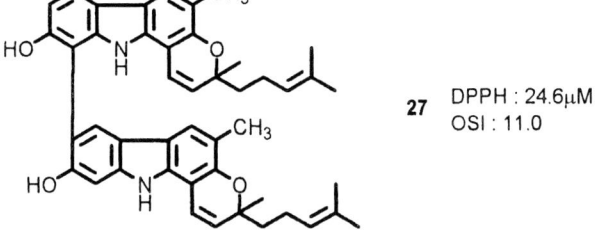

Figure 4 Structure of carbazoles isolated from curry leaf and their DPPH radical-scavenging and antioxidant activities. DPPH radical-scavenging activity of each carbazole was expressed in IC_{50} (μM). (DPPH conc. 100 μM. α-tocopherol: 27.7 μM, BHT: 83.2 μM). OSI was obtained at the sample concentration of 0.2 $\mu mol/g$ oil at $110^{\circ}C$. α-Tocopherol and BHT showed 4.4 and 3.4 of OSI under the same condition, respectively.

Curry leaf is usually used for fried and deep-fried cooking to add a distinct curry-like odor. Furthermore, curry leaf may play the role of prevention of lipid oxidation in the process of cooking.

Table II. DPPH Radical-scavenging and Antioxidant Activities of the *Murraya* Extracts

Sample	DPPH[a]	OSI[b]
CH$_2$Cl$_2$ Extract	89	4.9
EtOAc-sol. Fraction	91	4.3
H$_2$O-sol. Fraction	43	0.9
α-Tocopherol	90	5.4
BHT	82	4.8

[a] DPPH radical-scavenging activity: Sample concentration 100μg/mL; DPPH conc.100μM.
[b] OSI of oil with each sample/OSI of oil without any additives (sample conc. 0.02%, w/w; 110 °C)

Conclusion

Various structural types of antioxidants were obtained from some tropical spices. Their antioxidant properties were dependent on their own chemical structures and the physicochemical state of the lipid substrate.

Nowadays, it is generally acceptable that antioxidants play an important role not only in food industry, increasing the shelf life of food products and improving the stability of lipid-containing foods, but also in preventing radical-induced diseases such as cancer and atherosclerosis. From nutraceutical viewpoint, these tropical spices might be one of the important foodstuffs which provide human health benefits.

Acknowledgements

This work was supported by the Program for Promotion of Basic Research Activity for Innovative Biosciences (BRAIN) of Japan.

References

1. Chipault, J.R.; Mizuno, G.R.; Hawkins, J.M.; Lundberg, W.O. *Food Res.* **1952**, *17*, 46-54.
2. Nakatani, N. In *Food Phytochemicals for Cancer Prevention II, Teas, Spices, and Herbs.* Ho, C.-T.; Osawa, T.; Huang, M.T.; Rosen, R.T., Eds., Washington, DC: ACS Symp. Series 547, American Chemical Society: Washington, D.C., 1994, pp 144-153.
3. Nakatani, N.; Kikuzaki, H. In *Quality Management of Nutraceuticals.* Ho, C.-T.; Zheng, Q.Y., Eds., ACS Symp. Series 803, American Chemical Society: Washington, D.C., 2002, pp 230-240.
4. Kikuzaki, H.; Hara, S.; Kawai, Y.; Nakatani, N. *Photochemistry* **1999**, *52*, 1307-1312.
5. Kikuzaki, H.; Sato, A.; Mayahara, Y.; Nakatani, N. *J. Nat. Prod.* **2000**, *63*, 749-752.
6. Nakatani, N.; Tachibana, Y.; Kikuzaki, H. *JAOCS* **2001**, *78*, 19-23.
7. Blois, M.S. *Nature* **1958**, *181*, 1199-1200.
8. Osawa, T.; Namiki, M. *Agric. Biol. Chem.* **1981**, *45*, 735-739.
9. de Guzman, C.C.; Siemonsma, J.S. Eds. *Plant Resources of South-East Asia, No 13, Spices.* Bogor Indonesia: PROSEA, 1999, pp139-143.
10. Kikuzaki, H.; Kawai, Y.; Nakatani, N. *J. Nutr. Sci. Vitaminol.* **2001**, *47*, 167-171.
11. Martin, T.S.; Kikuzaki, H.; Hisamoto, M.; Nakatani, N. *JAOCS* **2000**, *77*, 667-673.
12. Cuvelier, M.E.; Richard, H,; Berset, C. *Biosci. Biotech. Biochem.* **1992**, *56*, 324-325.
13. Kikuzaki, H. In *Functional Foods & Nutraceuticals Series - Herbs, Botanicals & Teas.* Mazza, C.; Oomah, B.D., Eds., Technomic: Lancaster, 2000, pp 75-105.
14. Kikuzaki, H.; Usuguchi, J.; Nakatani, N. *Chem. Pharm. Bull.* **1991**, *39*, 120-122.
15. Kikuzaki, H.; Kobayashi, M.; Nakatani, N. *Phytochemistry* **1991**, *30*, 3647-3651.
16. Nakatani, N.; Kikuzaki, H. *Chem. Express* **1992**, *7*, 221-224.
17. Kikuzaki, H.; Tsai, S.M.; Nakatani, N. *Phytochemistry* **1992**, *31*, 1783-1786.
18. Kikuzaki, H.; Kawasaki, Y.; Nakatani, N. In *Food Phytochemicals for Cancer Prevention II, Teas, Spices, and Herbs.* Ho, C.-T.; Osawa, T.;

Huang, M.T.; Rosen, R.T., Eds., ACS Symp. Series 547, American Chemical Society: Washington, D.C., 1994, pp 237-243.
19. Kikuzaki, H.; Nakatani, N. *Phytochemistry* **1996**, *43*, 273-277.
20. Chakrabarty, M.; Nath, A.C.; Khasnobis, S.; Chakrabarty, M.; Konda, Y.; Harigaya, Y.; Komiyama, K. *Phytochemistry* **1997**, *46*, 751-755.
21. Tachibana, Y.; Kikuzaki, H.; Lajis, N.H.; Nakatani, N. *J. Agric. Food Chem.* **2001**, *49*, 5589-5594.

Chapter 13

Antioxidant Capacity of Berry Crops and Herbs

Shiow Y. Wang

Fruit Laboratory, Beltsville Agricultural Research Center, Agricultural Research Service, U.S. Department of Agriculture, Beltsville, MD 20705

Berry fruits and herbs are good sources of natural antioxidants. In addition to common nutrients such as vitamins and minerals, extracts of berries and herbs are also rich in anthocyanins, flavonoids and phenolic acids. Berry fruits and herbs have shown a remarkably high scavenging activity toward chemically generated radicals, thus making them effective in inhibiting oxidation of human low-density lipoproteins and preventing various human diseases. The different species of berries and herbs have varying abilities to inhibit different active oxygen species (peroxyl radicals, superoxide radicals, hydrogen peroxide, hydroxyl radicals, and singlet oxygen). Berries and herbs contain numerous phytochemicals including luteolin, naringin, rosmarinic acid, rosmanol, rutoside, caffeic acid, hispidulin, cirsimaritin, apigenin, vanillic acid, ellagic acid, chlorogenic acid, *p*-coumaric acid, kaempferol, quercetin, myricetin, cyanidin, delphinidin, petunidin, malvidin, pelargonidin, and peonidin. These phytochemicals are effective natural antioxidants. Therefore, a balanced diet with berries and herbs should have beneficial health effects.

Introduction

Active oxygen species are generated as by-products of normal metabolism (*1*). Increased levels of these active oxygen species or free radicals create oxidative stress, which leads to a variety of biochemical and physiological injuries often resulting in impairment of metabolism, and eventually cell death (*2,3,4*). Berry crops and herbs are rich in anthocyanins, flavonoids and phenolic acids (*2,5-8*), which exhibit remarkably high scavenging activities toward chemically generated radicals, thus making them effective in inhibiting oxidation of human low-density lipoproteins and preventing various human diseases (*9-12*).

Antioxidant Capacity in Berries

Many berries are good sources of natural antioxidants (*2,7,13,14*). Antioxidants have long been used for food preservation, but there has been concern that synthetic antioxidants used for preservation such as butylated hydroxyanisole (BHA) and butylated hydroxytoluene (BHT) may cause liver damage and cancer (*15*). Thus, the interest in natural antioxidants has increased considerably (*16*). In addition, high levels of natural antioxidant consumption have been shown to have multiple benefits to human health (*9,12*). Berry fruits have high antioxidant capacity against peroxyl radical (ROO$^{\bullet}$). Among the berry fruits, chokeberrries, lingonberries, blackberries, blueberries, black raspberries have higher antioxidant activities (oxygen radical absorbance capacity, ORAC) compared to red raspberries. Cranberries and strawberries generally have the lowest values of antioxidant activity against ROO$^{\bullet}$ (Table I) (*7*). The antioxidant activities of different species vary significantly (Table I). The total anthocyanin, phenolic content and ORAC $_{ROO^{\bullet}}$ values in chokeberries are 4.28 mg of cyanidin 3-glucoside equivalents/g fresh weight, 25.56 mg of gallic acid (GAE) equivalents/g fresh weight and 160.2 µmol of Trolox equivalents/g fresh weight, respectively. Previous research showed that a linear relationship exists between total phenolic or anthocyanin content and ORAC in various berry crops (*7,13,17*). This indicates that the antioxidant activity of fruit is mainly derived from the contribution of phenolic and anthocyanin compounds in fruits.

Berries also possess antioxidant activities against radicals other than ROO$^{\bullet}$ such as superoxide radicals ($O_2^{\bullet -}$), hydrogen peroxide (H_2O_2), hydroxyl radicals (OH$^{\bullet}$), and singlet oxygen ($^1O_2$) (*18*). Different species and cultivars show varying degrees of scavenging capacity on different active oxygen species ($O_2^{\bullet -}$, H_2O_2, OH$^{\bullet}$, and $^1O_2$) (*18*). The scavenging capacity among the berries range

Table I. ORAC Value, Anthocyanin Content and Total Phenolic Content of Various Berries and Herbs[a]

Various Berries & Herbs	Botanical Name	ORAC[b] (μmol TE/g)	Anthocyanin[c] (mg/100g)	Total phenolic[d] (mg/100g)
blackberry (Chester Thornless)	Rubus sp.	22.2±0.9	153.3±10.6	226±4.5
blueberry (Sierra)	Vaccinium corymbosum	28.9±1.2	119.8±2.5	412±6.7
chokeberry (wild)	Aronia melanocarpa E	160.2±9.5	428.3±11.7	2556±23.4
cranberry (Ben Lear)	Vaccinium macrocarpon	18.5±0.8	32.4±1.5	315±8.2
lingonberry (Amber Land)	Vaccinium vitis-idaea	38.1±1.6	45.3±0.7	652±14.1
black raspberry (Jewel)	Rubus occidentalis L.	28.2±1.4	197.2±1.4	267±4.3
red raspberry (Sentry)	Rubus idaeus L.	18.2±0.3	52.8±2.4	227±6.0
strawberry (Delmarvel)	Fragaria x ananassa D.	13.2±0.1	26.9±0.8	119±2.3
garden sage	Salvia officinalis	13.3±0.4	—	134±5.2
garden thyme	Thymus vulgaris	19.5±0.2	—	213±11.0
peppermint	Mentha x piperita	15.8±0.4	—	226±16.1
chives	Allium schoenoprasum	9.2±0.3	—	105±5.4
dill	Anethum graveolens	29.1±0.3	—	312±6.2
Greek mountain oregano	Origanum vulgare ssp. hirtum	64.7±1.1	—	1180±36.1
Mexican oregano	Poliomintha longiflora	92.2±0.7	—	1751±22.3
hard sweet marjoram	Origanum x majoricum	71.6±1.25	—	1165±29.0
rosemary	Rosmarinus officinalis	19.2±0.6	—	219±15.4
sweet basil	Ocimum basilicum	14.3±0.5	—	223±15.2
coriander	Polygonum odoratum	22.3±0.7	—	309±12.6

[a]Data expressed as mean ± SEM. [b]Data expressed as micromoles of Trolox equivalents per gram of fresh weight. [c]Data expressed as mg of cyanidin 3-glucoside equivalents per 100 g of fresh weight. [d]Data expressed as mg of gallic acid equivalents per 100 g of fresh weight.

from 40.8 to 72.0 % for $O_2^{\cdot-}$, from 50.7 to 73.9% for H_2O_2, from 52.4 to 77.3% for OH·, and from 6.3 to 17.4% for $^1O_2$. Among the small fruits of blackberry, strawberry, cranberry, raspberry, and blueberry, blackberries have the highest antioxidant capacity against $O_2^{\cdot-}$, H_2O_2 and OH·. Meanwhile strawberry is lower in antioxidant capacities for these same free radicals ($O_2^{\cdot-}$, H_2O_2 and OH·). With regards to $^1O_2$ scavenging activity, strawberry has the highest activity, followed by blackberry, compared to other fruits assayed. Among the least effective scavengers are raspberry for $O_2^{\cdot-}$, cranberry for H_2O_2, OH· and $^1O_2$ (*18*).

Antioxidant Compounds in Berry Crops

Berries contain a wide range of phenolic acids, flavones and flavonoids. The main flavonoid subgroups in berries are anthocyanins, proanthocyanins, flavonols, and catechins. Phenolic acids present in berries are glycosides of hydroxylated derivatives of benzoic acid and cinnamic acid (*19,20*). Acidic compounds incorporating phenolic groups have been repeatedly implicated as active antioxidants. Phenolic acids such as caffeic acid, chlorogenic acid, *p*-coumaric acid and vanillic acid are widely distributed in berries as natural antioxidants (*19,21*). Their antioxidant activities are associated with the number of hydroxyl groups in their molecule structure to some extent. For example, in a comparison of the content of two hydroxycinnamic acids, caffeic acid and *p*-coumaric acid, the antioxidant activity of the former is higher than that of the latter. It is likely that dihydroxylation in the 3, 4 position could enhance antioxidant potency due to being more available as hydrogen donors (*21*). Caffeic acid and its derivative have been found to be two major phenolic acids in chokeberry and both have a high proportion of antioxidant activity with 20.6% and 17.6%, respectively (Table II).

Chlorogenic acid is the most abundant phenolic acid in blueberry extract and also the most active antioxidant. A 1.2×10^{-5} M solution of chlorogenic acid inhibits over 80% of peroxide formation in a linoleic acid test system (*22*). We found that chlorogenic acid was effective in inhibiting OH·, $O_2^{\cdot-}$ and H_2O_2 free radicals (*18*). Vanillic acid, a benzoic acid derivative, was found in cranberries (49.3 µg/g fresh weight) and contributed 4.4% of total antioxidant activity in cranberries (Table II). The activity of cinnamic acid derivatives with two hydroxyl groups is superior to other phenolic acids with only one free OH group (*22,23*). *p*-Coumaric acid and *p*-coumaroyl glucose occur naturally in lingonberries and strawberries and also have antioxidant activity. Ellagic acid, a putative anticarcinogenic compound, was detected in many berry crops (*24,25*) and contributed 9.8 % of total antioxidant activity in blackberries. *p*-Coumaroyl-glucose and ellagic acid together constituted 12.7% of the total antioxidant activity in strawberries.

Table II. Contribution of Important Phenolic Acids and Flavones to Antioxidant Activity in Blackberries (Chester thornless), Blueberries (Sierra), Chokeberries (Wild), Cranberries (Ben Lear), Lingonberries (Amber land), Black Raspberries (Jewel) and Strawberries (Delmarvel)

Compound	\multicolumn{7}{c}{Percent (%) of antioxidant activity}
	blackberry	blueberry	chokeberry	cranberry	lingonberry	raspberry	strawberry
chlorogenic acid	—	20.9	—	—	—	—	—
caffeic acid derivative	—	—	17.6	—	—	—	—
vanillic acid	—	—	—	6.3	—	—	—
caffeic acid	—	—	20.6	6.4	6.9	—	—
p-coumaric acid	—	—	—	—	4.1	—	10.5
myricetin 3-arabinoside	—	5.4	—	5.9	—	—	—
quercetin 3-galactoside	—	6.2	4.3	10.2	9.0	—	—
quercetin 3-glucoside	6.7	2.0	4.5	—	—	—	6.9
quercetin 3-arabinoside	—	2.9	—	5.0	3.1	—	—
quercetin 3-glucuronide	—	—	—	—	—	6.9	—
quercetin derivative	—	4.2	—	—	5.6	—	—
quercetin 3-rhamnoside	4.3	—	—	6.6	9.8	—	—

The most important single group of phenolics in berries is flavonoids which consist mainly of proanthocyanidins, anthocyanidins, flavones, flavonols and their glycosides (*19*). Proanthocyanidins are polyflavonoid in nature, consisting of chains of flavan-3-ol units. Proanthocyanidins have relatively high molecular weights and have the ability to complex strongly with carbohydrates and proteins. Flavones (quercetin, myricetin and kaempherol) show high antioxidant activity and have a structure which allows them to have more effective antioxidant activity than that of anthocyanins. The 2, 3 double bond in conjunction with a 4-oxo function in the C ring of quercetin allowed electron delocalization from the B ring and show extensive resonance, thereby resulting in significant effectiveness for radical scavenging (*26*). An additional OH group at C5=of quercetin, as in myricetin, increases the ORAC value. Myricetin has higher antioxidant activity than quercetin when comparing their ORAC values (4.32 vs 3.2) (*27*). Kaempferol, which is a related structure of quercetin, has only a single 4'-OH group in the B ring, and has much lower antioxidant activity compared to quercetin. The antioxidant capacities (ORAC value) for quercetin and kaempferol are 3.29 and 2.67, respectively (*28*).

The occurrence of *p*-coumaroyl glucose, dihydroflavonol, quercetin 3-glucoside 3-glucuronide, kaempferol 3-glucoside and kaempferol 3-glucuronide have been detected in berry fruits and are effective antioxidants (*19,23,29-32*). Kaempferol 3-glucoside and kaempferol 3-glucuronide constitute 4.6 and 4.3% of the total antioxidant activity in strawberries. Quercetin contribute 15.3%, 21.9%, 8.8% and 27.5% of total antioxidant activity in blueberries, cranberries, chokeberries and lingonberries, respectively (Table II). Kaempferol and quercetin are potent quenchers of $ROO^{\cdot}$, $O_2^{\cdot-}$ and $^1O_2$ (*22*). Quercetin and other polyphenols have been shown to play a protective role in carcinogenesis by reducing bioavailability of carcinogens (*33*).

Anthocyanins are probably the largest group of phenolic compounds in the human diet. Anthocyanins have been used for several therapeutic purposes including the treatment of diabetic retinopathy, fibrocystic disease, and visual disorders (*34,35*). Berry fruits contain a large of number anthocyanins (cyanidin, delphinidin, pelargonidin, peonidin and malvidin). The total ORAC value of 11 identified anthocyanins in blueberry accounts for 56.3% of the total ORAC value (Table III). Pelargonidin 3-glucoside is the predominant anthocyanin in strawberries and contributes 27.3% of total antioxidant activity, while pelargonidin 3-rutinoside constitutes 8.3%. Peonidin 3-galactoside comprises 20.8% of the total ORAC value in cranberry extract (Table III). Cyanidin 3-galactoside is the predominant anthocyanin in chokeberry and lingonberry and 43.5% of antioxidant capacity in lingonberry is derived from cyanidin 3-galactoside. Andersen (*36*) identified cyanidin 3-galactoside (88%) as the main anthocyanin in Norwegian lingonberry. In addition to cyanidin 3-galactoside, cyanidin 3-arabinoside is also a major constituent in chokeberry and has a

Table III. Contribution of Important Anthocyanins to Antioxidant Activity in Blackberries (*Chester thornless*), Blueberries (Sierra), Chokeberries (Wild), Cranberries (Ben Lear), Lingonberries (Amber land), Black Raspberries (Jewel) and Strawberries (Delmarvel)

Compound	\multicolumn{7}{c	}{Percent (%) of antioxidant activity}
	blackberry	blueberry	chokeberry	cranberry	lingonberry	raspberry	strawberry
delphinidin 3-galactoside	—	9.1	—	—	—	—	—
delphinidin 3-glucoside	—	6.2	—	—	—	—	—
delphinidin 3-arabinoside	—	4.4	—	—	—	—	—
cyanidin 3-glucoside	70.8	2.0	0.3	1.2	1.6	15.5	10.6
cyanidin 3-arabinoside	—	—	18.4	6.4	6.0	—	—
cyanidin 3-galactoside	—	3.8	28.5	11.1	43.5	—	—
cyanidin 3-rutinoside	—	—	—	—	—	67.9	—
pelargonidin 3-glucoside	—	—	—	—	—	—	27.3
pelargonidin 3-rutinoside	—	—	—	—	—	—	8.3
pelargonidin 3-glucoside-succinate	—	—	—	—	—	—	13.3
cyanidin 3-xyloside	2.5	—	5.9	—	—	—	—
petunidin 3-glucoside	—	3.4	—	—	—	—	—
petunidin 3-galactoside	—	5.8	—	—	—	—	—
petunidin 3-arabinoside	—	5.1	—	—	—	—	—
malvidin 3-galactoside	—	6.3	—	—	—	—	—
malvidin 3-glucoside	—	5.1	—	—	—	—	—
malvidin 3-arabinoside	—	5.0	—	—	—	—	—
peonidin 3-galactoside	—	—	—	20.8	—	—	—
peonidin 3-arabinoside	—	—	—	10.3	—	—	—

remarkably high ORAC value (17.49 µmol of TE/g) (Wang, personal communication).

It has been shown that anthocyanidins are strong antioxidants with free radical scavenging properties attributed to the phenolic hydroxyl groups attached to ring structures (3,26,27). The hydroxyl radical scavenging activities of flavonoids increase with the number of hydroxyl groups substituted on the B-ring, especially at C-3' (37). A single hydroxy substituent generates little or no antioxidant. All flavonoids, such as cyanidin with 3',4'-dihydroxy substitution in the B ring and conjugation between the A- and B-rings, possess antioxidant activity (38) and have antioxidant potentials four times that of Trolox (26). Cyanidin shows higher antioxidant activity and the order of antioxidant potency defined by ORAC values is as follows: cyanidin > delphinidin > malvidin . peonidin . petunidin > pelargonidin (4,27). The results from our experiments showed that chokeberries, and lingonberries had higher antioxidant activities than cranberries and that strawberries generally had lower antioxidant activity (Table I). Chokeberries and lingonberries contain high amounts of cyanidin, strong antioxidant, whereas strawberries are rich in pelargonidin and ascorbic acid, which are weak antioxidants (19). The antioxidant capacities (ORAC value) for cyanidin 3-glucoside and pelargonidin 3-glucoside were found to be 2.24 and 1.54, respectively (27,28). The anthocyanin content in berries was found to correlate with oxygen radical absorbance capacity against $ROO^{\cdot}$, $O_2^{\cdot-}$, H_2O_2, $OH^{\cdot}$ and $^1O_2$ with correlation coefficients from 0.855 to 0.980 (20).

Berry crops show remarkably high scavenging activity of chemically generated active oxygen species. For example, with 100 g of 'Earliglow' strawberries, the antioxidant capacities against $ROO^{\cdot}$, $O_2^{\cdot-}$, H_2O_2, $OH^{\cdot}$ and $^1O_2$ are equal to 375.6 mg of Trolox, 188.9 mg of α-tocopherol, 44.2 mg of ascorbic acid, 155.7 mg of chlorogenic acid and 33.8 mg of β-carotene, respectively (20). Clearly, consumption of berry crops should be beneficial to our health.

Antioxidant Activity and Antioxidant Compounds of Essential Herbs

Herbs are rich sources of antioxidants. Each herb contains different phenolic compounds and each of these compounds possess differing amounts of antioxidant activity (39). There is a positive linear correlation between the phenolic content and antioxidant capacity of the herbs. The antioxidant activity in many herbs are higher than those of berries, fruits and vegetables when compared by weight (Table I) (5-8). Oregano (*Poliomintha longiflora*, *Origanum vulgare ssp. hirtum* and *Origanum* x *majoricum*) has extremely high contents of total phenolics and ORAC values (Table I). On a per gram fresh weight basis, oregano has 42 times more antioxidant activity than apples, 30 times more than potatoes, 12 times more than oranges and 4 times more than blueberries. The antioxidant activity of oregano is higher than α-tocopherol against linoleic acid oxidation (*40*). The ORAC values for *Poliomintha longiflora*, *Origanum vulgare ssp. hirtum* and *Origanum* x *majoricum* are 92.18, 64.71 and 71.64 μmol of Trolox equivalents/g fresh weight, respectively. Oregano species show high content of rosmarinic acid (124.8~154.6 mg/100g fresh weight) and other hydroxycinnamic acid compounds (*39*). Rosmarinic acid and hydroycinnamic acid compounds have been demonstrated to possess strong antioxidant activity (*22,41*). The antioxidant activity of rosmarinic acid is much higher than α-tocopherol and BHT (*41*).

The garden sage (*Salvia officinalis*) has been used as both a culinary and medicinal herb (*42*). Sage has been purported to cure all manner of maladies from epilepsy to warts and worms, and sage oil has astringent, antiseptic, and estrogenic properties (*42,43*). Rosmarinic acid (117.8 mg/100g fresh weight) and luteolin (33.4 mg/100g fresh weight) are the most abundant phenolic constituents in the extract of sage (*39*). Other compounds such as vanillic acid, caffeic acid, apigenin, hispidulin, and cirsimarin are also present, but in small amount. In addition, many volatile constituents of sage had been studied such as 1,8-cineole, thujone, isothujione and camphor (*44*). The ORAC value of sage was 13.28 μmol of Trolox equivalents/g fresh weight. Carnosol, rosmarinic acid and carnosic acid have the most antioxidant activity in sage followed by caffeic acid and cirsimaritin (*45*). Vanillic acid has only half of the antioxidant activity seen in caffeic acid and the ratio of antioxidant activity among caffeic acid, luteolin and apigenin are 1.3, 2.1, and 1.5 mM Trolox equivalent activity (TEAC) (*46*).

The herbs, thyme and rosemary, have been known to have high antioxidant capacity. The ORAC for thyme and rosemary were 19.5 and 19.2 μmol of Trolox equivalents/g fresh weight, respectively (Table I). The essential oils of thyme, thymol and carvacrol are recognized as major components that showed high antioxidant and antimicrobial activity (*47,48*). A biphenyl compound (3,4,3′,4′-tetrahydroxy-5,5′-diisopropyl-2,2′-dimethylbiphenyl) and a flavonoid (eriodicytol) have also been isolated from thyme as extremely potent antioxidants inhibiting superoxide anion production in the xanthine/xanthine oxidase system and mitochondrial and microsomal lipid peroxidation (*49*). The biphenyls, dimers of thymol and flavonoids isolated from thyme showed antioxidant activity as strong as BHT (*40*). In our study, high contents of rosmarinic acid (91.8mg/100g fresh weight) and luteolin (39.5 mg/100 g fresh weight) was found in the extract of thyme (*Thymus vulgaris*). A number of phenolic compounds such as rosmanol (124.1 mg/100 g fresh weight), rosmarinic acid (32.8 mg/100 g fresh weight), naringin (53.1 mg/100 g fresh weight), cirsimaritin (24.4 mg/100 g fresh weight) and carnosic acid (126.6 mg/100 g fresh weight) are found in rosemary (*39*). Similar to sage herb, these phenolic compounds in rosemary extract have been known to be very potent antioxidants and utilized in food products. Rosmanol is an active antioxidant and shows more activity than α-tocopherol or BHT (*40*). Rosmarinic acid has been shown to posses more antioxidant activity than rosmanol (*45*). There were also some antioxidant activities in herbs which apparently were from unidentified substances or synergistic interactions.

Other common herbs also appear to contain significant amounts of antioxidants. Among the more familiar, are dill (*Anethum graveolens*), coriander (*Polygonum odoratum*), peppermint (*Mentha x piperita*), sweet basil (*Ocimum basilicum*), parsley (*Petroselinum crispum*), and chives (*Allium schoenoprasum*) and their respective antioxidant activity ORAC values are 29.12, 22.30, 15.84, 14.27, 11.03, and 9.15 µmol of Trolox equivalents/g fresh weight (*39*).

In conclusion, berries and herbs contain large numbers of flavonoids and phenolics with antioxidant properties. Different flavonoids and phenolics contribute to antioxidant activity depending on their chemical structure and concentration. This study revealed that berries and herbs are effective potential sources of natural antioxidants. Consumption of berry crops and herbs indeed is beneficial to our health. The relevance of this information to the human diet is summarized by Rice-Evans and Miller (*3*) who state that total antioxidant potential of fruits and vegetables is more important than levels of any individual specific antioxidant constituent. Therefore, supplementing fruits, vegetables and herbs with a balanced diet should have beneficial health effects.

References

1. Elstner, E. F. *Ann. Rev. Plant Physiol.* **1982**, *33*, 73-82.
2. Heinonen, I. M.; Meyer, A. S.; Frankel, E. N. *J. Agric. Food Chem.* **1998**, *46*, 4107-4112.
3. Rice-Evans, C. A.; Miller, N. J. *Biochem. Soc. Trans.* **1996**, *24*, 790-795.
4. Satué-Gracia, M. T.; Heinonen, I. M.; Frankel, E. N. *J. Agric. Food Chem.* **1997**, *45*, 3362-3367.
5. Cao, G.; Sofic, E.; Prior, R.L. *J. Agric. Food Chem.* **1996**, *44*, 3426-3441.
6. Wang, H.; Cao, G.; Prior. R. L. *J .Agric Food Chem.* **1996**, *44*, 701-705.
7. Wang, S.Y.; Lin, H. S. *J. Agric Food Chem.* **2000**, *48*, 140-146.
8. Zheng, W., Wang, S. Y. *J. Agric. Food Chem.* **2001**, *49*, 5165-5170.
9. Ames, B. M.; Shigena, M.K.; Hagen, T. M. *Proc. Natl. Acad. Sci. U.S.A.* **1993**, *90*, 7915-7922.
10. Ascherio, A.; Rimm, E. B.; Giovannucci, E. L.; Colditz, G.A.; Rosner, B.; Willett, W. C.; Sacks, F.; Stampfer, M.J. *Circulation* **1992**. *86,* 1475-1484.
11. Gey, K. F. *Biochem. Soc. Trans.* **1990**, *18,* 1041-1045.
12. Steinberg, D. *Biochem. Soc. Trans.* **1991.** *24*, 790-795.

13. Prior, R. L.; Cao, G.; Martin, A.; Sofic, E.; McEwen, J.; O'Brien, C.; Lischner, N.; Ehlenfeldt, M.; Kalt, W.; Krewer, G.; Mainland, C. M. *J. Agric. Food Chem.* **1998**, *46,* 2686-2693.
14. Wang, M.; Goldman, I. L. *J. Amer. Soc. Hort. Sci.* **1996**, *121*, 1040-1042.
15. Ito, N.; Fukushima, S.; Hasegawa, A.; Shibata, M.; Ogiso, T. *J. Natl. Cancer Inst.* **1983**, *70,* 343-347.
16. Löliger, J. In: Arouma, O. I. and B. Halliwell (eds.). Free Radicals and Food Additives. Taylor and Francis. London. 1991, pp. 121-150.
17. Kalt, W.; Forney, C. F.; Martin, A.; Prior, R. L. *J. Agric. Food Chem.* **1999**, *47,* 4638- 4644.
18. Wang, S. Y.; Jiao, H. *J. Agric. Food Chem.* **2000**, *48,* 5677-5684.
19. Macheix, J. J.; Fleuriet, A.; Billot, J. *Fruit Phenolics*; CRC Press: Boca Raton, FL, 1990.
20. Wang, S. Y.; Zheng, W. *J. Agric. Food Chem.* **2001**, *49*, 4977-4982.
21. Shahidi, F.; Wanasundara, P. K. J. *Crit. Rev. Food Sci. Nutr.* **1992**, *32,* 67-103.
22. Larson, R. A. *Phytochemistry* **1988**, *27*, 969-978.
23. Pratt, D. *Amer. Chem. Soc. Symposiun* (507) 1992 ; pp54- 71.
24. Maas, J. L.; Wang, S. Y.; Galletta, G. J. *HortScience,* **1991**, *26*, 66-68.
25. Wang, S. Y.; Maas, J. L.; Payne, J. A.; Galletta, G. J. *J. Small Fruit Vit.* **1994**, *2,* 39-49.
26. Rice-Evans, C. A.; Miller, N. J.; Bolwell, P. G.; Bramley, P. M.; Pridham, J. B. *Free Radical Res.* **1995**, *22*, 375-383.
27. Wang, H.; Cao, G.; Prior, R.L. *J. Agric. Food Chem.* **1997**, *45*, 304-309.
28. Cao, G.; Sofic, E.; Prior, R.L. *Free Radicals Biol. Med.* **1997**, *22*, 749-760.
29. Bakker, J,; Bridle, P.; Bellworthy, S. J. *J. Sci. Food Agric.* **1994**, *64*, 31-37.
30. Gil, M. I.; Holcroft, D. M.; Kader, A. A. *J. Agric. Food Chem.* **1997**, *45*, 1662-1667.
31. Mabry, T. J.; Markham, K. R.; Thomas, M. B. The aglycone and sugar analysis of flavonoid glycosides. In *The Systematic Identification of Flavonoids*; Springer-Verlag: New York, 1970.
32. Van Buren, J. In *The Biochemistry of Fruits and their Products*; Hulme, A. C. Ed.; Academic Press: New York, 1970; pp 269-304.
33. Starvic, B.; Matula, T. I.; Klassen, R.; Downie, R. H.; Wood, R. J. In *Phenolic Compounds in Food and Their Effects on Health* II. *Antioxidants & Cancer Prevention*; Huang, M. T., Ho, C. T. and Lee, C. Y. (Eds.). ACS Symposium Series 507, Washington, DC 1992; pp 239-249.
34. Leonardi, M. *Minerva Ginecol.* **1993**, *45*, 617-621.
35. Scharrer, A.; Ober, M. *Klin. Monatsbl. Augenheikd.* **1981**, *178*, 386-389.
36. Andersen, O. M. *J. Food Sci.* **1985**, *50*, 1230-1232.
37. Ratty, A. K.; Das, N. P. *Biochem. Med. Metabolic Biology.* **1988**, *39*, 69-79.

38. Dziedzic, S. Z.; Hudson, B. J. F. *Food Chemistry*, **1983** *12*, 205-212.
39. Zheng. W.; Wang, S. Y. *J. Agric. Food Chem*. **2001,** *49*, 5165-5170.
40. Nakatani, N. In *Phenolic Compounds in Food and Their Effects on Health II*; Huang, M.T.; Ho, C.-T.; Lee, C.Y. Eds.; ACS Symp. Ser. 507: Washington, DC, 1992; pp 72-86.
41. Chen, J. H.; Ho, C. T. *J. Agric. Food Chem.* **1997,** *45*, 2374-2378.
42. Areias, F.; Valentão, P.; Andrade, P. B.; Ferreres, F.; Seabra, R. M. *J. Agric. Food Chem.* **2000,** *48,* 6081-6084.
43. Okamura, N.; Fujimoto, Y.; Kuwabara, S.; Yagi, A. *J. Chromatogr. A* **1994,** *679*, 381-386.
44. Pietta, P.; Simonetti, P.; Mauri, P. *J. Agric. Food Chem.* **1998,** *46,* 4487-4490.
45. Cuvelier, M..E.; Richard, H.; Berset, C. *J. Am. Oil. Chem. Soc.* **1996,** *73*, 645-652.
46. Robards, K.; Prenzler, P. D.; Tucker, G.; Swatsitang, P.; Glover, W. *Food Chem.* **1999,** *66*, 401-436.
47. Kähkönen, M. P.; Hopia, A. I.; Vuorela, H. J.; Rauha, J.; Pihlaja, K.; Kujala, T. S.; Heinonen, M. *J. Agric. Food Chem.* **1999,** *47*, 3954-3962.
48. Schwarz, K.; Ernst, H.; Ternes, W. *J. Sci. Food Agric.* **1996,** *70*, 217-223.
49. Haraguchi, H.; Saito, T.; Lshikawa, H.; Date, H.; Kataoka, S.; Tamura, Y.; Mizutani, K. *Planta Med.* **1996,** *62,* 217-221.

Chapter 14

Antioxidant Properties of Hsian-tsao (*Mesona procumbens* Hemsl.)

Gow-Chin Yen, Chien-Ya Hung, and Yen-Ju Chen

Department of Food Science, National Chung Hsing University, 250 Kuokuang Road, Taichung 40227, Taiwan, Republic of China

The reactive oxygen species scavenging activity of Hsian-tsao (*Mesona procumbens* Hemsl.) and its protective effect against oxidative damage in Chang liver cells were investigated. Water extracts of Hsian-tsao (WEHT) had antioxidant activity in linoleic acid peroxidation system. WEHT exhibited a positive concentration-dependent scavenging effect on DPPH radical, superoxide anion, hydrogen peroxide, nitric oxide and peroxyl radical. WEHT inhibited lipid peroxidation in Chang liver cells induced by hydrogen peroxide in a concentration-dependent manner. WEHT could inhibit 89% of oxidative DNA damage induced by hydrogen peroxide in Chang liver cells, under a concentration of 100 µg/mL, and the phenolic compounds of Hsian-tsao had a 18-41% inhibitory effect. Seven phenolic compounds (kaempferol, apigenin, caffeic acid, protocatechuic acid, syringic acid, vanillic acid and *p*-hydrobenzoic acid) were isolated and identified from the extracts of Hsian-tsao. Caffeic acid, kaempferol and protocatechuic acid had marked inhibitory effects on lipid peroxidation in Chang liver cells induced by hydrogen peroxide. The water extracts of Hsian-tsao reduced, by 78%, intracellular reactive oxygen species in Chang liver cells at a concentration of 100 µg/mL, and these species were decreased by 31 and 52% by 10 µM caffeic acid and kaempferol, respectivity. Thus, caffeic acid and kaempferol, which had the highest contents of phenolic compounds in the extracts of Hsian-tsao, could be the active components responsible for the antioxidant activity of Hsian-tsao.

Introduction

Reactive oxygen species (ROS), such as superoxide anion radical ($O_2^{\bullet-}$), singlet oxygen ($^1O_2$), hydroxyl radical ($^{\bullet}OH$), peroxyl radical ($ROO^{\bullet}$), hydrogen peroxide, and nitric oxide ($NO^{\bullet}$), play important roles in certain clinical diseases and the process of aging (1). They are generated by normal metabolisms or induced by environmental pollution and living styles (2). The generation of ROS and the antioxidant defense systems are normally approximately in balance, but when this balance is tipped, oxidative stress occurs and disease results. The accumulation of excess ROS can cause damage in DNA, lipids, and proteins, and can cause the function of cell metabolism to be lost. An important etiological mechanism of these diseases may be a causal relationship between the presence of oxidants and the generation of lipid hydroperoxides. Recent epidemiological studies indicated that oxidative stress can be partially decreased by increasing the dietary intake of natural antioxidants (3,4). Therefore, it is desirable to find safer, more efficient and more economical dietary antioxidants which can diminish ROS and may prevent aging and related diseases (5).

The herb *Mesona procumbens* Hemsl., called Hsian-tsao in China, is consumed in an herbal drink and jello-type dessert in the Orient. It is also used as an herbal remedy in Chinese folk medicine and is effective against heat-shock, hypertension, diabetes, hepatitis, muscle pain and joint pain. Most studies on this product have focused on its gelation property and proximate composition. In Japan, Hsian-tsao is an ingredient in an SOD (superoxide dismutase) beverage. It has been reported that the herb contains hypoglycemic and hypertension substances, including β-sitosterol, stigmasterol, α– and β–amyrin, oleanolic acid, maslinic and β-sitosterol glycoside (6). Hung and Yen (7) reported that WEHT have an antioxidant activity equal to that of trolox and butylated hydroxyanisole (BHA). Yen et al. (8) indicated that WEHT could reduce UV-C and H_2O_2-induced DNA damage, and that the protective effect might be due to the fact that it contains polyphenol compounds and/or other active components. Recently, Hung and Yen (9) isolated and identified seven phenolic compounds (kaempferol, apigenin, caffeic acid, protocatechuic acid, syringic acid, vanillic acid and *p*-hydroxybenzoic acid) from the extracts of Hsian-tsao. Although Hsian-tsao is used as an herbal remedy for hepatitis in Chinese folk medicine, scientific data concerning the protective effect of Hsian-tsao on liver cells against oxidative damage and the contribution of its active compounds are not available. The objectives of this study were to investigate the ROS scavenging activity of

WEHT and to evaluate the protective effect of WEHT and its active phenolic compounds on oxidative damage in Chang liver cells.

Materials and Methods

Reagents and Cell Culture

Sun-dried Hsian-tsao (*Mesona procumbens* Hemsl.) (harvested in 1995) was obtained from a local market in Hua-lian county, Taiwan. The Chang liver cell line (CCRC 60024) was obtained from the Food Industry Research and Development Institute (Hsinchu, Taiwan). Protocatechuic acid, *p*-hydroxybenzoic acid, caffeic acid, vanillic acid, syringic acid, kaempferol and apigenin were purchased from Sigma Chemical Co. (St. Louis, MO). All other chemicals and cultural medium were of reagent grade.

Preparation of Water Extracts from Hsian-tsao (WEHT)

The dried Hsian-tsao was cut into small pieces and ground into a fine power using a mill (RT-08, Rong Tsong, Taichung, Taiwan). The powder was passed through an 80-mesh sieve, collected and sealed in a polyethylene plastic bag, and then stored at 0-4° C for further use. Hsian-tsao (20 g) was extracted with boiling water (400 mL) for 30 min. The extract was filtered through Whatman No. 1 filter paper, and the filtrate was freeze-dried. The extract was dissolved in sterilized water and phenolic compounds standard in 0.1% DMSO/sterilized water for further study.

Determination of Antioxidative Activity

The antioxidative activity of Hsian-tsao extracts was determined using the thiocyanate method (*10*). The percentage of inhibition of peroxidation (%) was calculated as follows: inhibition of peroxidation (%) (capacity to inhibit peroxide formation in linoleic acid at 72 h) = $[1 - (A_{sample\ at\ 500\ nm} / A_{control\ at\ 500\ nm})] \times 100$.

Scavenging Effects on DPPH Radical, Superoxide Anion, Hydrogen Peroxide and Nitric Oxide

The effect of WEHT on DPPH radical was estimated according to the method of Shimada et at. (*11*) by measuring spectrophotometrically at 517 nm. The influence of WEHT on the generation of superoxide in a nonenzymic system was measured by means of spectrophotometric measurement of the product on reduction of nitro blue tetrazolium (*12*). The method used to detect hydrogen peroxide was based on the peroxidase assay system. Horseradish peroxidase (HRPase) was used, which reacts with hydrogen peroxide to oxidize phenol red into a purplish-red product, as described by Rhinkus and Taylor (*13*). The method of Marcocciet et al. (*14*) was used to determine the scavenging effect on nitrite oxide. The reaction mixture was mixed with sulfanilamide and N-(1-naphthyl)-ethylenediamine dihydrochloride, and the absorbance was measured at 570 nm.

Cell Viability

Cell viability was determined on the basis of mitochondrial-dependent reduction of 3-(4,5-dimethylthiazol-2-yl)-2,5-diphenyltetrazolium bromide (MTT) to formazan.

Determination of Lipid Peroxidation in Cells

The lipid peroxidation was determined by measuring thiobarbituric acid-reactive substances (TBARS) using a modified method of Ramanathan et al. (*15*).

Alkaline Single-Cell Gel Electrophoresis (Comet Assay)

Comet assay was performed under alkaline conditions following to the methods of Yen et al. (*8*). The slides were observed using a fluorescent microscope attached to a CCD camera connected to a personal computer based image analysis system (Komet 3.0; Kinetic Imaging Ltd.) For each analysis, 50 individual cells were calculated, and in most cases three separate experiments were conducted for each series. Single cells were analyzed under the fluorescent microscope. The DNA damage was expressed as % Tail DNA = [Tail DNA/(Head DNA + Tail DNA) × 100. A higher % Tail DNA meant a higher level of DNA damage.

Reactive Oxygen Species (ROS) Levels

Intracellular ROS was measured using a fluorescent probe, DCFH-DA (*16*). A reaction mixture containing 0.889 mL Chang liver cells (1×10^5 cells/mL) and 100 μL water extracts of Hsian-tsao (final concentrations 10, 50 and 100 μg) or phenolic compounds (final concentration 10 μM) was incubated in 24 wells for 20 h. Then, 10 μL hydrogen peroxide (10 mM) was added to the cells simultaneously with 1 μL DCFH-DA (final concentration 5 mM), and the cells were then incubated at 37° C. The DCF fluorescence intensity was measured using a FLUO star-galaxy fluorescence plate reader (BMG Labtechnologies GmbH Inc., Offenburg, Germany) with the temperature maintained at 37° C. The excitation filter was set at 485 ± 10 nm, and the emission filter was set at 530 ± 12.5 nm. Data points were taken every 5 min for 30 min, and the data were exported to Excel (Microsoft, Seattle, WA) spreadsheet software for analysis.

Results and Discussion

Antioxidant Activity of WEHT

The antioxidant activity of WEHT increased significantly as the concentration increased from 1 to 50 μg/mL; then the antioxidant activity increased more slowly when the concentration exceeded 50 μg/mL. At 200 μg/mL, the capacity to inhibit the peroxidation of linoleic acid was 97%, which was equal to that of BHA (99%). In our previous experiment (*17*), polyphenols (185 mg/g WEHT), and a lower content of ascorbic acid, α-tocopherol, and β-carotene were determined in WEHT. Recently, Hung and Yen (*9*) isolated and identified seven phenolic compounds (kaempferol, apigenin, caffeic acid, protocatechuic acid, syringic acid, vanillic acid and *p*-hydrobenzoic acid) from the extracts of Hsian-tsao. High correlation ($r^2 = 0.92$) was found between the antioxidant activity and the total polyphenolic content of WEHT. Therefore, polyphenols could be the major antioxidant components responsible for the total antioxidant activity of WEHT.

Scavenging Effects of WEHT on DPPH Radical, Superoxide Anion, Hydrogen Peroxide and Nitric Oxide

Figure 1 shows the scavenging effects of WEHT on DPPH radical, superoxide anion, hydrogen peroxide and nitric oxide. The scavenging activity of WEHT on DPPH radical increased as the concentration of WEHT increased and when IC_{50} was 18.7 µg/mL. The correlation coefficients between DPPH radical scavenging activity and antioxidant activity and the content of total polyphenolics in WEHT were $r^2 = 0.89$ and $r^2 = 0.83$, respectively. The scavenging effect of WEHT on superoxide anion radical increased linearly from 54 to 90% as the concentration increased from 25 to 125 µg/mL when the IC_{50} was 16.4 µg/mL. The correlation coefficients between superoxide anion scavenging activity and antioxidant activity and the content of total polyphenolics in WEHT were $r^2 = 0.90$ and $r^2 = 0.96$, respectively. WEHT showed 71% scavenging ability on hydrogen peroxide at a concentration of 400

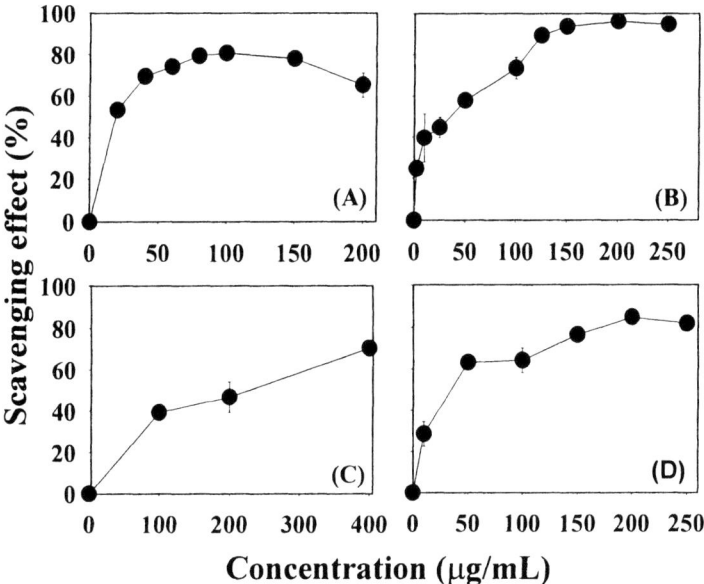

Figure 1. Scavenging effect of water extract from Hsian-tsao on DPPH radical (A), superoxide anion radical (B), hydroxygen peroxide (C) and nitric oxide (D) with different concentration. Each value is mean±standard derivation of three replicate analyses.

µg/mL with the IC$_{50}$ at 227 µg/mL. The correlation coefficient between the scavenging activity of hydrogen peroxide and antioxidant activity was r^2 = 0.60. The scavenging ability of WEHT on nitric oxide was 61 and 85% when the concentration was 50 and 200 µg/mL, respectively. IC$_{50}$ was 35 µg/mL, indicating that WEHT was a good scavenger on nitric oxide. The correlation coefficients in our study between nitric oxide scavenging activity, and antioxidant activity and the content of total polyphenolics were r^2 = 0.82 and r^2 = 0.87, respectively.

Cytotoxicity of WEHT and Isolated Phenolic Compounds in Chang Liver Cells

The cytotoxicity of WEHT in Chang liver cells was evaluated based on its effect on cell growth (MTT test). Cell viability was greater than 95% when WEHT (0-1000 µg/mL) was incubated with cells at 37° C for 24 h. In addition, no cytotoxicity was found when Chang liver cells were treated with isolated phenolic compounds of Hsian-tsao under a concentration range of 0-100 µM (data not shown).

Effects of WEHT and Isolated Phenolic Compounds on Hydrogen Peroxide-Induced Lipid Peroxidation

As shown in Table I, the inhibitory effect of WEHT on lipid peroxidation in Chang liver cells increased from 41 to 61% when the concentration increased from 100 mg/mL to 400 mg/mL. This result suggests that WEHT is a good antioxidant, possessing a protective effect that can reduce the oxidative damage on Chang liver cells induced by H$_2$O$_2$. Hung and Yen (9) isolated and identified seven phenolic compounds (kaempferol, apigenin, caffeic acid, protocatechuic acid, syringic acid, vanillic acid and *p*-hydrobenzoic acid) from the extracts of Hsian-tsao. Therefore, these phenolic compounds in WEHT might play important roles in modulating H$_2$O$_2$-induced lipid peroxidation in Chang liver cells. At a concentration of 10 µM, the inhibitory effects of caffeic acid, kaempferol and protocatechuic acid on lipid peroxidation in cells were 44, 31 and 22%, respectively, and the effects of the other compounds were less then 20% (Figure 2). The results indicated that WEHT and its phenolic compounds had protective effects inhibiting lipid peroxidation in cells induced by hydrogen peroxide.

Table I. Effect of Water Extract of Hsian-tsao on H_2O_2-induced Production of Lipid Oxidation Formation in Chang Liver Cells

Hsian-tsao water extract (μg/mL)	TBARS (nmol/mg protein)[a]	Inhibition of peroxidation (IP %)[b]
0	1.77 ± 0.46 [ac]	0
10	1.65 ± 0.07 [b]	16.46 ± 3.17 [d]
50	1.39 ± 0.10 [bc]	29.45 ± 5.03 [c]
100	1.15 ± 0.06 [cd]	41.09 ± 2.97 [b]
200	0.95 ± 0.08 [cd]	46.75 ± 4.34 [b]
300	0.88 ± 0.02 [cd]	50.36 ± 1.28 [b]
400	0.70 ± 0.09 [d]	60.83 ± 4.85 [a]

[a] Values are means±SD of triplicate determinations of malondialdehyde. The concentration of the lipid peroxides was calculated by using a molar absorption coefficient of 1.56×10^5 M^{-1} cm^{-1}.

[b] Inhibition of peroxidation (IP %) = $[1-(A_{sample}/A_{control})] \times 100$. The control containing no added sample represents 100 % lipid peroxidation. The hydrogen peroxide added concentration was 50 μM.

[c] Values in a column with the different superscripts are significantly different ($p<0.05$).

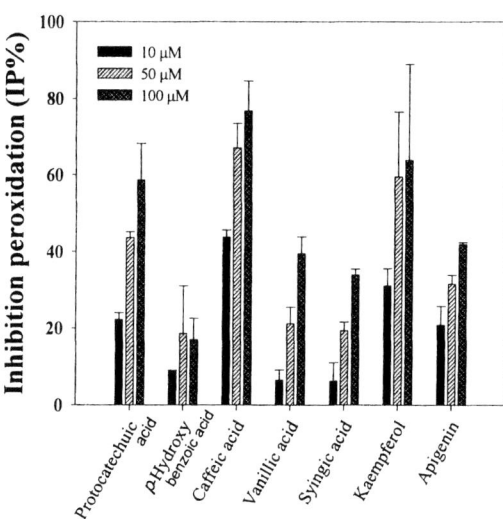

Figure 2. Inhibitory effect of phenolic compounds on H_2O_2-induced lipid oxidation of Chang liver cells. A control containing no added sample represents 100 % lipid peroxidation.

Effects of WEHT and Isolated Phenolic Compounds on Hydrogen Peroxide-Induced DNA Damage

No significant ($p < 0.05$) DNA damage was found when Chang liver cells were treated with WEHT under a concentration range of from 0 to 100 µg/mL. The results indicated that WEHT did not cause DNA damage under this concentration range as compared with the control group. H_2O_2-induced DNA damage in cells has been thought to occur through the Fenton reaction, which produces hydroxyl radicals that attack DNA, resulting in damage. As the results in Figure 3 show, DNA damage in Chang liver cells was reduced by treatment with WEHT. WEHT could inhibit 21, 78 and 89% oxidative DNA damage in Chang liver cells induced by hydrogen peroxide under concentrations of 10, 50 and 100 µg/mL, respectively. Yen et al. (*8*) reported that WEHT had a protective effect on UV-C and H_2O_2-induced DNA damage in human lymphocytes, and that this effect may be due to the fact that it contained polyphenol compounds and/or other active compounds. Many researches showed that phenolic compounds could reduce oxidative DNA damage induced by hydrogen peroxides in cells (*18,19*). Fenech et al. (*20*) reported that the phenolic compounds of wine reduced oxidative DNA damage induced by H_2O_2 in human lymphocytes. Thus, the protective effect of WEHT on Chang liver cells might be related to its phenolic compounds.

As the results in Figure 4 show, when the % tail DNA was distributed between 0 and 14%, there is no significant difference between the phenolic compounds (10 µM) and the control group. The results indicated that the phenolic compounds in Hsian-tsao did not cause DNA damage at a concentration of 10 µM. When the Chang liver cells were treated with hydrogen peroxide and then 10 µM different phenolic compounds were added, the % tail DNA was reduced from 51% (control group) to 30-42% (Figure 5). These results indicated that the phenolic compounds could inhibit 18-41 % of oxidative DNA damage induced by hydrogen peroxide at a concentration of 10 µM. Therefore, the phenolic compounds in WEHT played an important role in modulation of the H_2O_2-induced DNA oxidative damage in Chang liver cells.

Thus, WEHT and its phenolic compounds had a protective effect against DNA oxidative damage in Chang liver cells.

Reductive Effects of WEHT and Isolated Phenolic Compounds on Intracellular ROS Level in H_2O_2-Treated Cells

The concentration of intracellular ROS was evaluated based on changes in DCF fluorescence intensity. As the results in Figure 6 show, WEHT effectively

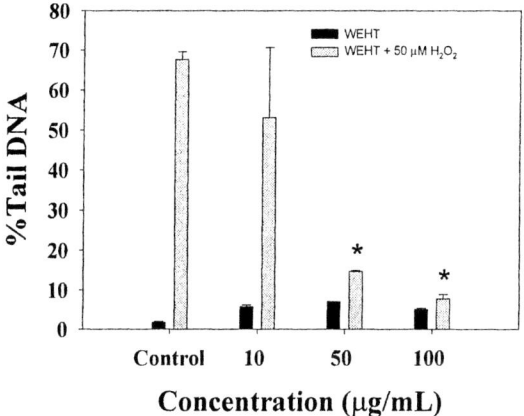

*Figure 3. Effect of water extract of Hsian-tsao (WEHT) on H_2O_2 -induced DNA damage of Chang liver cells. *$p < 0.05$ when compared with the control.*

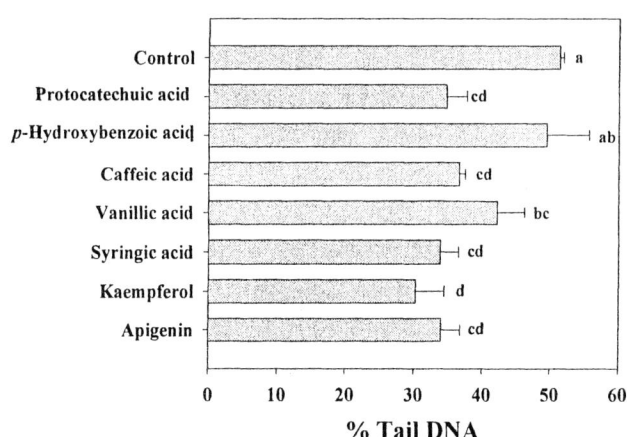

Figure 4. Genotoxicity of Chang liver cells treated with phenolic compounds. DNA strand breaks were detected by the comet assay. The concentration of sample was 10 μM. [a]Values in each column with different letters are significantly different ($p < 0.05$).

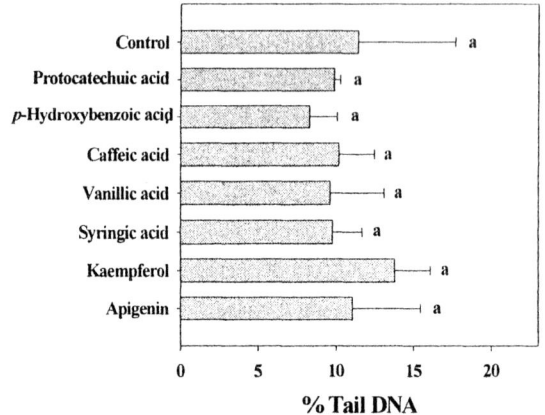

Figure 5. Effect of phenolic compounds on H_2O_2 –induced DNA damage of Chang liver cells. DNA strand breaks were detected by the comet assay. The concentration of phenolic compounds was 10 µM. [a-d] Values in each column with different letters are significantly different ($p < 0.05$).

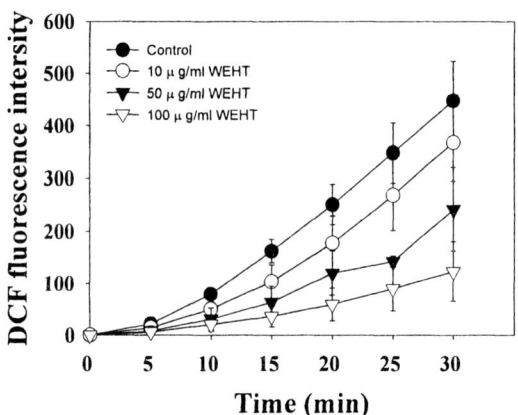

Figure 6. Effect of water extract of Hsian-tsao (WEHT) pretreatment on hydrogen peroxide-induced intracellular ROS level in Chang liver cells. Chang liver cells was preincubated with sample for 20 h before analysis.

reduced the intracellular ROS level in H$_2$O$_2$-treated cells in a dose-dependent manner. Yang et al. (21) indicated that Ebselen (2-phenyl-1,2-benzisoselenazol-3(2H)-one) had a scavenging effect on ROS, protecting against cytotoxicity, DNA oxidative damage and lipid peroxidation in HepG2 cells caused by hydrogen peroxide. Figure 7 also indicates that intracellular ROS in Chang liver cells was decreased by 31 and 52% by caffeic acid and kaempferol, respectively, at a concentration of 10 μM, indicating that their scavenging ability was equal to that of 10 and 50 μg WEHT, respectively. Therefore, the results indicated that the phenolic compounds isolated from WEHT could reduce intracellular ROS and had protective effects against lipid peroxide and oxidative DNA damage in Chang liver cells.

Conclusion

WEHT showed strong antioxidant activity and scavenging activity on DPPH radical, superoxide anion radical, and nitric oxide. The antioxidant activity and the scavenging capability on radicals correlated highly with polyphenolic content. The findings also indicated that WEHT and its phenolic compounds possessed no cytotoxicity and were capable of inhibiting H$_2$O$_2$-induced oxidative DNA damage and lipid peroxidation by reducing the intracellular reactive oxygen species (ROS) in Chang liver cells. The results suggest that WEHT might have

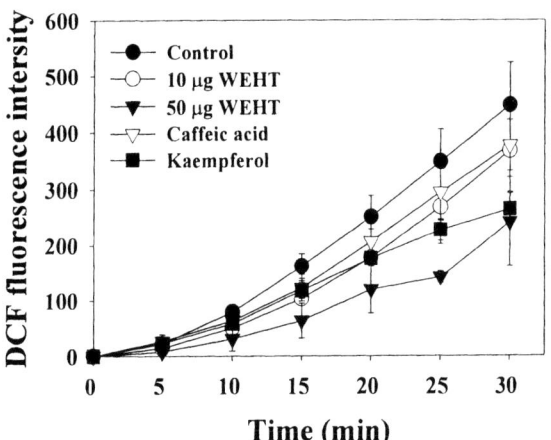

Figure 7. Effect of Hsian-tsao water extract (WEHT), caffeic acid and kaempferol pretreatment on hydrogen peroxide-induced intracellular reactive oxygen species (ROS) level in Chang liver cells. Chang liver cells was preincubated with sample for 20 h before analysis. The concentration of caffeic acid and kaempferol was 10 μM.

protective effects in human liver cells. Caffeic acid and kaempferol, which were the most abundant and had the highest levels of antioxidant activity among the phenolic compounds in Hsian-tsao, were the important antioxidant compounds inhibiting lipid peroxidation and oxidative DNA damage, and decreasing the intracellular reactive oxygen species in Chang liver cells. Thus, the water extract of Hsian-tsao is a healthy beverage.

References

1. Bergendi, L.; Beneš, L.; Ďuračková,F.; Ferenčik, M. *Life Sci.* **1999**, *65*, 1865-1874.
2. Gracy, R. W.; Talent, J. M.; Kong, Y.; Conrad, C. C. *Mutat. Res.* **1999**, *428*, 17-22.
3. Hertog, M. G. L.; Feskens, E. J. M.; Hollman, P. C. H.; Katan, M. B. Kromhout, O. *Lancet* **1993**, *342*, 1007-1011.
4. Coccetta, R. A. A.; Croft, K. D.; Beilin, L. J.; Puddey, I. B. *Am. J. Clin. Nutr.* **2000**, *71*, 67-74.
5. Culter, R. G. *Am. J. Clin. Nutr.* **1991**, *53*, 373s-379s.
6. Sheu, S. Y.; Liu, C.; Chiang H. C. *T'ai-wan K'o Hsueh* **1984**, *38*, 26-31.
7. Hung, C. Y.; Yen, G. C. *Lebensm.-Wiss. U.-Technol.* **2001**, *34*, 306-311.
8. Yen, G. C.; Hung, Y. L.; Hsien, C. L. *Food Chem. Toxic.* **2000**, *38*, 747-754.
9. Hung, C. Y.; Yen, G. C. *J. Agric. Food Chem.* **2002**, *50*, 2993-2997.
10. Mitsuda, H.; Yasumoto, K.; Iwami, K. *Eiyo to Shokuryo* **1996**, *19*, 210-214.
11. Shimada. K.; Fujikawa. K.; Yahara, K.; Nakamura, T. *J. Agric. Food Chem.* **1992**, *40*, 945-948.
12. Robak, J.; Gryglewski, I. R. *Biochem. Pharma.* **1988**, *37*, 837-841.
13. Rhinkus, S. J.; Taylor, R. T. *Food Chem.Toxic.* **1990**, *28*, 323-331.
14. Marcocci, L.; Maguire, J. J.; Droy-Lefaix, M. T.; Packer, L. *Biochem. Biophys. Res. Commun.* **1994**, *201*, 748-755.
15. Ramanathan, A.; Das, N. P.; Tan, C. H. *Free Radic. Biol. Med.* **1994**, *16*, 43-48.
16. Shen, H. M.; Shi, C. Y.; Shen, Y.; Ong, C. N. *Free Radic. Biol. Med.* **1996**, *21*, 139-146.
17. Yen, G. C.; Hung, C. Y. *Food Res. Internal.* **2000**, *33*, 487-492.
18. Duthie, S. J.; Collins, A. R.; Duthie, G. G.; Dobson, V. L. *Mutat. Res.* **1997**, *393*, 223-231.
19. Noroozi, M.; Anderson, W. J.; Lean M. E. *Am. J. Clin. Nutr.* **1998**, *67*, 1210-1218.
20. Fenech, M.; Stockley, C.; Aitken, C. *Mutat. Res.* **1997**, Supplement pp. S173.
21. Yang, C. F.; Fhen, H. M.; Ong, C. N. *Biochem. Pharm.* **1999**, *57*, 273-279.

Chapter 15

Effect of Different Heating Processes on Cytotoxic and Free Radical Scavenging Properties of Onion Powder

Hui-Yin Fu[1] and Tzou-Chi Huang[2]

[1]Department of Food Science and Technology, Tajen Institute of Technology, 907, Pingtung, Taiwan, Republic of China
[2]Department of Food Science and Technology, National Pingtung University of Science and Technology, 912, Pingtung, Taiwan, Republic of China

Drying method affects the quercetin composition of onion subsequently and significantly affecting the antioxidative activity. Onion powder dried by hot-air (60 °C), vacuum (35 °C) and lyophilization (35 °C) were used to prepare methanol extract. DPPH radical and peroxide radical scavenging activities of crude extracts from dried onion powders were evaluated. The results showed that hot air dried onion had higher radical scavenging activities in both DPPH and peroxide radicals than that of the lyophilized and vacuum dried onion. HPLC analyses showed that lyophilized and vacuum dried onion contained more quercetin glycosides, whereas hot air dried onion dominated in aglycone. The inhibitory effects of onion on the growth of human leukemia cell lines CCFF-CEM, U937, K562, P3H-1 and Raji *in vitro* were investigated. A strong cell proliferation inhibition activity in hot-air dried onion was observed for cell lines CEM and U937, whereas freeze dried and vacuum dried onion gave comparatively moderate inhibition. Low growth reduction was seen in cell lines K562, Raji, and no inhibition was observed in P3H-1 for both onion samples.

Introduction

Onion (*Allium cepa* L.) is a major source of flavonoids which are recognized as bioactive substances with potential health effects (*1*). Previous pharmacological testing showed that onion could be used as an anti-asthmatic, anti-thrombotic, anti-hypertensive, anti-hyperglycemic, anti-hyperlipemic and anti-tumor agent (*2,3*). A simultaneous case-referent study of a high epidemic area in Jiangsu Province, China, along with the results of nine out of 11 epidemiological studies, including those of a cohort study, have suggested a decreased risk of both stomach and esophageal cancer with increased consumption of onion, garlic or related Allium vegetables (*4*).

Dehydrated onion has great commercial value due to its culinary and medicinal properties as a nutraceutical (*5*). In modern times, onion powder is sold as a nutraceutical and as a dietary supplement. Several experiments have been conducted to investigate the effect of processing on the composition of quercetins in onion. During cooking we found an overall loss for quercetin-3,4'-diglucoside (Qdg) and quercetin-4'-monoglucoside (Qmg) of 11.3% and 2.6%, respectively. A significant proportion, 18% of quercetin-3,4'-diglucoside and 19% of quercetin-4'-monoglucoside, of each conjugates was leached, unchanged, from the onion tissue into the cooking water (*6*). Hirota et al. (*7*) found similar results and attributed the decrease of quercetin conjugates to the release of quercetin-3,4'-diglucoside and quercetin-4'-monoglucoside into cooking water and their oxidation by peroxidase. In studying the effect of processing on major flavonoids in processed onions, Ewald et al. (*8*) reported that the greatest loss of flavonoids in onions took place during the pre-processing step where the onion was peeled, trimmed and chopped before blanching. Boiling for 60 min caused overall flavonol losses of 20.6 % in onion (*9*). Very few papers have been published concerning the effect of drying on the nutraceutical changes in onion.

Although, according to Polish recommendations, sliced onions should be dried at 58-60° C (*10*), a vacuum dryer (Lab-Lone Duo-Vac oven) and a freeze dryer (Labconco Freezone Plus 6) were utilized to dehydrate onion slices at 30°C to study the effect of drying temperature on nutraceutical properties of onion. The fresh onions harvested in February, 2001 from a local farm (Herchung, Pingtung) were used in this study. The onion bulbs were hand-peeled and mechanically sliced into 3-3.5 mm thick slices, and dehydrated to a final moisture content of approximately 5%, moisture-free basis or less, as follows: Freeze-drying: freezing temperature −30° C, drying temperature 35° C, vacuum 0.15-0.20 mm Hg, drying time 16 h. Vacuum dryer: drying temperature 35+2° C, vacuum 510 mm Hg, and total drying time 16 h. Hot air shelf-drying: drying temperature 60 +2 °, and total drying time 20 h.

Moisture sorption isotherms were determined by the method described by Hutchinson and Otten (*11*). Sorption isotherm curves of these samples are shown in Figure 1. The isotherms had sigmoidal shapes, typical for dehydrated vegetables. This is in accordance with the results of Debnath et al. (*5*) for dehydrated onions. Moisture content at equilibrium was slightly higher in the hot air-dried sample than that of either the freeze-dried or cold air-dried ones. Some of the propectic substance was transformed into water soluble pectin. The isotherm curves for both freeze-dried (30 °C) and vacuum-dried (30 °C) onion are the typical sigmoidal shape. They indicate that equilibrium moisture increases very slowly, with an increase in Aw up to 0.43, beyond which there is a steep rise in moisture in all three samples. Moisture-sorption data showed that onion powder belongs to a class of food powders that is highly hygroscopic by nature. Polyuronides, with a high level of the galactan side chain, were characterized in dietary fiber of onion tissue (*12*). These chains bring about the solubility of pectin. This data indicates that enzymatic autolysis may occur during the hot air-drying process rather than in either vacuum or freeze-drying.

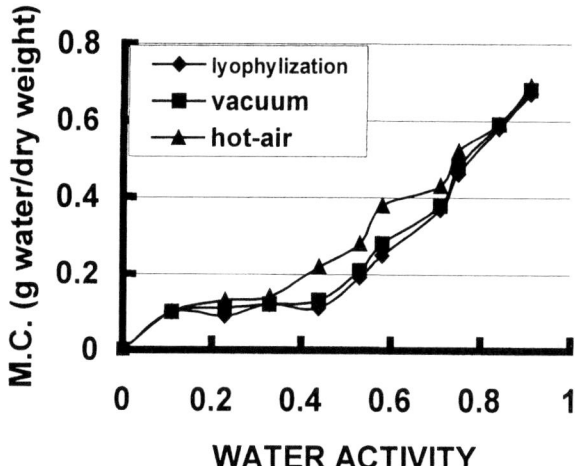

Figure 1. Moisture sorptions of onion powder.

The levels of the individual quercetin glucosides and quercetin aglycone were determined for the ethanol extracts from all the three onion powder samples. Changes in the composition of the quercetin conjugates due to drying are shown in Table I. The proportion of Qdg, Qmg and quercetin in the hot air-dried onion powder was 47%, 39% and 12%, respectively. Both freeze-drying and vacuum-drying at low temperature (35° C) showed a lesser extend of transformation of Qdg into Qmg or aglycone quercetin as compared with that in hot air-dried onion powder. Flavonoid composition in the edible portion of the onion bulb is made up primarily of two quercetin conjugates, quercetin-3,4'-diglucoside and quercetin-4'-monoglucoside (13). The quercetin conjugates present in the onion bulb are resistant to degradation during low temperature drying operations and there is only a small amount of free quercetin present in the ethanol extract. Peeling, trimming and chopping processes may lead to some extend of autolysis in the broken onion tissues.

Table 1. Changes of quercetin glycosides in onion powder by different heating treatments

Heating process	Concentration (μg/g dry weight) (%)			
	Qdg	Qmg	Quercetin	Total
Freeze-dried	2834^a (62%)	1375^a (30%)	365^a (8%)	4574^a
Vacuum-dried	2770^a (63%)	1294^a (29%)	352^a (8%)	4466^a
Hot air-dried	2112^b (47%)	1723^b (39%)	539^b (12%)	4474^a

Qdg = quercetin-3,4'-diglucoside, Qmg = quercetin-4'-monoglucoside; Data indicate is means of triplicate trials; ***Within the same column, means with no letter in common are significantly different (p<0.05).

In addition to leaching and thermal degradation during the drying process, enzymatic degradation of quercetin glycosides should be taken into consideration. In commercial mass production, it takes at least 16 h to complete the process. β-Glucosidase (1,4-β-D-glucosidase glucohydrolase; EC 3.2.1.21) is defined as an enzyme that hydrolyzes compounds containing β-glucosidic linkages by splitting off the terminal β-glucose residue (14). β-Glucosidase, which exists in a different cellular compartment from that of the glucosides, comes into contact with the glucosides and hydrolyzes the quercetin conjugates to release aglycone (15). High levels of β-glucosidase activity have been shown in wheat soon after the germination stage (16). β-Glucosidase activity was detected in onion tissue and crude enzyme was prepared in this experiment. Optimal temperature for the hydrolysis of p-nitrophenyl-β-glucopyranoside (PNPG) was found to be about 50° C as shown in Figure 2. This is in good agreement with the optimal temperature of maize β-glucosidase (17). We hypothesized that the β-glucosidase in onion tissue may be attributed to the partial degradation of the quercetin-3,4'-diglucoside in onion bulbs.

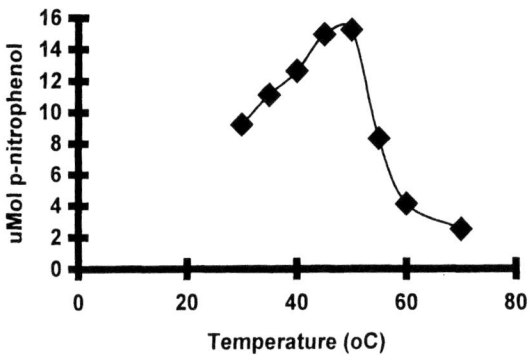

Figure 2. Effect of temperature on the β-glucosidase activity in onion.

Scavenging Effect on DPPH Radicals

Both water and methanolic extracts were investigated in DPPH radicals scavenging activity. The methanolic extract showed the higher DPPH scavenging capacity than that of water extracts of onion products, as indicated in Table II. The order of DPPH scavenging capacity of the methanolic extracts was in the decreasing order: hot air dried >freeze dried >cold air dried. Water extracts of onion products showed a similar trend. Antioxidant activities of quercetin and its glycosides has been compared. Ioku et al. (*18*) compared the relative antioxidation activity among quercetin and quercetin monoglucoside in both solution and phospholipid bilayers systems. They found that quercetin acts as a chain-breaking antioxidant more effectively than quercetin monoglucosides on the oxidation of phospholipid bilayers induced by aqueous oxygen radicals. The binding of glucose moiety to hydroxyl group of quercetin lower the peroxy radical scavenging activity irrespective of the position of hydroxyl group. In study the radical scavenging compounds in onion skin, Suh et al. (*19*) found that quercetin exhibited higher radical scavenging activity toward DPPH and the superoxide anion and hydroxy radical than that of quercetin-4'-glucoside. DPPH method has been used to evaluate various natural extracts (*20,21*). Free radical scavenging is a generally accepted mechanism for phenolic antioxidants to inhibit lipid oxidation (*22*). The antioxidative activity of phenolic acids is generally governed by their chemical structure, the activity increases with increasing the number of hydroxyl (OH) groups (*23*). Pekkaainen et al. (*24*) attributed the antioxidant activity of phenolic acids in a bulk lipid system to their DPPH radical scavenging activity.

Table II. DPPH Scavenging Capacity of different Heating Methods

Onion Products[*]	DPPH Scavenging Capacity[**]	
	Water extract	Methanol extract[***]
Freeze-dried	23.85[a]	71.31[a]
Vacuum-dried	24.08[a]	72.58[a]
Hot air-dried	31.61[a]	76.06[b]

[*]Freeze-dried (35° C), vacuum-dried (35° C), hot air-dried (60° C); [**]Means of triplicate trials; [***]2.5% methanol extract of onion powder.
Within the same column, means with no letter in common are significantly different ($p<0.05$).

Evaluation of Antioxidative Activity Using a Thermal Oxidative Stability Test

Oxidative Stability Index offers a simple, reproducible method for evaluating flavonoid antioxidant properties and is accepted as the AOAC Standard Method (Cd 12b-92). The effect of methanolic onion extracts on retarding lard oxidation after heat treatment was investigated by the Oxidative Stability Test (25). The potency of antioxidative activity is shown in Table III. Hot-air dried onion showed the highest activity among the three dried onion powders. The higher antioxidative activity in hot-air dried onion is attributed to the generation of some hydrolyzed quercetin derivatives during the prolong heating process.

Table III. Induction Time of Lipid Oxidation Measured by Oxidative Stability Index

Onion products[*]	Induction times (hrs)[**]
Control	6.7[a]
Freeze dried	10.4[b]
Vacuum dried	9.8[b]
Hot air dried	13.9[c]

[*]Freeze-dried (35 °C), Cold air-dried (35 °C), Hot air-dried (60 °C); [**]Means of triplicate trials. Within the same column, means with no letter in common are significantly different ($p<0.05$).

Cell Proliferation Inhibition Assay

Five human leukemia cell lines were obtained from the American Type Culture Collection (ATCC), namely CCRF-CEM, K562, P3HR-1, Raji and U937 cell lines. These cell lines were cultured in RPMI 1640 medium with 10% fetal bovine serum (Gibco, BRL), supplemented with 5 mm L-glutamine and 50 μg/mL of antibiotics (Penicillin/Streptomycin, Gibco, BRL) at 37 °C in a humidified 5% CO_2 incubator (26). Cells were seeded in 96-wells plates at an initial density of 1.5×105/mL in 180 μl RPMi-1640 medium per well. Twenty microliters of various concentrations of test samples and reference compound (5-Fluorouracil, 5-FU) were following added to triplicate culture wells to a final concentration. As a control vehicle, 0.1% of DMSO was added to the cells. After the cells were incubated for 48h, a cell proliferation kit (Roche Molecular Biochemical, Germany) was used to perform the cell proliferation or cytotoxic test according to the manufacturer's instructions. After 4-8 h of incubation at 37° C, all samples were measured their absorbance at 450 nm with a multiskan EX ELISA reader (Labsystems).

Onions dried by different heating process showed different inhibition properties on various cell lines. Among the tested cell lines, hot-air dried onion showed good inhibition on CCRF-CEM and U937 cell lines in dose dependent manner. The EC50 values on human leukemia cell lines CEM and U937 were 285.7 and 270.72 mg/mL, respectively. For authentic quercetin aglycone, the EC50 on CEM and U937 were 1.39 and 3.31 mg/mL, respectively. Quercetin-induced apoptosis has been observed in the monoblastoid cell line U937 *in vitro* as well (26). The EC50 values of cold air and freeze dried onions were between 335-628 mg/mL. Quercetin inhibits the growth of human leukemic cell lines in a dose-dependent manner and is able to enhance the antiproliferative activity of cytosine arabinoside. Rong et al. (26) postulated that quercetin can act as a cytostatic agent for leukemic cells by modulating the production of TCF-β-1. The growth inhibitory effect of quercetin was proposed to be the result of an arrest in the G1 phase of the cell cycle (27). Uddin and Choudhry (28) reported that quercetin exhibits a dose-dependent inhibition of DNA synthesis in the human leukemia cell, HL-60.

Low inhibition was observed in the investigation of K562, P3HR-1 and Raji leukemia cell lines. But, the hot air-dried onion still gave better inhibition than other dried onion. Hot-air dried samples in three different concentrations; 250, 500 and 1000 μg/mL, of Raji cell lines have a significant inhibition rate: 49.8, 48.5 and 36.7%, respectively. It is postulated that quercetin may inhibit tumor promotion through inhibiting effects on phosphorylase C and protein kinase. The biological activity of quercetin is predicted to be highly dependent on the structure, particularly the availability of a hydroxyl group (29). The high proportion of quercetin and quercetin-4'-glucoside with a high level of free hydroxy groups may play an important role in antiproliferative activity against human leukemia cell lines.

References

1. Day, A.J.; Williamson, G. *Brit. J. Nutri.* **2001**, *86*, 105-110.
2. Kleijnen, J.; Knipschld, P.; Terriet, G. *Brit. J. Clinic. Pharma.* **1989**, *28*, 535-544.
3. Formica, J.V.; Regelson,W. *Food Chem. Toxic.* **1995**, *33*, 1061-1080.
4. Gao, C.M.; Takexaki, T.; Ding, J.H.; Li, M.H.;Tajima, K. *Jpn. J. Cancer Res.* **1996**, *90*, 614-621.
5. Debnath, S.; Hemavathy, J.; Bhat, K.K. *Food Chem.* **2002**, *78*, 479-482.
6. Price, K.R.; Bacon, J.R.; Rhodes, M.J.C. *J. Agric, Food Chem.* **1997**, *45*, 938-942.
7. Hirota, S.; Shimoda, T.; Takahama, U. *J. Agric. Food Chem.* **1998**, *46*, 3497-3502.
8. Ewald, C.; Fjelkner-Modig, S.; Jahansson, K.; Sjoholm, I.; Akesson, B. *Food Chem.* **1999**, *64*, 231-235.
9. Makris, D.P.; Rossiter, J.T. *J. Agric. Food Chem.* **2001**, *49*, 3216-3222.
10. Liwicki, P.P.; Witrowa-Rajchert, D.; Nowark, D. *Drying Technol.* **2001**, *16*, 83-100.
11. Hutchinson, D.H.; Otten, L. *Cereal Chem.* **1984**, *61*,155-158.
12. Jaime, L.; Molla, E.; Fernadez, A.; Martin-Carejas, M.A.; Lopez-Andreu, F.J.; Esteban, R.M. *J. Agric, Food Chem.* **2002**, *50*, 122-128.
13. Rhodes, M.J.C.; Price, K.R. *Food Chem.* **1996**, *57*, 113-117.
14. Esen, A. In *β-Glucosidases. Biochemistry and Molecular Biology.* ACS Symp. Ser. 533. American Chemical Society, Washington, DC, 1993, pp.1-14.
15. Esen, A.; Stetler, D.A. *Crop News Lett.* **1993**, *67*, 19-20.
16. Sue, M.; Ishibara, A.; Iwamura, H. *Planta* **2000**, *210*, 432-438.
17. Esen, A. *Plant Physiol.* **1992**, *98*, 174-182.
18. Ioku, K.; Tsushida, T.; Takei, Y.; Nakatani, N.; Terao, J. *Biochim. Biophy. Acta.* **1995**, *1234*, 99-104.
19. Sue, M.; Ishibara, A.; Iwamura, H. *Planta* **2000**, *210*, 432-438.
20. Yamaguchi, T.; Takamura, H.; Matoba, T.; Terao, J. *Biosci. Biotechnol. Biochem.* **1998**, *62*, 1201-1204.
21. Yen, G.C.; Chen, H.Y. *J. Agric. Food Chem.* **1995**, *43*, 27-32.
22. Bors, W.; Saran, M. *Free Radical Res. Commun.* **1987**, *2*, 289-294.
23. Dziedric, S.Z.; Hudson, B.J.F. *Food Chem.* **1984**, *14*, 45-51.
24. Pekkarinen, S.S.; Stockmann, H.; Schwarz, K.; Heinonen, I.M.; Hopia, A.I. *J. Agric. Food Chem.* **1999**, *47*, 3036-3043.

25. Chen, C.W.; Ho, C.-T. *J. Food Lipids* **1995**, *2*, 35-46.
26. Rong, Y.;Yang, E.B.; Zhang, K.; Mack, P. *Anticancer Res.* **2000**, *20*, 4339-4345.
27. Yoshida, M.; Sakai, T.; Hosokawa, N.; Marui, N.; Matsumoto, K.; Fujioka, A.; Nishino, H.; Aoike, A. *FEBS Letters* **1990**, *260*, 10-13.
28. Uddin, S.; Choudhry, M.A. *Biochem. Molecul. Biolog. Intl.* **1995**, *36*, 545-550.
29. Rice-Evans, C.A.; Miller, N.J.; Paganga, G. *Free Radic. Biol. Med.* **1996**, *20*, 933-956.

Chapter 16

Identification of Antioxidants from Du-Zhong (*Eucommia ulmoides* Oliver) Directed by DPPH Free Radical-Scavenging Activity

Yong Chen, Nanqun Zhu, and Chi-Tang Ho

Department of Food Science, Rutgers, The State University of New Jersey, 65 Dudley Road, New Brunswick, NJ 08901–8520

A free radical 2,2-diphenyl-picryhydrazyl (DPPH) scavenging activity-directed fractionation and purification process was used to identify the antioxidative components of Du-Zhong (*Eucommia ulmoides* Oliver). Dried bark of Du-Zhong was extracted with 95% ethanol and then separated into water, ethyl acetate, butanol and hexane fractions. Among these the butanol phase showed relatively strong antioxidant activity by DPPH testing when compared with water, hexane and ethyl acetate phases. The butanol fraction was then subjected to separation and purification using silica gel column chromatography and Sephadex LH-20 chromatography. Two compounds showed good antioxidant activity as identified by spectral methods (^{1}H-NMR, ^{13}C-NMR and MS) and by comparison with authentic samples of (+)-pinoresinol-*O*-β-D-glucose and syringic acid.

Eucommia ulmodies Oliver is a type of deciduous tree growing to about 20 meters tall. Known as Du-Zhong in traditional Chinese medicine, its bark and leaves have been used as tonic, analgesic and antihypertensive agents as well as for other medicinal purposes. In recent years, this plant has evoked various research interests and a number of biological functions have been reported, including recuperative effects for hypercholesterolemia and fatty liver (*1*), antifungal activity (*2*), and other pharmacological effects (*3*). Studies have shown that its functions are presumably attributed to the presence of certain chemical constituents. Identified in this plant, for instance, pinoresinol diglucoside is an antihypertensive component (*4*); loliolide from the chloroform extract of its leaves had immunosuppressive activity (*5*); geniposidic acid and aucubin from its leaves facilitated collagen synthesis, possibly related its reported pharmacological effects on healing organs and strengthening bone and muscle (*6*).

Du-Zhong has also been found to possess antioxidant activity. Using different methods, Yen and Hsieh (*7,8*) compared the antioxidant activities of its leaves, roasted cortex and raw cortex on oxidative damage in biomolecules and in lipid peroxidation model systems and discovered that these leaves exhibit the most potent antioxidant capacity, followed by the roasted cortex and the raw cortex. Their following study (*8*) revealed that the leaves had marked free radical or reactive oxygen species (ROS) scavenging activity while the roasted and raw cortex showed only modest and weak activity, respectively, and concluded that the scavenging abilities of Du-zhong extracts were correlated to their protocatechuic acid content.

In this study, we used a DPPH free radical scavenging activity directed identification process to pinpoint the antioxidants from the dried bark of Du-zhong and to elucidate their chemical structures as well as to compare their antioxidant capacities.

Materials and Methods

General Procedures

^{1}H-NMR and ^{13}C-NMR spectra were obtained on a VXR-200 instrument and MS analysis was performed on a Micromass Platform II system (Micromass Co., MA) equipped with a Digital DECPc XL560 computer for data analysis. Positive-ion mass spectra were obtained using a heated nebulizer atmospheric-pressure chemical-ionization (APCI) interface. The ion-source temperature was set at 150°C, and the sample cone voltage was 10 V. Thin-layer chromatograghy was performed on Sigma-Aldrich silica gel TLC plates (250 µm thickness, 2-25 µm particle size), with compounds visualized by spraying with 5% (v/v) H_2SO_4

in an ethanol solution. Silica gel (130-270 mesh) and Sephadex LH-20 (Sigma Chemical Co., St. Louis, MO) were used for column chromatography. Both 2,2-diphenyl-picryhydrazyl (DPPH) and silica gel (130-230 mesh) were purchased from Aldrich Chemical Co. (Milwaukee, WI). All solvents used for chromatographic isolation were analytical grade and purchased from Fisher Scientific (Springfield, NJ).

Plant Material

Dried bark of Du-Zhong was purchased from a local Oriental store and it was imported from China.

Extraction and Isolation Procedures

The sliced dried bark of Du-Zhong (1362 g) was immersed into 95% ethanol (10 L) and extracted continuously, at room temperature for 4 weeks. The plant material was filtered off and the EtOH extracts were combined and concentrated under reduced pressure by a rotary evaporator (Rotavapor R-114, Buchi, Switzerland). The obtained dry extract (113 g) was fractionated into hexane, ethyl acetate, butanol and water phases as shown in Figure 1. Each fraction was subjected to DPPH testing and the result indicated that all fractions had relatively weak DPPH scavenging activity, although the butanol fraction had the highest activity, as shown in Figure 2. Consequently, the separation and identification work was focused on butanol fraction. The dry butanol extract (23 g) was then subjected to column chromatography on silica gel and eluted with a solvent mixture of chloroform-methanol with increasing methanol content (30:1, 20:1, 10:1, 8:1, 5:1, 1:1, each 1000 mL), and collected into 16 fractions, which were subjected to DPPH testing again and the result is shown in Figure 3. After a series of separations using different solvent systems, about 38 mg of compound A was obtained from combined fractions of 7, 8 and 9, and about 20 mg of compound B was purified from combined fractions of 3 and 4.

Structure Determination of Isolated Compounds

(+)-Pinoresinol-O-β-D-glucose (A): Amorphous powder; APCI m/z: 519 [M-1]$^+$. ^{13}C NMR (50 MHz, CD$_3$OD): δ137.7 (s, C-1), 111.8^a (d, C-2), 147.7 (s, C-3), 151.2 (s, C-4), 111.2^a (d, C-5), 120.3 (d, C-6), 87.8 (d, C-7), 55.8^b (d, C-8), 71.6 (t, C-9), 134.0 (s, C-1'), 111.8 (d, C-2'), 147.7 (s, C-3'), 151.2 (s, C-4'), 116.3 (d, C-5'), 118.2 (d, C-5'), 87.3 (d, C-7'), 55.6^b (d, C-8'), 71.6 (t, C-9'), 103.1 (d, C-1''), 75.2 (d, c-2''), 78.1 (d, C-3''), 73 (d, C-4''), 70.5 (d, C-5''), 62.8 (t, C-6'') identical with literature (9), (a,b may be interchangeable).

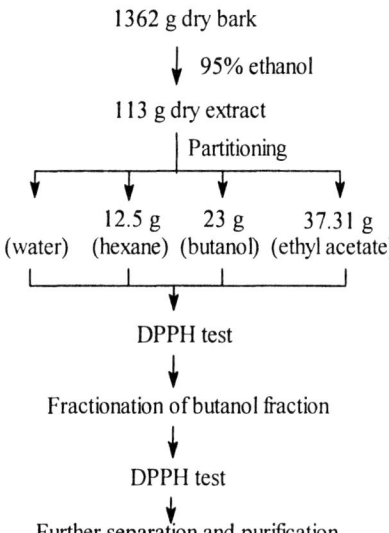

Figure 1. Extraction, separation and identification procedures.

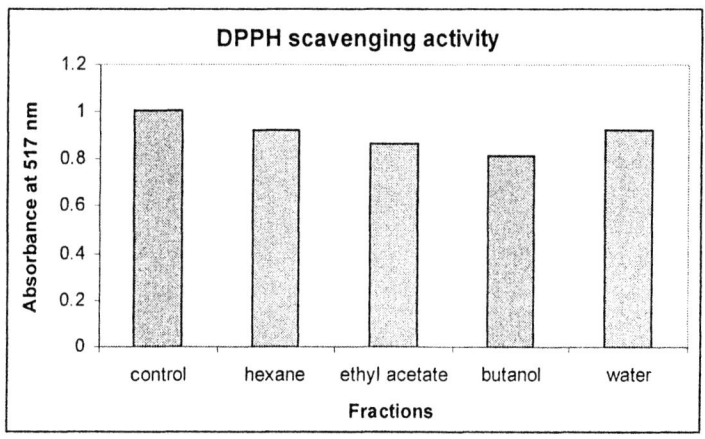

Figure 2. DPPH test for the 4 fractions (hexane, ethyl acetate, butanol, water) of crude alcohol extract

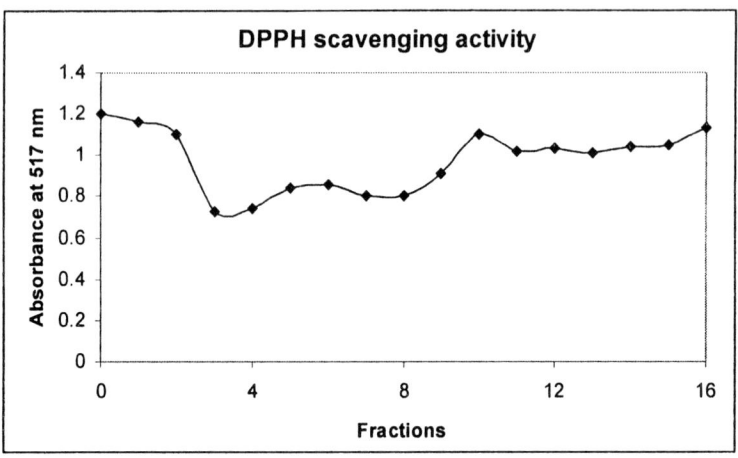

Figure 3. DPPH scavenging activity of the 15 sub-fractions of the butanol fraction.

Syringic acid (B): APCI m/z: 197 [M-1]$^+$. ^{1}H-NMR (200 MHz, CD$_3$OD): δ 7.33 (2H, s, H-2, 6), δ 3.88 (6H, s, H-OCH$_3$ at 3, 5) identical with literature (*10,11*).

Determination of the Scavenging Effect on DPPH Radicals

This method was adapted from Chen and Ho (*12*). The tested compounds were added into a 1.0 x 10^{-4} M ethanol solution of DPPH to make a final concentration at 30 μM. After mixing thoroughly, the solutions were kept in the dark for 30 minutes. Thereafter, the absorbency of the samples was measured using a spectrophotometer (Milton Roy, Model 301) at 517 nm against ethanol without DPPH as a blank reference. The test for each sample was duplicated and the values were averaged. For the determination of IC$_{50}$, each of the purified compounds was made into 8 different concentrations for DPPH testing. IC$_{50}$ was obtained by extrapolation from linear regression analysis.

Results and Discussion

The free radical DPPH (2,2-diphenyl-1-picrylhydrazyl) has been widely used to test free radical scavenging ability (*13*). For this study, it was also used to guide our separation and identification process. In this experiment, the crude ethanol extract from the dried bark was separated into 4 fractions and DPPH test indicated the butanol fraction had relatively higher free radical scavenging activity. Our work was then concentrated on the butanol fraction, which was further eluted into 15 sub-fractions that subsequently went through a thorough DPPH testing. The result is shown in Figure 3. Fractions 3 and 4 were combined and subjected to further separation and purification. About 20 mg of compound B was obtained and detected to be syringic acid, a gallic acid analogue. About 38 mg of compound B was purified and isolated from fractions 7, 8 and 9 and determined to be (+)-pinoresinol-*O*-β-D-glucose. Their structures, as shown in Figure 4, were determined by NMR and confirmed by literature. (+)-pinoresinol-*O*-β-D-glucose, regarded as an antioxidative component, is a lignan found in many plants including sesame seeds, extra-virgin olive oil (*14*) and valerian (*15*). Pinoresinol from Todopon Puok (*Fagraea racemosa*), a medicinal plant from Borneo, exhibited analgesic properties in a mice model system (*16*). Pinoresinol isolated from the stem bark of *Eucalyptus globulus* showed significant inhibitory action toward lipid peroxidation in rat liver microsome (*17*). In addition, it was also found to possess anti-inflammatory effects by inhibiting TNF-alpha production (*18*). On the other hand, syringic acid was identified as a strong DPPH scavenging agent from adlay hulls (*19*) and was one of the phenolic compounds isolated from *Mesona procumbens* Hemsl showing antioxidant activity (*10*). We conducted a study to compare their IC$_{50}$ in scavenging DPPH radical with those of BHA, BHT, vitamin E and trolox, as

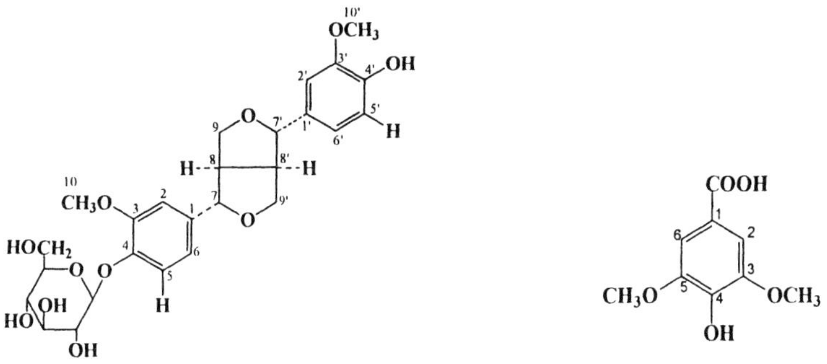

A: (+)-pinoresinol-*O*-β-D-glucose   B: Syringic acid

Figure 4. The structures of two identified compounds.

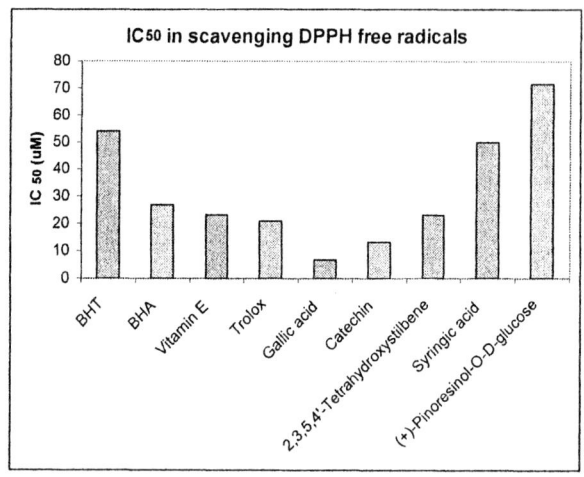

Figure 5. IC$_{50}$ of DPPH scavenging activities of 9 tested compounds.

well as gallic acid, catechin and 2,3,5,4'-tetrahydroxystilbene-2-O-β-D-glucoside, three antioxidants we previously identified from *Polygonum multiflorum* Thunb (*20*). The results, shown in Figure 5, indicate that the free radical scavenging activity of syringic was equivalent to that of BHT while (+)-pinoresinol-*O*-β-D-glucose showed the weakest among all 9 tested compounds, correlating to the report that roasted cortex and raw cortex had only modest and weak free radical scavenging activity (*8*).

In summary, this DPPH directed fractionation and identification study resulted in the identification of two compounds from the dried bark of Du-Zhong, which show free radical scavenging activity.

References

1. Nakasa, T.; Yamaguchi, M.; Okinaka, O.; Metori, K.; Takahashi, S. *J. Japan Soc. Biosci. Biotechnol. Agrochem.* **1995**, *69*, 1491-1498.
2. Huang, R.H.; Xiang, Y.; Liu, X.Z.; Zhang, Y.; Hu, Z.; Wang, D.C. *Febs. Lett.* **2002**, *521*, 87-90.
3. Deyama, T.; Nishibe, S.; Nakazawa, Y. *ACTA Pharmacol. Sinica* **2001**, *22*, 1057-1070.
4. Tomoda, M.; Gonda, R.; Shimizu, N.; Kanari, M. *Phytochem* **1990**, *29*, 3091-3094.

5. Okada, N.; Shirata, K.; Niwano, M.; Koshino, H.; Uramoto, M. *Phytochem.* **1994**, *37*, 281-282.
6. Li, Y. M.; Sato, T.; Metori, K.; Koike, K.; Che, Q. M.; Takahashi, S. *Biol. Pharm. Bull.* **1998**, *21*, 1306-1310.
7. Yen, G.C.; Hsieh, C.L. *J. Agric. Food Chem.* **1998**, *46*, 3952-3957.
8. Yen, G.C.; Hsieh, C.L. *J. Agric. Food Chem.* **2000**, *48*, 3431-3436.
9. Deyama, T.; Ikawa, T.; Kitagawa, S.; Nishibe, S. *Chem. Pharm. Bull.* **1986**, *34*, 523-537.
10. Hung, C.Y.; Yen, G.C. *J. Agric. Food Chem.* **2002**, *50*, 2993-2997.
11. Russell, K.M.; Molan, P.L.; Wilkins, A.L.; Holland, P.T. *J. Agric. Food Chem.* **1990**, *38*, 10-13.
12. Chen, C.W.; Ho, C.-T. *J. Food Lipids* **1995**, *2*, 35-46.
13. Dinis, T.C.; Maderia, V.M.; Almeida, L.M. *Arch. Biochem. Biophys.* **1994**, *315*, 161-169.
14. Owen, R.W.; Giacosa, A.; Hull, W.E.; Haubner, R.; Wurtele, G.; Spiegehalder, B.; Bartsch, H. *Lancet Oncol.* **2000**, *1*, 107-111.
15. Schumacher, B.; Scholle, S.; Holzl, J.; Khudeir, N.; Hess, S.; Muller, C. E. *J. Nat. Prod.* **2002**, *65*, 1479-1485.
16. Okuyama, E.; Suzumura, K.; Yamazaki, M. *Chem. Pharm. Bull.(Tokyo).* **1995**, *43*, 2200-2204.
17. Yun, B.S.; Lee, I.K.; Kim, J.P.; Chung, S.H.; Shim, G.S. Yoo, I.D. *Arch. Pharm. Res.* **2000**, *23*, 147-150.
18. Cho, J.Y.; Kim, A.R.; Park, M.H. *Planta Med.* **2001**, *67*, 312-316.
19. Kuo, C.C.; Chiang, W.; Liu, G.P.; Chien, Y.L.; Chang, J.Y.; Lee, C.K.; Lo, J.M.; Huang, S.L.; Shih, M.C.; Kuo, Y.H. *J. Agric. Food Chem.* **2002**, *50*, 5850-5855.
20. Chen, Y.; Wang, M.F.; Rosen, R.; Ho, C.-T. *J. Agric. Food Chem.* **1999**, *6*, 2226-2228.

Phytochemistry

Chapter 17

Schisandra chinensis: Chemistry and Analysis

Mingfu Wang[1], Qing-Li Wu[1], Yaakov Tadmor[1], James E. Simon[1], Shengmin Sang[2], and Chi-Tang Ho[2]

[1]New Use Agriculture and Natural Plant Products Program, Department of Plant Biology and Pathology, Rutgers, The State University of New Jersey, 65 Dudley Road, New Brunswick, NJ 08901
[2]Department of Food Science, Rutgers, The State University of New Jersey, 65 Dudley Road, New Brunswick, NJ 08901-8520

Schisandra chinensis is a famous traditional Chinese medicine and has been used in China for thousands of years. Schisandra berries have mainly been used for the lungs and kidneys as an astringent tonic, also as anti-hepatotoxic, anti-asthmatic, antitussive, sedative and tonic medicine. Recently it was found to be a very good liver protective drug and can lower the level of serum piruvic transaninase in hepatitis patients and lignan type compounds are believed to be the active components. In order to control the quality and ensure the reliability and repeatability of clinical and pharmacological research, it is necessary to develop and validate analytical methods for active components in Schisandra products. This paper will focus on the chemistry of Schisandra and the analytical methods used to analyze compounds in this herb. We will discuss the application of HPLC, GC and GC-MS, LC-MS, TLC, capillary electrochromatography and micellar electrokinetic capillary chromatography to analyze compounds in Schisandra.

© 2003 American Chemical Society

Schisandra chinensis (turcz) Baill grows wild in northeastern China, Japan, Korea and eastern parts of Russia. Its stem can reach 10-15 m in length and usually is twisting around the trunks of trees. The leaf has a wedge-shaped based and flower is white or slightly cream-colored. The fruits, ripening in September and October, have bright red color, 5-8 mm in diameter and the Seeds are brownish-yellow (*1*). In China, Schisandra Berry is called Wu-Wei-Zi (literal translation: five taste fruit) and has been used in China as herbal medicine since antique time. The first recorded use of Schisandra is found in Chinese earliest record of herbal medicine "Shennong Bencao Jing" as a superior drug to prevent asthma. According to Chinese Medicine philosophy, Schisandra berry has sour and warm properties and enters the lung and kidney channels and stomach meridians. In Traditional Chinese Medicine (TCM), Schisandra berries have been used mainly for the lungs and kidneys as an astringent tonic, also as antihepatotoxic, antiasthmatic, antitussive, sedative and tonic medicine.

More recently in China, crude Schisandra berries, their preparation, and individual constituents are widely used as an effective liver protective drug. In the 1970s, Chinese scientists reported that in experimental models, The levels of glutamic piruvic transaninase (GPT which is an enzyme found primarily in the liver that is released into the bloodstream as the result of liver damages) activities induced by carbon tetrachloride or paracetamol in mice, by thiacetamide in rats, and ethinylestradio 3-cyclopentylether in rabbits were significant decreased by oral administration of ethanol extracts from the fruits of Schisandra prior to and after the administration of hepatotoxic agents (*1,2*). Schidandra was also found to induce liver microsomal cytochrome P450, to stimulate the biosynthesis of protein and liver glycogen (*2*), to protect against tacrine- and bis(7)-tacrine induced hepatotoxicity (*3*), to protect against menadione-induced hepatotoxicity (*4*), and to protect aflatoxin B1 and cadmium chloride-induced hepatotoxicity in rats (*5*).

In the 1980s, a clinical trial with more than 5000 cases of various types of hepatitis patients had been performed in China (*6*). All these patients were treated with a Schisandra fruits preparation, the elevated GPT level returned to normal in 75 percents of patients after they were treated 20 days with schisandra. Another controlled study (*7*) was also conducted in China with 189 chromic viral hepatitis B patients with elevated GPT levels. Tablets prepared with an ethanol extract of schisandra berries (with 20 mg of lignans inside corresponding to 1.5 crude Schisandra fruit) were given to 107 of these patients, and liver extract and vitamins were give to the control group (82 of these patients). In the treated group, normal GPT levels were observed in 73 patients or 68 percent while in the control group, normal GPT levels were observed in 36 patients, or 44 percent. These researches and other studies have led to the development of the antihepatotoxic drug DDB (dimethyl-4,4'-dimethoxy-5,6,5',6'-dimethylene-dioxybiphenyl-2,2'-dicarboxylate) in China which has been widely used to treat chronic viral and drug-induced hepatitis (*4*).

Meantime Schisandra herbs or herbal extracts have also been found to have antioxidant and detoxificant activities (*1*), anti-carcinogenic effects, antiviral (*1*), anticonvulsant (*8*), to improve physical performance, to act on the central nerve system and prevent and treat neurodegenerative disease (*1*), to inhibit chitin synthase II (*9*), as platelet activating factor antagonists (*10-11*), and to inhibit acetyl-Co A: cholesterol acetyltransferase (*12*).

Chemistry

S*hisandra chinensis* is a rich source of lignan compounds and total about 30 lignans have been purified from it with schisandrol A (other names: schisandrin, wuweizi alcohol A, wuweizichun A, 2-9%), schisandrin B (γ-schisandrin, wuweizisu B, 1-5%), schisandrol B (Gosimin A, wuweizi alcohol B, Wuweizichun B, 0.7-3%), schisandrin A (deoxyshisandrin, weiweizisu A, 0.2-1.1%), schisantherin A (gomisin C, wuweizi ester A) and schisantherin B (gomin B, wuweizi ester B) as the major ones inside (*13-34*). The sturcures of four major lignans in S*hisandra chinensis* are shown in Figure 1. Other lignans includes gomisin E, F, G, H, Angeloylgomisin H, Tigloygomisin H, gomisin J, K1, K2, K3 (schisanhenol), L1, L2, M1, M2, N, O, 6-*O*-benzoylgomisin O, benzoylisogomisin O, angeloylisogomisin O, epigomisin O, tigloylgomisin P, angeloylgomisin P, gomisin R, S and T, and schisandrin C (wuweizisu C) and they are minor lignans (below 0.2% for each of them) and the contents of them vary depending on the plant origin and the harvesting season (*13-34*). Most of these lignans have dibenzo[a,c]-cyclooctene skeleton, with two methyl group at 6, 7 positions and substituted groups at other positions vary depending on compounds, the substituted groups could be –OH, -OCH$_3$, -*O*-benzoyl, -*O*-angeloyl, -*O*-tigloyl. The fruits also contain tannins, sugar and essential oils (with copaene, alpha-farnesene, ansd alpha-cubebene as the major constituents) (*35*). Recently several other types of compounds including protocatechuic acid, quinic acid, 2-Me citrate, 5-hydroxymethyl-2-furancarboxaldehyde, zingerone glucoside, thymoquinol 2-glucoside, thymoquinol 5-glucosides and daucosterol were also isolated form the fruits of *schisandra chinensis* (*36*).

Analysis of Chemical Components in *Schisandra chinenesis*

High Performance Liquid Chromatography (HPLC)

As the lignans have been found to be the active components in *Schisanda chinensis*, they are used as marker compounds in Schisandra herbs and herbal preparations. Right now HPLC is the most useful approach to analyze lignans in Schisandra and can offer good separation, shorter analysis times and easier

237

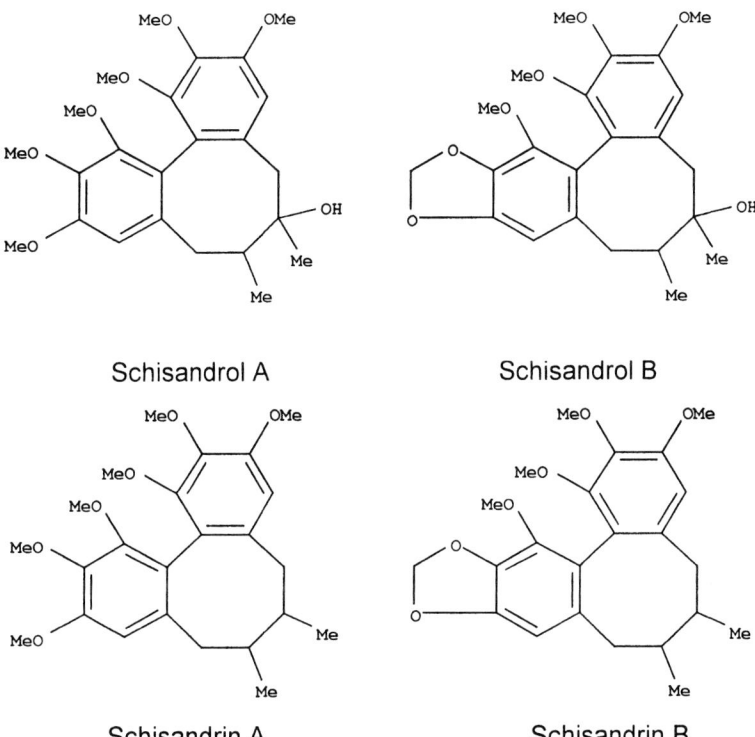

Figure 1. Structures for four major lignans in Schisandra chinensis.

quantification. Several HPLC methods have been developed to analyze lignans in Schisandra. Nakajima et al. (37) reported the first reliable HPLC method to analyze lignans in Schisandra collected from different regions (China, Korea and Japan) and they found schisandrol A and schisandrol B are the main components for the samples collected from central and northern parts of Korea or China while schisandrol A and schisandrin A were the main components from samples collected in Japan. In 1983, Wang et al. (38) also published a HPLC method to separate wuweizisu C, schisandrol A, schisandrol B, Schisandrin A and schisandrin B. By using an octadecylsilane column with 77% methanol as the mobile phase and detection wavelength at 254 nm, these lignans had good separation. Zhu et al. (39) reported in 1988 a HPLC method for assay of lignans of *Schisandra chinensis* in Shengmai San, by using a YWG C18 (10um) column with MeOH-H_2O (72:28, vol./vol.) as mobile phase and detection at 254 nm. Lignans including schisandrol A, shisandrol B, schisandrin A and schisandrin B, schisantherin A and Schisantherin B were separated and identified in Shengmai-San. A similar method was also reported in 1989 (40), but the authors were using 77% methanol as mobile phase. Zhang et al. (41) developed and validated in 1990 a simple, rapid and reproducible reverse phase HPLC method for detection of lignans of *Schisandra chinensis*, they were using a µ-Bondapak C18 (10 µm) column and two mobile phases, I: acetonitrile-methanol-water (11:11:16) and II: acetonitrile-methanol-water (10:10:10) for separation. The flow rate were set up at 1 mL/min and detection wavelength was at 254 nm. From 0-12 minutes, they were using mobile phase I and then used mobile phase II. The average recoveries for this method was 93.3% for Schisandrin A, 99.1% for Schisandrol B and 101% for Gomisin N. He et al. (13) developed in 1997 a gradient HPLC method using a Prodigy ODS, 5 µM, 150 x 2.0 mm column (Phenomenex, USA), mobile phase A, water; B, methanol; gradient elution was with 60 to 100% B in 15 minutes, and keep at 100% B for 5 minutes and change from 100% to 60% B in 5 minutes and the flow rate was at 0.2 mL/minute.

Recently we developed and validated a nice reverse phase PHLC method to analyze the lignans in Schisandra fruits. Analytical HPLC analyses were performed on a Hewlett-Pachard 1100 modular system equipped with an auto-sampler, a quaternary pump system, a photodiode array detector and a HP Chemstation data system. A pre-packed 250 × 4.6 mm (5 µM particle size) Luna C18 (2) column (Phenomenex, Torrance, CA) were selected for HPLC analysis. The absorption spectra were recorded from 200 to 400 nm for all peaks; quantification was carried out at a single wavelength of 255 nm.

The column temperature was ambient and the mobile phase included water (containing 0.1% formic acid, solvent A) and acetonitrile (solvent B) in the following gradient system: initial 45% B, linear gradient to 60% B in 12 minutes, hold at 60% B for 12 minutes, linear gradient to 90% B in 16 minutes and then linear gradient to 100% B in 5 minutes. The total running time was 45 minutes. The post running time was 12 minutes. The flow rate was set at 1 mL/minute.

The dried Schisandra fruits and fruit extracts (9 to 1) were gifts from Frutarom Inc (USA). The dried Schisandra fruits were ground to powder and accurately weighed about 400 mg or 200 mg of ground fruits powder or extract, respectively, into a 100 mL volumetric flask. Seventy ml of methanol were then added and samples were sonicated for 30 min. The flasks were allowed to cool to room temperature and then filled to volume with methanol. Using a disposable syringe and 0.45 µm filter, the samples were filtered into HPLC vials for HPLC analysis.

The four reference standards (schisandrol A, schisandrol B, schisandrin A and schisandrin B) used in this analysis were purified from a Schisandra fruit extract using column chromatography on silica gel, reverse phase silica gel and Sephadex LH-20, and they structures were validated by NMR and MS spectra.

This method showed excellent linearity, accuracy and precision. The profiles we got match with the reported profile for *Schisandra chinensis*. One representative HPLC chromatogram is shown in Figure 2.

Gas Chromatography (GC) and GC/MS

GC and GC/MS methods are widely acceptable methods for analysis of volatile and hydrophobic compounds. Sohn *et al.* reported in 1989 (*42*) using GC-MS method to identify the lignans in Schisandra. A total of 11 lignans including gomisin J, schisandrin A, gomisin N, schisandriol A, schisandrin C, schisandrol B, angeloylgomisin H, tigloygomisin H, angeloylgomisin Q, Schisantherin B and benzoylgomisin have been detected. They used a SPB-1 fused silica capillary column and the column temperature was 200-300°C at the rate 4°C/min, the mass ionization voltage was 70 eV (EI mode). They (*43*) also published a GC/FID method to determine lignans in Schisandra fruits using same column and condition, this time they only detected 7 lignans including schisandrin A, gomisin N, schisandrol A, schisandrin C, schisandrol B, angeloylgomisin H, and tigloylgomisin H by FID detector. The FID responses were maintained at 2-50 ppm for schisandrin A and schisandrin C, and 5-500 ppm for other compounds. Two GC-MS methods (*35,44*) were also published recently to analyze the essential oil of Schisandra, one detected 56 compounds and identify 49 compounds, another one detected 44 peaks and identified 27 of them representing 95.13% of the total contents and concluded that copanene, alpha-farnesene and alpha-cubenene were the main components in seed oil and the chemical components of seed oil were different from those of the oil of Schisandra fruit.

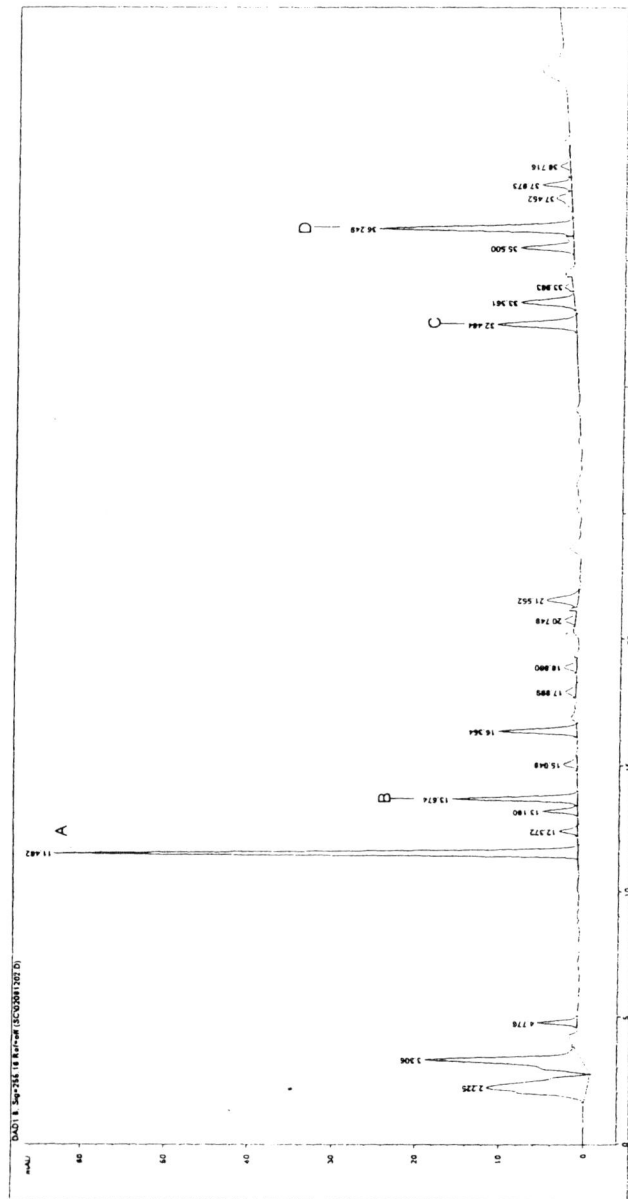

Figure 2. Representative HPLC chromatogram of Schisandra chinensis fruit (Peak A, schisandrol A; Peak B, schisandrol B; Peak C, schisandrin A and Peak D, Schisandrin B identified by comparison with authentic reference standards)

Thin Layer Chromatography Method

Suprunov et al. (45) developed a thin layer chromatography and colorimetric method to determine shisandrin and schisandrol in *Schisandra chinensis* fruit powders or extracts. The lignans were extracted using chloroform and separated on silica gel plate using ethyl acetate-petroleum ether (1:1, v/v) as mobile phase. The lignans were then extracted using chloroform, treated with concentrated H_2SO_4 and acetone, and measured using a dark violet filter. Zhu et al. reported in 1988 (46) a simple, rapid and accurate method for the TLC-densitometric detection of 2 active constituents of lignans, *i.e.* schisandrol A and B in Shengmai-San (a Chinese herbal formula with *Panax ginseng, Schidandra chinensis* and *Ophiopogon japanicus*). The lignans was extracted with ether at room temperature. The sample solution and that of mixed authentic standard lignans were spotted on silica gel plate, and developed by PhMe-EtOAc (6:4, v/v). The separated spots were scanned using Shimadzu TLC Scanner. Wang et al. (47) published in 1990 another TLC densitometry method to analyze lignans in Schisandra. In this research, the samples were extracted with hexane, the hexane extracts were then evaporated to dryness and the samples were re-dissolved in methanol and spotted on a silica gel GF_{254} plate, and developed with toluene-ethyl acetate (9:1) for schisandrin A, B and C, with toluene-ethyl acetate (4:6) for schisandrol A, schisandrol B, wuweizi ester and gomisin K3. A Shimadzu TLC model 910 was used for scanning. Zhao et al. (48) also reported another TLC method to determine Schisandrin B. The quantitative determination of this compound was achieved by first extracting with ether under reflux, drying to dryness and re-dissolving in methanol, passing the methanol extract through a neutral Al_2O_3 column and then spotting on a silica gel GF_{254} plate, developing with petroleum ether-ethyl formate-formic acid (15:5:1).

Capillary Electrochromatography and Micellar Electrokinetic Capillary Chromatography

Capillary electrophoresis has become a very popular method for botanical samples analysis. Kvasnickova et al. (49) reported recently a method to determine Schisandra lignans by capillary electrochromatography using polymer-based monolithic stationary phase. The column were prepared by co-polymerization of acrylamide, N,N'-methylenebisacrylamide, vinylsulfonic acid and lauryl acrylate in the presence of poly(ethylene glycol) as a porogenic agent. The column was successfully used to determine and quantify the major lignans in Schidandra in 35 minutes. The quantification for major lignans (schisandrol A, schisandrol B, gomisin N and schisandrin C) were in good agreement with those detected by the HPLC method. Micellar electrokinetic capillary chromatography (MEKC) has been developed as a promising method for analysis of botanical samples recently. Sterbova et al. (50) published this year a MEKC method to analyze schisandra lignans. In this method, the background electrolyte consisted of 40 mM SDS and 35% acetonitrile in 10 mM tetraborate buffer (pH 9.3), the applied voltage was at 28 kV (positive polarity) and the capillary temperature was optimized at 25 °C. By using these conditions, the lignans had very good separation and the results are similar to the results tested by HPLC or capillary electrochromatography.

High Performance Liquid Chromatography-Mass Spectrometry

LC-MS methods have been widely used for the analysis of natural products recently. HPLC-MS coupled with a photodiode-array detector enable us to obtain the UV and MS spectra for each peak almost at the same time. By careful design of experiments, we can get reliable and useful data to compare with reported literature, sometime with the spectra of reference standards to check the identification and purity of each peak. LC-MS is also very sensitive to certain types of compounds comparing with UV detector. LC-MS-MS is even more sensitive than LC-MS by only picking up certain daughter ion for a specific compound and discarding non-related compounds. He et al. (13) had reported the only LC-MS method for *Schisandra chinensis*. In this research, they used a HP1090 series HPLC coupled with a HP 5989 B quadrupole mass spectrometer without stream splitting, the flow rate was set at 0.2 mL/min. the drying N_2 was set at 40 mL/min and temperature was at 350 °C and nebulizing N_2 was 80 p.s.i. They identified total fifteen peaks, six of that were unambiguously identified as schisandrol A, schisantherin A, schisantherin B, gomisin K3, schisandrin A and schisandrin B based on their abundant $[M+1]^+$, $[M+Na]^+$ ions, UV spectra and retention times, compared with those data of reference compounds. Other 9 peaks were tentatively identified, based on their intense $[M+H]^+$ and $[M+Na]^+$ and UV spectra.

We analyzed one Schisandra fruit samples under a different condition using an Agilent 1100 Series LC/MSD system (Agilent Technologies, Germany) equipped with quaternary pump, diode array and multiple wavelength detector, thermostatted column compartment, degasser, MSD trap with an electrospay source and software of HP ChemStation, Bruker Daltonics 4.0 and Data analysis 4.0. The mobile phase we used was same as the HPLC method mentioned before, but with a 2 to 1 stream splitting for the MSD detector.

The electrospray mass spectrometer (ESI-MS) was operated under positive ion mode and optimized collision energy level of 20%, scanned from m/z 100 to 700. ESI was conducted using a needle voltage of 3.5 kV. High-purity nitrogen (99.999%) was used as dry gas and flow rate at 9 L/Min, capillary temperature at 325 °C. Helium was used as Nebulizer at 45 psi. The ESI interface and mass spectrometer parameters were optimized to obtain maximum sensitivity. The

243

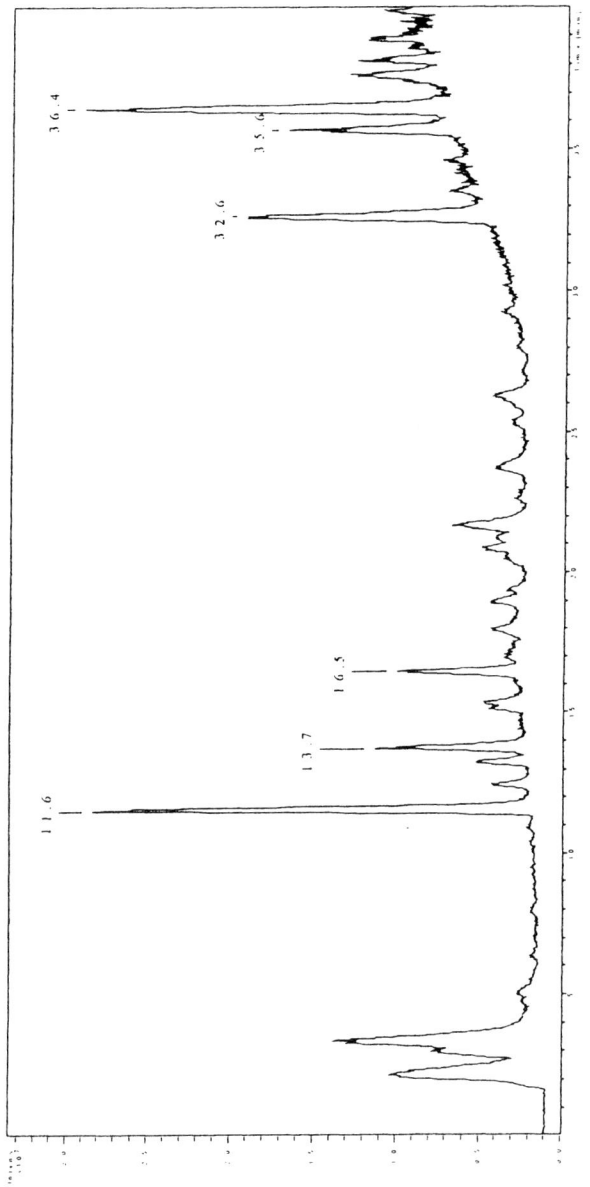

Figure 3. Representative TIC of Schisandra chinensis fruit (11.6 min, schisandrol A; 13.7 min schisandrol B; 16.5 min, angeloylgomisin H; 32.6 min, schisandrin A; 35.6 min, gomisin N and 35.4 schisandrin B, respectively).

Total Ion Chromatogram (TIC) was shown in Figure 3. Total 6 major lignans were tentatively identified by comparing the UV and MS spectra with the reference standards and by their $[M+1]^+$ and $[M+Na]^+$ ions.

Conclusion

Schisandra chinensis is a rich source of lignan compounds and their lignans have been analyzed by various methods including HPLC, GC, GC-MS, TLC, LC-MS and CE. Comparing with other methods, HPLC is widely used and more reliable method for Schisandra lignans analysis and LC/MS is very useful for identification of lignans in Schisandra fruits.

References

1. Hanche, J. L.; Burgos, R. A.; Ahumada, F. *Fitoterapia* **1999**,*70*, 451-471.
2. Liu, G.-T. *Emerging Drugs (Westbury, NY, United States)*, **2001**, *1(molecular Aspects of Asian Medicines)*, 461-478.
3. Pan, S. Y.; Han, Y.F.; Carlier, P.R.; Pang, Y.P.; Mak, D.H.F.; Lam, B.Y.H. Ko, K.M. *Planta Medica* **2002**, *68*, 217-220.
4. Ip, S.-P.; Yiu, H.-Y.; Ko, K.-M. *Molecular and Cellular Biochemisty* **2000**, *208*, 151-155.
5. Ip, S.-P.; Mak, D.H.F.; Li, P.C.; Poon, M. K.T.; Ko, K.M. *Pharmacology & Toxicology (Copenhagen)* **1996**, *78*, 413-416.
6. Chang, H.M., But, P.H.; *Pharmacology and Application of Chinese Materia Medica (Singapore: World Sci)*, **1986**, *vol.1*, 773.
7. Liu, G. T. *Acta Pharmaceutica Sinica* **1983**, *18*, 714-720.
8. Baek, N.-I.; Han, J.-T.; Ahn, E.-M.; Park, J.K.; Cho, S.W.; Jeon, S.-G.; Jang, J.-S.; Kim, C.K.; Choi, S.Y. *Agricultural Chemistry and Biotechnology* **2000** *43*, 72-77.
9. Hwang, E.I.; Kim, M.K.; Lee, H.B.; Kim, Y.K.; Kwon, B.M.; Bae, K.H.; Kim, S.U. *Yakhak Hoechi* **1999**, *43*, 509-515.
10. Lee, I.S.; Jung, K. Y.; Oh, S.R.; Park, S.H.; Ahn, K.S.; Lee, H.-K. *Biological & Pharmaceutical Bulletin*, **1999**, *22*, 265-267.
11. Jung, K. Y.; Lee, I.S.; Oh, S. R.; Kim, D. S.; Lee, H.K. *Phytomedicine* **1997**, *4*, 229-231.
12. Kwon, B.M.; Jung, H.J.; Lim, J.-H.; Kim, M.-K.; Kim, Y.-K., Bok, S.H.; Bae, K.-H.; Lee, I.-R. *Planta Medica* **1999**, *65*, 74-76.
13. He, X.-G.; Lian, L.-Z.; Lin, L.-Z. *Journal of Chromatography A*, **1997**, *757*, 81-87.

14. Song, W.Z.; Tong, Y.Y.; Cheng, L. S. *Tianran Chanwu Yanjiu Yu Kaifa*, **1990**, *2*, 51-8.
15. Taguchi, H.; Ikeya, Y. *Chem. Pharm. Bull.* **1975** *23*, 3296-3298.
16. Ikeya, Y.; Taguchi, H.; Litaka, Y. *Tetrahedron Lett.* **1976**, *17*, 1359-1362.
17. Chen, Y.-Y.; Li, L.N. *Huaxue Xuebao* **1976**, *34*, 45-52.
18. Taguchi, H. Ikeya, Y. *Chem. Pharm. Bull.* **1977**, *25*, 364-366.
19. Ikeya, Y.; Taguchi, H.; Yosioka, I. *Chem. Pharm. Bull.* **1978**, *26*, 328-331.
20. Ikeya, Y.; Taguchi, H.; Yosioka, I. *Chem. Pharm. Bull.* **1978**, *26*, 682-684.
21. Ikeya, Y.; Taguchi, H.; Yosioka, I.; Kobayashi, H. *Chem. Pharm. Bull.* **1978**, *26*, 3257-3260.
22. Ikeya, Y.; Taguchi, H.; Yosioka, I. Kobayashi, H. *Chem. Pharm. Bull.* **1979**, *27*, 1395-1401.
23. Ikeya, Y.; Taguchi, H.; Yosioka, I. Kobayashi, H. *Chem. Pharm. Bull.* **1979**, *27*, 1383-1394.
24. Ikeya, Y.; Taguchi, H.; Yosioka, I. Kobayashi, H. *Chem. Pharm. Bull.* **1979**, *27*, 1576-1582.
25. Ikeya, Y.; Taguchi, H.; Yosioka, I. Kobayashi, H. *Chem. Pharm. Bull.* **1979**, *27*, 1583-1588.
26. Ikeya, Y.; Taguchi, H.; Yosioka, I. Kobayashi, H. *Chem. Pharm. Bull.* **1979**, *27*, 2695-2709.
27. Ikeya, Y.; Taguchi, H.; Yosioka, I. *Chem. Pharm. Bull.* **1980**, *28*, 2422-2427.
28. Ikeya, Y.; Taguchi, H.; Yosioka, I. Kobayashi, H. *Chem. Pharm. Bull.* **1980**, *28*, 3357-3361.
29. Ikeya, Y.; Taguchi, H.; Yosioka, I. *Chem. Pharm. Bull.* **1982**, *30*, 132-139.
30. Ikeya, Y.; Ookawa, N. Taguchi, H.; Yosioka, I. *Chem. Pharm. Bull.* **1982**, *30*, 3202-3206.
31. Ikeya, Y.; Taguchi, H.; Mitsuhashi, H.; Takada, S.; Kase, Y.; Aburada, M. *Phytochemistry* **1988**, *27*, 569-573.
32. Ikeya, Y.; Kanatani, H.; Hakozaki, M.; Taguchi, H.; Mitsuhashi, H. *Chem. Pharm. Bull.* **1988**, *36*, 3974-3979.
33. Chen, C.-C.; Shen, C.-C.; Shih, Y.-Z.; Pan, T.-M. *J. Nat. Prod.* **1994**, *57*, 1164-1165.
34. Slanina, J.; Tabotska, E.; Lojkova, L. *Planta Medica* **1997**, *63*, 277-280.
35. Wang, Y.; Wang, J.; You, H.; Cui, Y. *Zhongguo Yaoxue Zazhi* **2001**, *36*, 91-92.
36. Dai, H.; Zhou, J.; Peng, Z.; Tan, N. *Tianran Chanwi Yanjiu Yu Kaifa*, **2001**, *13*, 24-26.
37. Nakajama, K.; Taguchi, H.; Ikeya, Y.; Endo, T.; Yosioka, I. *Yakugaku Zasshi* **1983**, *103*, 743-749.
38. Wang. M.; LI, B. *Yaoxue xuebao* **1983**, *18*, 209-214.

39. Zhu, Y.; Yan, K.; Wu, J.; Tu, G. *J. Chromatogr.* **1988**, *438*, 447-450.
40. Tong, Y.; Song, W. *Zhongguo Zhongyao Zazhi* **1989**, *14*, 611-614.
41. Zhang, Y.; Guo, Y.; Nakajima, K.; Ikeya, Y.; Mitsuhashi, H. *Yaowu Fenxi Zazhi* **1990**, *10*, 146-148.
42. Sohn, H. J.; Bock, J. Y. *Han'guk Nonghwa Hakhoechi* **1989**, *32*, 344-349.
43. Sohn, H. J.; Bock, J. Y.; Baik, S. O., Yong, M. *Han'guk Nonghwa Hakhoechi* **1989**, *32*, 350-356.
44. Li, X., Cui, H.; Song, Y.; Liang, Y. *Yaoxue Xuebao* **2001**, *36*, 215-219.
45. Suprunov, N.; Samoilenko, L. I. *Farmatsiya*, **1975**, *24*, 35-37.
46. Zhu, Y.; Yan, K.; Tu, G. *Yaowu Fenxi Zazhi* **1988**, *8*, 71-73.
47. Wang, K.; Tong, Y.Y.; Song, W. Z. *Yaoxue Xuebao* **1990**, *25*, 49-53.
48. Zhao, H.; Xu, J.; Liu, R. *Yaowu Fenxi Zazhi* **1995**, *15*, 33-35.
49. Kvasnickova L.; Glatz, Z.; Sterbova, H.; Kahle, V.; Slanina, J.; Musil, P. *Journal of Chromatography A* **2001**, *916*, 265-271.
50. Sterbova, H.; Sevcikova, P.; Kvasnickova, L.; Glatz, Z.; Slanina J. *Electrophoresis* **2002**, *23*, 253-258.

Chapter 18

Phytochemical and Biological Studies on *Evodia lepta*

Guolin Li[1], Dayuan Zhu[2], and Ravindra K. Pandey[1]

[1]PDT Center, Roswell Park Cancer Institute, Elm and Carlton Streets, Buffalo, NY 14263
[2]Department of Phytochemistry, Shanghai Institute of Materia Medica, Chinese Academy of Sciences, 294 Tai-Yuan Road, Shanghai, 200031, People's Republic of China

Chemical investigation of *Evodia lepta* resulted in the isolation and identification of twenty-five compounds with the basic unit of 2,2-dimethylchromene or 2,2-dimethylchroman, one semiterpine and two flavones. The biological assay of some of these compounds showed leptol A (**7**) has killing mosquito activity (mortality 30% at 2.0 ppm), and leptin A (**81**), evodione (**6**), ethylleptol A (**9**), leptol A (**7**) have weak anti-HIV activity.

Evodia lepta (Spreng.) Merr., which belongs to Rutaceae family, is a deciduous shrub or arbor, and distributes in the south of China. As a traditional Chinese herb medicine, this plant is widely used for treating sore throat, malaria, infectious jaundice, rheumatic ostalgia, eczema, dermatitis, ulcer, etc (*1*). Chemical investigation of this plant by Gunawardana et al gave three alkaloids, (-)-edulinine, (-)-ribalinine and (+)-isoplatydesmine (Figure 1) (*2*). To elucidate the effective constituents, we had investigated the aerial part of this plant which was collected from Hainan province, China. In this paper, we like to give a review of phytochemical and biological studies on *Evodia lepta* in our laboratory.

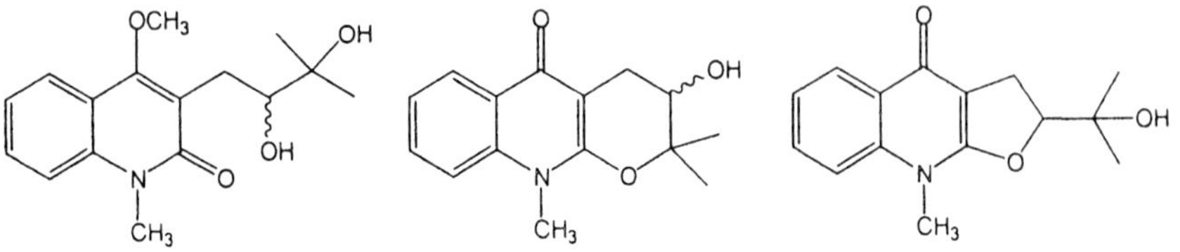

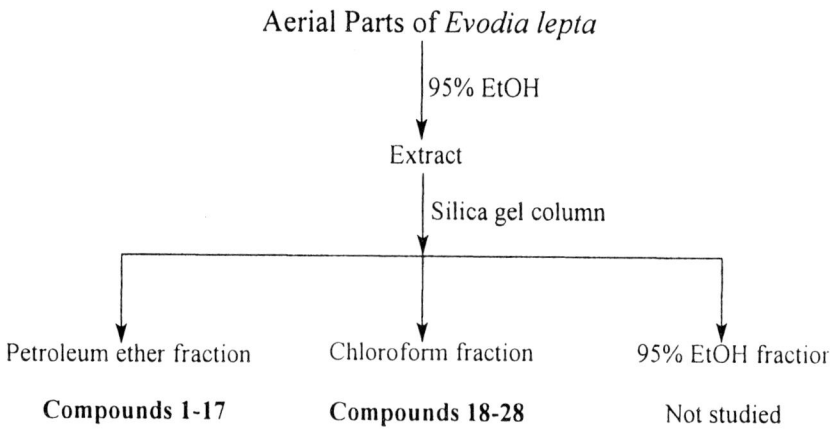

(-)-edulinine   (-)-ribalinine   (+)-isoplatydesmine

Figure 1. alkaloids from Evodia lepta.

Phytochemical Studies on *Evodia lepta*

From the aerial parts of *Evodia lepta*, we isolated and identified thirteen 2,2-dimethyl-chromenes, four dichromenes, eight 2,2-dimethyl-chromans, one sesquiterpene and two flavones. Figure 2 shows the extraction and isolation procedure.

Aerial Parts of *Evodia lepta*
↓ 95% EtOH
Extract
↓ Silica gel column

Petroleum ether fraction	Chloroform fraction	95% EtOH fraction
Compounds 1-17	**Compounds 18-28**	Not studied

Figure 2. Extraction and isolation procedure.

2,2-Dimethylchromenes (3,4,5)

All of the thirteen chromenes have the similar structures (Figure 3). The major difference is the substituted groups on their benzene rings. Methylevodionol (**1**) and isoevodionol (**11**) are known compounds, and their structures were determined by comparison their spectral data with those reported in the literature (6,7,8).

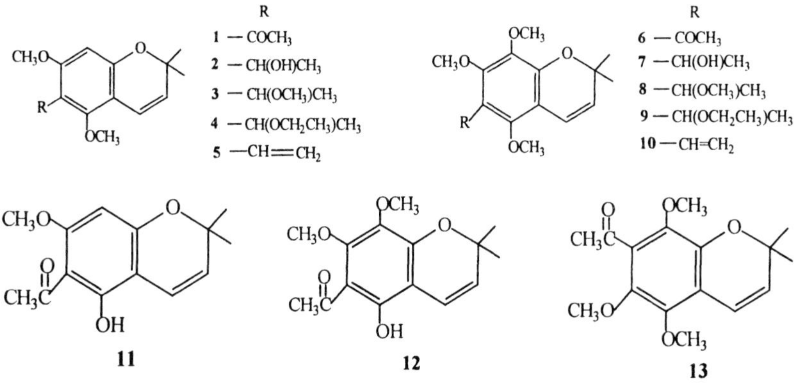

Figure 3. 2,2-Dimethyl-chromenes from Evodia lepta.

Evodione (**6**) is also a known natural product which had been isolated from *Evodia elleryana* (9,10). Its structure had been determined by chemical degeneration (11) and total synthesis (12) previously, but there was no ^{1}H NMR and ^{13}C NMR spectral data available in literature. To confirm that compound **6** was evodione, isoevodionol (**11**) was transformed to evodione in two steps of reactions. First, **11** was converted to **11a** via Elb reaction (13), then methylation of **11a** with $(CH_3)_2SO_4/K_2CO_3$ gave evodione (Figure 4). The spectral data (^{1}H NMR and EIMS) of the synthetic evodione is exactly same to compound **6**.

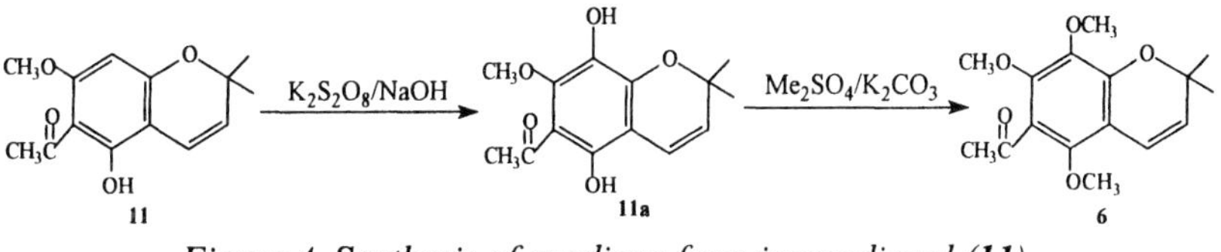

Figure 4. Synthesis of evodione from isoevodionol (11).

The structures of chromenes **2**, **3**, **4**, **5**, **7**, **8**, **9**, **10** were also determined by chemical correlation method (Figure 5).

For compound **12**, the most outstanding issue was to determine the position of hydroxyl group. Arnone, A., et al had studied a serial of chromenes with hydroxyl group on benzene ring, and they found that H-4 resonance has an upfield shift about 0.3-0.4 ppm and H-3 resonance has an downfield shift about 0.1 ppm in ^{1}H NMR spectra if HO-5 was acetylated, while acetylation of the hydroxyl groups at other positions had very little effect on H-3 and H-4 (14).

Compound **12** was treated with acetic anhydride in pyridine to give the acetylated derivative **12a** (Figure 6). The ^{1}H NMR resonances of H-4 and H-3 of compound **12a** appear at 6.23 ppm and 5.60 ppm, respectively, while those of compound **12** appear at 6.65 ppm and 5.49, respectively. The upfield shift of 0.42 ppm for H-4 and downfield shift of 0.11 ppm for H-3 indicated that the hydroxyl group in compound **12** should be at position 5.

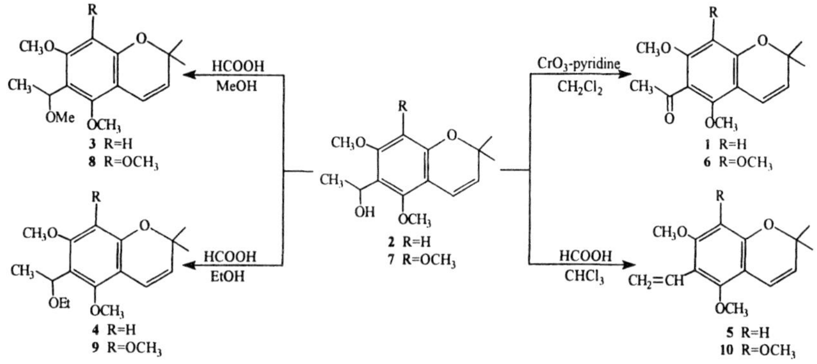

Figure 5. Chemical correlation among chromenes.

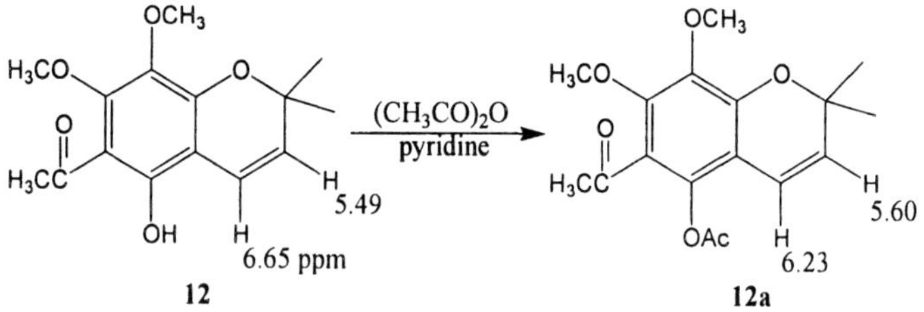

Figure 6. Acetylation of compound 12.

Dichromenes (*15,16*)

Four dichromenes were isolated from *Evodia lepta* (Figure 7). Their structures were determined by extensive 2D NMR studies (H-H COSY, HMQC and HMBC) and analysis of their mass spectra. Interestingly, compounds **15** an **16** have the same planar structures. Both of compounds have two chiral centers, so four isomers are possible (RR, SS, RS and SR). RS and SR are meso isomers because there is a plane of symmetry in this molecule. The configurations of two chiral centers in compound **15** should be RS and in compound **16** should be RR or SS owing the fact that the optical rotation of compound **15** is 0.00° (25 °C, acetone, *c* 0.61) and that compound **16** is -13.7° (20 °C, acetone, *c* 1.2).

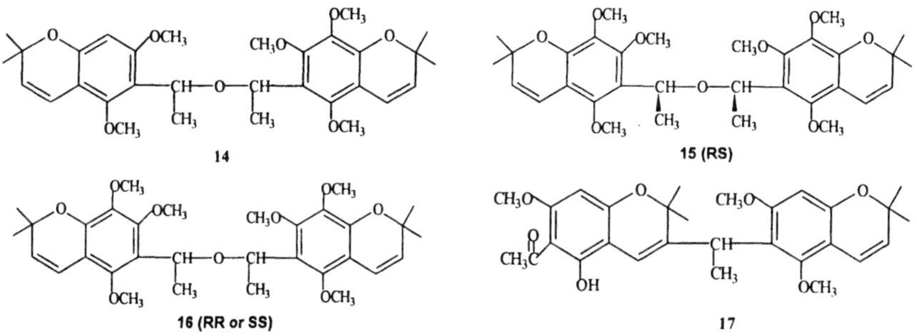

Figure 7. Dichromenes from Evodia lepta.

2,2-Dimethylchromans (*17,18*)

Eight 2,2-dimethylchromans were isolated (Figure 8). Their structures were elucidated by chemical correlations and 2D NMR studies. For example, treating isoevodionol with KMnO$_4$-NaOH gave compound **18** (*19*), so the structure of compound **18** was determined as shown in Figure 9, and the diol at positions 3 and 4 has *cis* configuration. It is well known that oxidation of olefin with H$_2$O$_2$-HCOOH method gives a diol compound with *trans* configuration (*20*). But when isoevodionol was treated with H$_2$O$_2$-HCOOH, a mixture of compounds **18** and **19** was obtained (Figure 9). This could be explained by the fact that HO-4 in compound **19** was at α position of a benzene ring, and its configuration could be reverted under acidic condition. This was confirmed when compound **19** was treated with HCOOH in dichloromethane at room temperature for overnight, some of compounds **19** was converted to **18** (Figure 9). So the diol group at positions 3 and 4 in **19** should have *trans* configuration. Using the same method, the structures of compounds **21** and **24** was identified. The structures of compounds **20, 22, 23** and **25** were elucidated by 2D NMR (H-H COSY, HMQC and HMBC) studies.

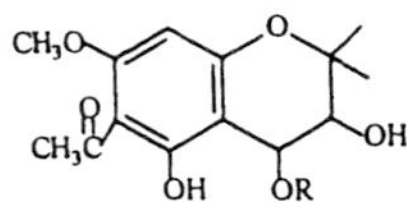

Figure 8. Compounds 18-25.

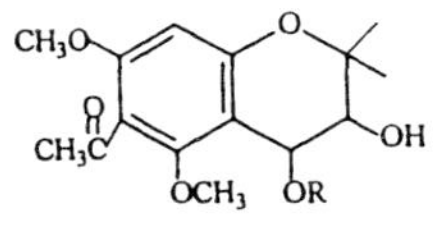

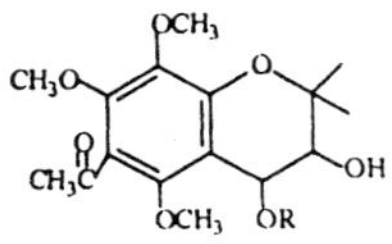

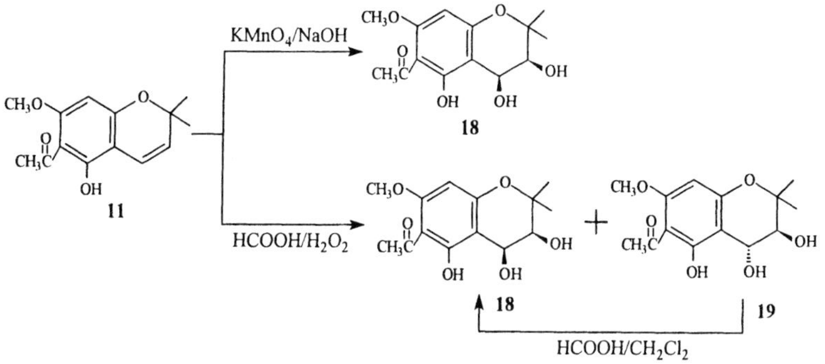

Figure 9. Structure determination of Compound 18.

Other Compounds from *Evodia lepta* (5)

Besides the chromene and chroman compounds, one known sesquiterpene (clovandiol, **26**) (*21*) and two known flavones (7,4-dihydroxy-3,5,3'-trimethoxyflavone, **27**, (*22*) and 3, 7-dimethylkaempferol, **28** (*23*)) were also isolated (Figure 10).

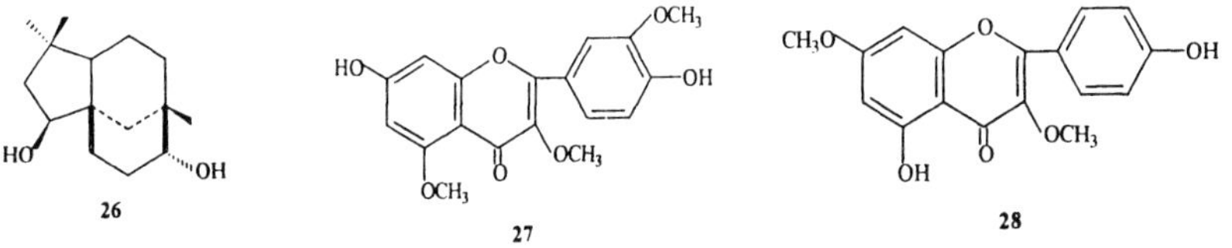

Figure 10. Other compounds from Evodia lepta

Biological Tests for Some of the Compounds from Evodia lepta

Some of the compounds from *Evodia lepta* had been tested for insecticide activity, anti-fungus activity, anti-tumor activity and anti-HIV activity. The results will be discussed as follows.

Insecticide Activity Tests

Precocene I and precocene II are two natural products isolated from *Ageratum houstonnianum* (Figure 11) (*24*). Studies showed these compounds possess antijuvenile hormone activity, and they are able to induce precocious metamorphosis, cause sterilization, and/or force diapause in certain insects.

Thus, it is possible that such natural products could form a basis for development of new generation of insecticide chemicals (24). Furthermore, cis- and trans-3,4-Dihydroxy-precocene-II are two metabolites of precocene II with highly active fat body monooxygenases of the cabbage looper (*trichoplusia ni* (Hübner)) (Figure 11) (25). Owing to the fact that the structures of compounds **1-17** are similar to those of precocene I and precocene II and the structures of compounds **18-25** are similar to those of cis- and trans-3,4-Dihydroxy-precocene-II. Compounds **6, 7, 9** and **11** which were isolated in large amount (0.3 - 2.5 g) were selected for insecticide activity test (Table I). The results showed compound **7** has killing mosquito activity (30 % mortality at 2 ppm).

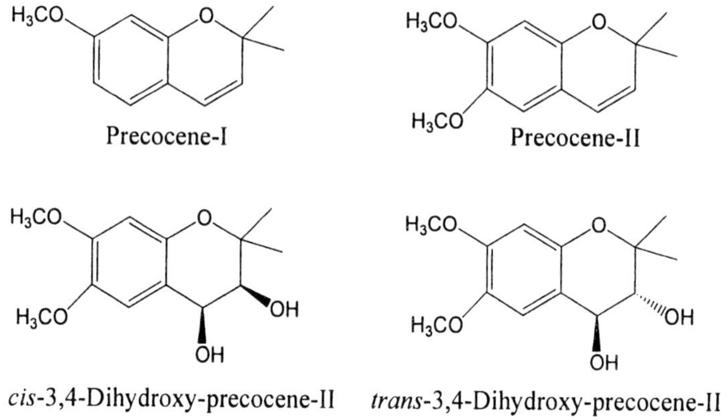

Precocene-I

Precocene-II

cis-3,4-Dihydroxy-precocene-II

trans-3,4-Dihydroxy-precocene-II

Figure 11. Precocene I and II and their metabolites.

Table I. Insecticide activity tests of compounds 6, 7, 9 and 11.

Insect	Concentration (ppm)	Mortality (%)			
		6	7	9	11
Mosquito	2	0	30	0	0
Armyworm	200	0	0	0	0
Bean aphid	200	0	0	0	0
Tribolium castaneum	50	0	0	0	0
Red mite	200	0	0	0	0
Corn borer	200	0	0	0	0

Anti-fungus Activity Tests

Compounds **1, 2, 5, 6, 7, 9, 11, 18, 19, 20, 24** and **25** had been tested for antifungus activity on *Candida albicans, Cryptococcus neoformas, Aspergillus fumigatus* and no activity indicated.

Anti-tumor Activity Tests

Compounds **2, 5, 6, 7, 9, 11, 18, 19, 20, 22, 24, 25, 27, 28** had been tested for anti-tumor activity tests on Lung cancer 7721, Stomach Cancer MKN-28, Mouse leukemia P388, and no activity was found.

Anti-HIV Activity Tests

Seven compounds had been tested for anti-HIV activity, and these compounds have very limited activity (Table II).

Table II. Anti-HIV activity tests

Compound No.	IC_{50} ($\mu g/mL$)	EC_{50} ($\mu g/mL$)	Therapeutic Index
5	>100	no	suppression
6	28.2	7.9	3.6
7	>100	25.5	>3.9
9	24.0	12.1	2.0
11	7.5	no	suppression
18	>100	30.5	>3.3
24	>100	no	suppression
25	>100	no	suppression

Conclusion

Total twenty-eight compounds were isolated and identified from petroleum ether and chloroform fractions of the aerial parts of *Evodia lepta*, and twenty-five of them are 2,2-dimethylchromene and 2,2-dimethylchroman derivatives. Biological tests of some of these compounds showed compound 7 has killing mosquito activity (mortality 30% at 2 ppm), and some compounds have weak anti-HIV activity.

Acknowledgments

We thank Prof. Kuo-Hsiung Lee (University of North Carolina at Chapel Hill) for anti-HIV tests. We also thank Prof. Hongrong Zhang's group (Shanghai, China) for the anti-fugal activity tests, Prof. Jian Ding's group (Shanghai, China) anti-tumor activity tests and State Key Laboratory of Elemental Organic Chemistry (Nankai University, China) for insecticide activity tests.

References

1. *The Dictionary of Chinese Herb*, Jiangsu New Medical College, China, **1986**, *Vol. 1*, p 68.
2. Gunawardana, Y. A. G. P.; Cordell, G. A. *J. Sci. Soc. Thailand* **1987**, *13*, 107.
3. Li, G.; Zeng, J.; Song, C.; Zhu, D. *Phytochemistry* **1997**, *44*, 1175-1177.
4. Li, G.; Zeng, J.; Zhu, D. *Acta Pharmaceutica Sinica* **1997**, *32*, 682.
5. Li, G.; Zhu, D. *Phytochemistry* **1998**, *48*, 1051.
6. Allan, R. D.; Correll, R. L.; Well, R. J. *Tetrahedron Lett.* **1969**, *10*, 4673-4674.
7. Baldwin, M. E.; Bick, I. R. C.; Komzak, A. A.; Price, J. R. *Tetrahedron* **1961**, *16*, 206-211.
8. Hlubucek, J.; Ritchie, E.; Tayler, W. C. *Aust. J. Chem.* **1971**, *24*, 2347.
9. Buckingham, J. Dictionary of Natural Products. Chapman & Hall, London, **1994**, *2*, 2340.
10. Barnes, C. S.; Occolowitz, J. L. *Aust. J. Chem.* **1964**, *17*, 75.
11. Wight, S. E. *J. Chem. Soc.* **1948**, 2005.
12. Huls, R.; Brunelle, S. *Bull. Soc. Chim. Belg.* **1959**, *68*, 325.
13. Sethna, S. M. *Chemical Reviews* **1951**, *49*, 51.
14. Arnone, A.; Cardillo, G.; Merlini, L.; Mondelli, R. *Tetrahedron Lett.* **1967**, *8*, 4201-4206.
15. Li, G.; Zhu, D. *J. Nat. Prod.* **1998**, *61*, 390-391.
16. Li, G.; Zhu, D. *J. Asian Nat. Prod. Res.*, **1999**, *1*, 337.
17. Li, G.; Zhu, D. *Acta Botanica Sinica*, **1997**, *39*, 670-674.
18. Li, G.; Zeng, J.; Zhu, D. *Phytochemistry*, **1998**, *47*, 101-104.
19. Wiberg, K. B.; Saegebarth, K. A. *J. Am. Chem. Soc.* **1957**, *79*, 2882.
20. Roebuck, A.; Adkins, H. *Org. Synth.* **1955**, *3*, 217.
21. Delgado, G.; Cárdenas, H.; Peláez, G., et al. *J. Nat. Prod.* **1984**, *47*, 1042.
22. Monache, G. D.; De Rosa, M. C.; Scurria, R.; Monacelli, B.; Pasqua, G.; Dall'Olio, G.; Bruno Botta, B. *Phytochemistry*, **1991**, *30*, 1849-1854.

23. Wang, Y.; Hamburger, M.; Gueho, J., et al. *Phytochemistry*, **1989**, *28*, 2323-2327.
24. Bowers, W. S.; Ohta, T.; Cleere, J. S.; Marsella, P. A. *Science*, **1976**, *193*, 542.
25. Soderlund, D. M.; Messeguer, A.; Bowers, W. S. *J. Agric. Food Chem.* **1980**, *28*, 724-731.

Chapter 19

Unique Chemistry of Aged Garlic Extract

Kenjiro Ryu[1,2] and Robert T. Rosen[1]

[1]Center for Advanced Food Technology, Cook College, Rutgers, The State University of New Jersey, New Brunswick, NJ 08901-8520
[2]Healthcare Institute, Wakunaga Pharmaceutical Company, Ltd., 1624 Shimo-kotachi, Kodacho, Takatagun, Hiroshima, 739-1195, Japan

Recent progress on the chemistry of the aged garlic extract (AGE) is discussed. AGE is one of the unique preparations for garlic, which is manufactured by soaking sliced garlic in aqueous ethanol for more than ten months. AGE has been reported to have various biological activities including lipid lowering, anti-cancer and antioxidant effects. S-allyl-L-cysteine, fructosylarginine and 1,2,3,4-tetrahydro-β-carboline-3-carboxylic acids were isolated and identified from AGE as the unique bioactive constituents. Chromatographic experiments including a LC/MS method were directed to determine the fluctuation of these compounds during the long-term extraction (aging) process. The data demonstrated that all of these compounds were not present or were in extremely lower concentration in raw garlic, but they formed and increased during process. In this chapter, we also discuss the biological properties and formation mechanisms of these compounds. The results indicate that the aging process provides additional unique health benefits to the garlic without odor.

Garlic (*Allium sativum* L.) is one of the oldest herbs in the world, and has been reported to have biological activities, which include cardiovascular, anticancer, immune enhancement, antimicrobial and antioxidant effects (*1,2*). The chemistry of garlic has enormous diversity. Many compounds including alliin, vinyldithiins, ajoenes, *S*-allyl-L-cysteine (SAC) and *S*-allylmercapto-L-cysteine have been reported from garlic and its preparations (*3*). It is well-known that allicin, the pungent principle of garlic, is formed from alliin, the odorless amino acid in intact garlic, by the enzyme when garlic is chopped (*4*). Because allicin is a highly reactive compound, it easily transforms into other organosulfur compounds as shown in Figure 1 (*5*) and disappear during the process. Thus, no garlic products contain allicin (*6*).

The enzymatic formation and the lability of allicin make the chemistry of garlic complicated. The chemical composition will be quite different if its processing way is different. For example, when garlic is simply crushed or chopped, allicin is formed. However, once garlic is boiled or heated, allicin will never be found in these preparations. There are some interesting reports on the relationship between processing and chemical composition. Iberl *et al.* reported that diallyl trisulfide was a dominant constituent as an allicin-derived compound in crushed garlic when it is stored in ethanol for 9 days. Meanwhile 2-vinyl-(4*H*)-1,3-dithiin was dominant when crushed garlic was stored in linseed oil (*7*). Kubec *et al.* reported that diallyl sulfides were generated by heating alliin without enzymatic conversion (*8*). Yu *et al.* reported the profile of various volatile compounds, which are generated from garlic when it is fried, baked, blanched and microwaved (*9,10*). Changes in allicin content and taste of pickled garlic during aging were determined by Kim *et al.* (*11*).

Aged garlic extract (AGE) is one of the unique preparations of garlic, which is manufactured by soaking garlic in aqueous ethanol for more than 10 months. The biological effects of AGE have been extensively studied. Many clinical and basic studies were conducted with AGE as a modulator of cardiovascular risk factors. The risk factors include blood pressure, total cholesterol, low density lipoprotein (LDL) (*12*), platelet aggregation and adhesion (*13*), microcirculation (*14*), oxidative damage caused by smoking-induced free radicals (*15*), oxidation resistance of LDL (*16*) and atherogenesis (*17*). Since cardiovascular diseases are not caused by only one risk factor, it is very important to suppress various risk factors at once for a long period of time. AGE has also been reported to have antioxidant activity in various types of assays such as radical scavenging, hydroperoxide scavenging and anti-lipid peroxidation (*18,19,20*). Important to note, it is reported that AGE inhibited the *t*-butyl hydroperoxide-induced lipid peroxidation in a chemiluminescence assay while other garlic preparations enhanced the lipid peroxidation (*18,21*). This result indicated that the garlic extract gained antioxidant activity during the aging process. Recently, AGE is also reported to reduce the number of Heinz bodies, a marker of sickle cell anemia in a clinical study (22). In addition to these biological activities, it has been reported that AGE showed antitumor-promoting (23), liver-protective (24,25), anti-aging (26,27), anti-stress (28) and immunomodulatory effects (29).

Safety is also well-established based upon various toxicological studies (30, 31).

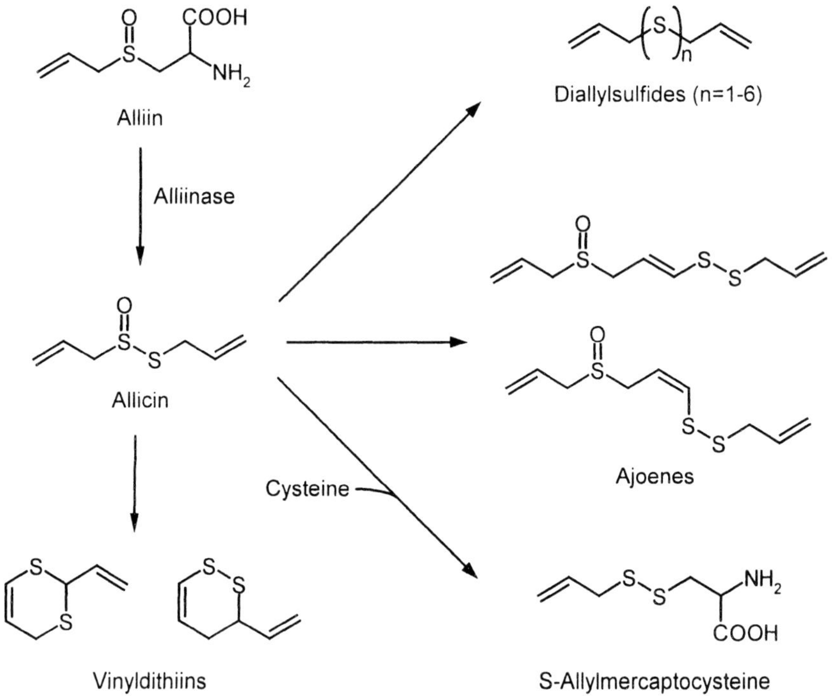

Figure 1. Organosulfur compounds derived from allicin.

In addition to the intensive research for the biological effects and safety of AGE, a few studies on the chemistry of AGE have been also conducted. It has been reported that AGE contains alliin, cycloalliin, SAC, S-1-propenyl-L-cysteine, S-allylmercapto-L-cysteine, dialk(en)yl sulfides and ethyl-2-propenesulfinate as organosulfur compounds (32,33,34). Besides these compounds, steroid saponins and sapogenins were isolated from AGE and were speculated to be the responsible constituents for the cholesterol lowering effect in addition to various organosulfur compounds (35). Also, fructosylarginine (Fru-Arg) has been reported to be an antioxidant component in AGE (36). Recently, four 1,2,3,4-tetrahydro-β-carboline-3-carboxylic acids were isolated and identified from AGE (37). Although AGE contains various chemical components, in this chapter, we will focus on the unique bioactive compounds in AGE, such as SAC, Fru-Arg and tetrahydro-β-carbolines, and will review their chemical and biological characteristics. In addition, the formation mechanisms of these compounds are also discussed.

S-Allyl-L-cysteine

S-Allyl-L-cysteine (SAC) has been isolated and identified from garlic as a precursor of alliin (38). In 1989, Nakagawa et al. reported that SAC is one of the major organosulfur components of AGE (24). The compositional change of SAC in garlic extract during the aging process was determined using a HPLC method, which is described in the U.S. Pharmacopeia (39). The results are shown in Table I. SAC increases gradually until 10 months and is stable in garlic extract after its formation. The proposed mechanism of SAC formation is shown in Figure 2. It is assumed that SAC is derived from the precursor γ-glutamyl-S-allyl-L-cysteine, which is a major dipeptide in fresh garlic, by the enzymatic conversion. The responsible enzyme for this conversion is γ-glutamyl transpeptidase. The activity of this enzyme was measured during aging period. The data indicated that the enzyme remains active after at least ten month of aging process (40).

As previously reported, S-1-trans-propenyl-L-cysteine can be generated from γ-glutamyl-S-1-trans-propenyl-L-cysteine by the same enzymatic conversion (34). The pharmacokinetics of SAC in humans has been investigated after oral administration. The half-life and excretion time were more than 10 h and 30 h, respectively (41). In experimental animal models, the bioavailability was 98.2, 103.0 and 87.2% in rats, mice and dogs, respectively (42). The biological activities of SAC are summarized in Table II.

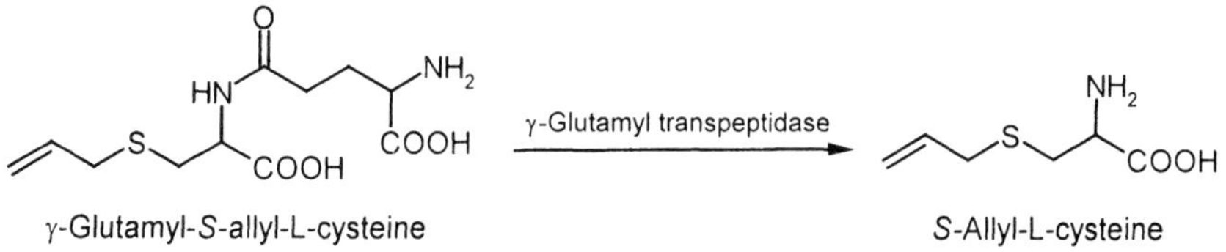

Figure 2. Mechanism of S-Allyl-L-cysteine formation.

Table I. Changes in Concentrations of Bioactive Constituents with Aging Process[a]

Aging Period (months)	SAC (mM)	Fru-Arg (mM)	Tetrahydro-β-carbolines (mM)
0	0.10	ND[b]	ND[c]
1	1.67 (±0.04)	ND[b]	0.13 (±0.00)
4	2.37 (±0.02)	ND[b]	0.15 (±0.01)
10	3.75 (±0.14)	0.60 (±0.06)	0.30 (±0.02)
22	3.82 (±0.17)	1.20 (±0.11)	0.36 (±0.01)

SAC, S-allyl-L-cysteine; Fru-Arg, fructosylarginine
[a] Values are means ±SD for three batches except 0 month (fresh garlic). These data were corrected by adjusting the solid extract content 28%.
[b, c] ND: Not detected (less than detection limit 0.24 mM[b] and 0.01 mM[c] respectively)

Fructosylarginine

Fructosylarginine (Fru-Arg) is an Amadori compound, which is generated in early step of a Maillard reaction. This compound was isolated and identified from AGE and has antioxidant properties (36). Using a LC/MS method, its fluctuation during the aging process was determined (62). ^{13}C-labeled Fru-Arg was used as internal standard for accurate and rapid quantification. The results are shown in Table I. Fru-Arg gradually increased during the aging process and is not present in fresh garlic. Interestingly, Fru-Arg was not detected in heated garlic juice, although it is a Maillard reaction product between arginine and glucose. In general, this kind of Maillard reaction can occur quite easily by heating amino acids with reducing sugars. Glucose contents in AGE and heated garlic juice were measured. It was revealed that lower glucose level in heated garlic juice might be the reason why Fru-Arg was not formed by heating garlic. When garlic juice was heated with additional glucose, Fru-Arg was detected in this preparation (36). The possible mechanism of Fru-Arg formation is shown in Figure 3. The fructan, which is a major carbohydrate in garlic, is hydrolyzed into free glucose in AGE (63). This hydrolysis is assumed to be enzymatic, because it was not observed when heat-treated garlic was aged. The released glucose may react with arginine, which is the most abundant free amino acid in garlic, to form the addition compound. Following the conversion to Schiff base and enol form compound, Amadori compounds including Fru-Arg are generated (64).

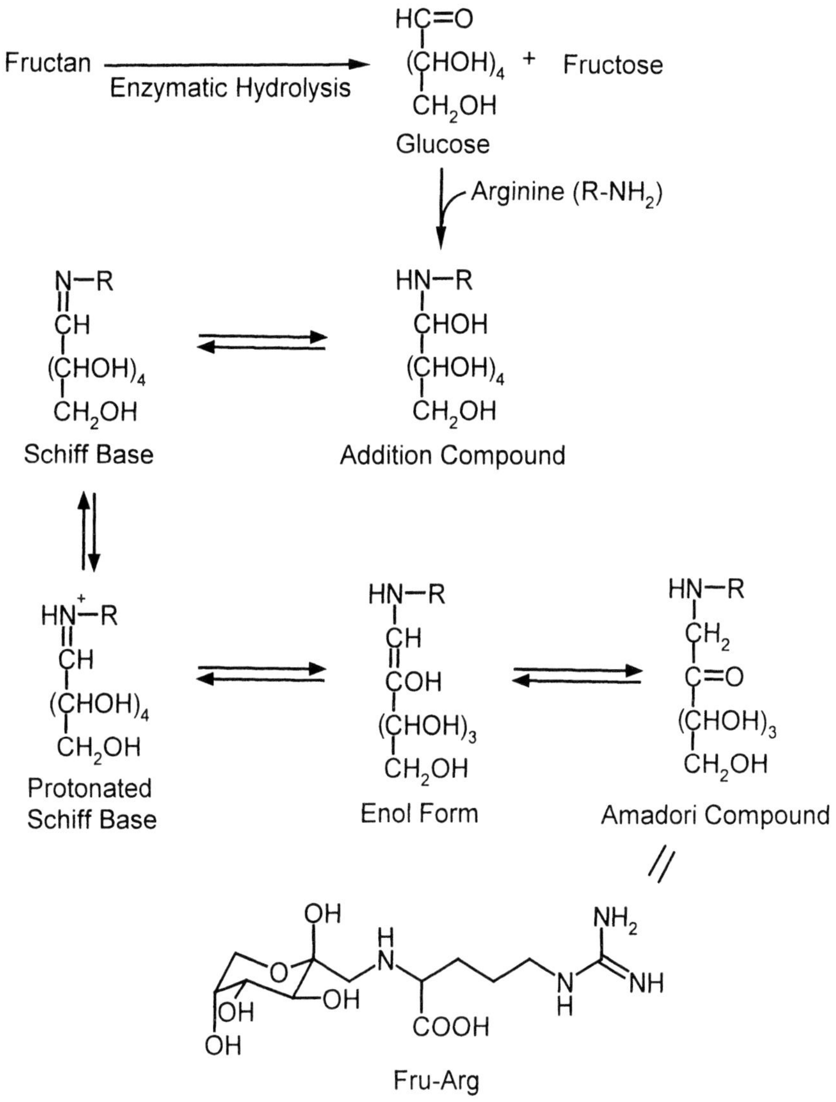

Figure 3. Possible mechanism of Fru-Arg formation.

Table II. Biological Activities of S-Allyl-L-cysteine

Effects	Reference
in vivo Anti-aging	*(43)*
Anti-carcinogenesis	*(44,45,46)*
Enhancement of circulatory antioxidants	*(47)*
Liver protective effect	*(24 48)*
Protective effect in brain ischemia	*(49,50)*
Enhancement of noradrenaline secretion	*(51)*
in vitro Anti-carcinogenesis	*(52)*
Hydrogen peroxide scavenging effect	*(53)*
Inhibition of Cu^{2+}-induced LDL oxidation	*(19)*
Inhibition of NF-κB activation	*(54,55)*
Protection from oxidized LDL-induced cell damage	*(20)*
Inhibition of cholesterol biosynthesis	*(56)*
Inhibition of human squalene monooxygenase	*(57)*
Inhibition of HMG-CoA reductase	*(58)*
Inhibition of dense cell formation in sickle cell disease model	*(59,60)*
Neurotrophic effect	*(61)*

LDL, low-density lipoprotein; NF-κB, nuclear factor kappa B; HMG-CoA, 3-hydroxy-3-methylglutaryl CoA

Fru-Arg was also isolated and identified from Korean red ginseng by Matsuura *et al.* (*65*) and reported to be detected in rat plasma after oral administration (*66*).

There are several reports on the biological activity of Fru-Arg. It was demonstrated that Fru-Arg attenuated Cu^{2+}-induced LDL oxidation and protected endothelial cells from the oxidant-induced membrane damage (*67*). Fru-Arg has been also reported to attenuate noradrenaline-induced hypertension and ameliorate microcirculation in an *in vivo* study (*68*). Yoo *et al.* reported the inhibitory effect of Fru-Arg on protein-arginine *N*-methyltransferase. However, its inhibitory effect was almost same as arginine. It was concluded that amino-group was critical for its activity (*69*). It is also reported that Fru-Arg might have the beneficial effects in lessening the incidence of anemia in sickle cell disease (*59,60*).

1,2,3,4-Tetrahydro-β-carboline-3-carboxylic acids

Recently, four compounds were isolated and identified from AGE as antioxidant constituents (*37*). Structure elucidation using NMR and MS techniques revealed that these compounds were 1-methyl-1,2,3,4-tetrahydro-β-carboline-3-carboxylic acid (MTCC, **1** in Figure 4), 1-methyl-1,2,3,4-tetrahydro-β-carboline-1,3-dicarboxylic acid (MTCdiC, **2** in Figure 4) and their diastereoisomers at position 1 (**3** and **4** in Figure 4).

The changes in tetrahydro-β-carbolines concentration during the aging process were determined by a LC/MS analysis. As shown in Table I, tetrahydro-β-carbolines increased during the aging process. Meanwhile, they were not detected in fresh garlic. The proposed mechanisms of their formations are shown in Figure 4. It is known that tetrahydro-β-carboline derivatives are formed by Pictet-Spengler condensation of tryptophan with various aldehydes or α-oxo-acids (*37,70*).

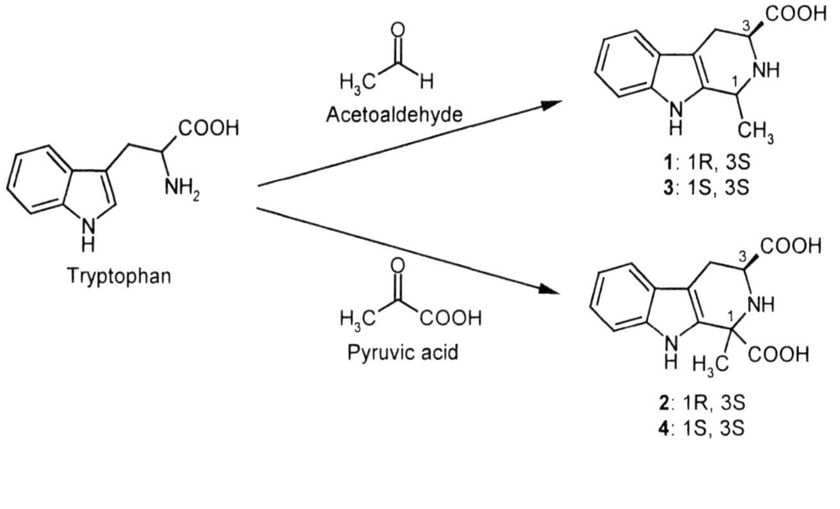

Figure 4. Possible mechanism of tetrahydro-β-carbolines formation.

MTCC has been isolated and identified from *Allium macrostemon* (*71*). Also, MTCC has been reported to be contained in various types of foods, such as alcoholic beverages, fermented products, flour, milk, fruit products and chocolate (*72,73,74*). MTCdiC has been also reported from seasoning sauces, soy sauces, yeast extracts, wine, vinegar, beer and fruit syrup (*75*). There are several reports on the biological activity of tetrahydro-β-carbolines. It was demonstrated that these compounds have strong antioxidant activity in several kinds of *in vitro* assays including inhibitory effect of 2,2'-azobis(2-amidinopropane)hydrochloride-induced lipid peroxidation and LPS-induced nitrite production (*37*). In addition, it has been reported that MTCC showed an inhibitory effect on platelet aggregation (*71*) and antioxidant activity (*75*). Recently, Pari *et al.* suggested the important protective role of tetrahydro-β-carbolines against oxidative stress in human lens (*76*).

Conclusions

In this chapter, the chemical constituents of AGE have been discussed. SAC, Fru-Arg and tetrahydro-β-carbolines were identified as principal bioactive components of AGE. Traditionally, it has been believed that thiosulfinates, allicin and its derivatives or degradation products were the main bioactive principles of garlic and its products. However, the compounds reported here are not thiosulfinates or their degradation products but more stable, present in the preparation, and most importantly bioactive.

Using HPLC and LC/MS, it was determined that SAC, Fru-Arg and tetrahydro-β-carbolines increased during the aging process and were not present or were in extremely lower concentration in raw garlic. The mechanisms of their formation are proposed. It was revealed that the aging process of garlic is a very complicated sequence of reactions, which include enzymatic and non-enzymatic conversions.

These data indicated that the aging process provides additional unique health benefits to the garlic extract.

Acknowledgements

We would like to thank Dr. Chi-Tang Ho (Rutgers University) for his helpful suggestions and advice.

References

1. Nagourney, R.A. *J. Med. Food* **1998**, *1* (1), 13-28.
2. Amagase, H.; Petesch, B.L.; Matsuura, H.; Kasuga, S.; Itakura, Y. *J. Nutr.* **2001**, *131*, 955S-962S.

3. Agarwal, K.C. *Med. Res. Rev.* **1996**, *16* (1), 111-124.
4. Block, E. *Angew. Chem. Int. Ed. Engl.* **1992**, *31*, 1135-1178.
5. Matsuura H. In *Nutraceuticals: Designer Foods III Garlic, Soy and Licorice*; Lachance, P.A., Ed.; Food and Nutrition Press, Inc., Trumbull, CT, 1997; pp 55-69.
6. Freeman, F.; Kodera, Y. *J. Agric. Food Chem.* **1995**, *43*, 2332-2338.
7. Iberl, B.; Winkler, G.; Knobloch, K. *Planta Med.* **1990**, *56*, 202-211.
8. Kubec, R.; Velíšek, J.; Doležal, M.; Kubelka, V. *J. Agric. Food Chem.* **1997**, *45*, 3580-3585.
9. Yu, T.-H.; Wu, C.-M.; Ho, C.-T. *J. Agric. Food Chem.* **1993**, *41*, 800-805.
10. Yu, T.-H.; Lin, L.-Y.; Ho, C.-T. *J. Agric. Food Chem.* **1994**, *42*, 1342-1347.
11. Kim, M.-R.; Yun, J.-H.; Sok, D.-E. *J. Korean Soc. Food Nutr.* **1994**, *23* (5), 805-810.
12. Steiner, M.; Khan, A.H.; Holbert, D.; Lin, R.I. *Am. J. Clin. Nutr.* **1996**, *64*, 866-870.
13. Steiner, M.; Lin, R.S. *J. Cardiovasc. Pharmacol.* **1998**, *31*, 904-908.
14. Moriguchi, T.; Takasugi, N.; Itakura, Y. *J. Nutr.* **2001**, *131*, 1016S-1019S.
15. Dillon, S.A.; Lowe, G.M.; Billington, D.; Rahman, K. *J. Nutr.* **2002**, *132*, 168-171.
16. Munday, J.S.; James, K.A.; Fray, L.M.; Kirkwood, S.W.; Thompson, K.G. *Atherosclerosis* **1999**, *143*, 399-404.
17. Efendy, J.L.; Simmons, D.L.; Campbell, G.R.; Campbell, J.H. *Atherosclerosis* **1997**, *132*, 37-42.
18. Imai, J.; Ide, N.; Nagae, S.; Moriguchi, T.; Matsuura, H.; and Itakura, Y. *Planta Med.* **1994**, *60*, 417-420.
19. Ide, N.; Nelson, A.B.; Lau, B.H.S. *Planta Med.* **1997**, *63*, 263-264.
20. Ide, N.; Lau, B.H.S. *J. Pharm. Pharmacol.* **1997**, *49*, 908-911.
21. Borek, C. *J. Nutr.* **2001**, *131*, 1010S-1015S.
22. Takasu, J.; Uykimpang, R.; Sunga, M.A.; Amagase, H.; Niihara, Y.; *BMC Blood Disord.* **2002**, *2*(1), 3-6.
23. Nishino, H.; Iwashima, A.; Itakura, Y.; Matsuura, H.; Fuwa, T. *Oncology* **1989**, *46*, 277-280.
24. Nakagawa, S.; Kasuga, S.; Matsuura, H. *Phytother. Res.* **1989**, *3* (2), 50-53.
25. Wang B.H.; Zuzel, K.A.; Rahman, K.; Billington, D. Toxicology 1999, 132, 215-225.
26. Moriguchi, T.; Takashina, K.; Chu, P.-J.; Saito, H.; Nishiyama, N. *Biol. Pharm. Bull.* **1994**, *17*(12), 1589-1594.
27. Moriguchi, T.; Saito, H.; Nishiyama, N. *Biol. Pharm. Bull.* **1996**, *19*(2), 305-307.
28. Kyo, E.; Uda, N.; Ushijima, M.; Kasuga, S.; Itakura, Y. *Phytomedicine* **1999**, *6*(5), 325-330.

29. Kyo, E.; Uda, N.; Suzuki, A.; Kakimoto, M.; Ushijima, M.; Kasuga, S; Itakura, Y. *Phytomedicine* **1998**, *5*(4), 259-267.
30. Nakagawa, S.; Masamoto, K.; Sumiyoshi, H.; Kunihiro, K.; Fuwa, T. *J. Toxicol. Sci.* **1980**, *5*, 91-112.
31. Sumiyoshi, H.; Kanezawa, A.; Masamoto, K.; Harada, H.; Nakagami, S.; Yokota, A.; Nishikawa, M.; Nakagawa, S. *J. Toxicol. Sci.* **1984**, *9*, 61-75
32. Horie, T.; Awazu, S.; Itakura, Y.; Fuwa, T. *Planta Med.* **1992**, *58*, 468-469
33. Weinberg, D.S.; Manier, M.L.; Richardson, M.D.; Haibach, F.G.; Rogers, T.S. *J. High Resol. Chromatog.* **1992**, *15*, 641-654
34. Lawson, L.D. In *Phytomedicines of Europe*; Lawson L.D., Bauer R., Eds.; *ACS Symposium Series 691*; American Chemical Society: Washington, D.C., 1998, pp176-209
35. Matsuura H. *J. Nutr.* **2001**, *131*, 1000S-1005S
36. Ryu, K.; Ide, N.; Matsuura, H.; Itakura, Y. *J. Nutr.* **2001**, *131*, 972S-976S
37. Ide, N.; Ichikawa, M.; Ryu, K.; Yoshida, J.; Sasaoka T.; Sumi, S.; Sumiyoshi, H. In *Food Factors for Health Promotion*; Ho, C.-T.; Shahidi, F., Eds.; *ACS Symposium Series*; American Chemical Society: Washington, D.C., *in press*
38. Suzuki, T.; Sugii, M.; Kakimoto T.; Tsuboi N. *Chem. Pharm. Bull.* **1961**, *9* (3), 251-252
39. *United States Pharmacopeia XXV National Formulary 20*; The United States Pharmacopeial Convention, Inc., ed.; United States Pharmacopeial Convention, Inc., Rockville, MD, 2002; pp 2553
40. Kodera Y.; Matsuura, H.; Sumiyoshi, H.; Sumi, S. In *Food Factors for Health Promotion*; Ho, C.-T.; Shahidi, F., Eds.; *ACS Symposium Series*; American Chemical Society: Washington, D.C., *in press*
41. Kodera Y.; Suzuki, A.; Imada, O.; Kasuga, S.; Sumioka, I.; Kanezawa, A.; Taru N.; Fujikawa, M.; Nagae, S.; Masamoto, K.; Maeshige, K.; Ono, K. *J. Agric. Food Chem.* **2002**, *50*, 622-632
42. Nagae, S.; Ushijima M.; Hatono S.; Imai J.; Kasuga, S.; Matsuura H.; Itakura, Y.; Higashi, Y. *Planta Med.* **1994**, *60*, 214-217
43. Nishiyama, N.; Moriguchi, T.; Morihara, N.; Saito, H. *J. Nutr.* **2001**, *131*, 1093S-1095S
44. Amagase, H.; Milner, J.A. *Carcinogenesis* **1993**, *14*(8), 1627-1631
45. Hatono, S.; Jimenez, A.; Wargovich, M.J. *Carcinogenesis* **1996**, *17*(5), 1041-1044
46. Balasenthil, S.; Nagini, S. *Oral Oncol.* **2000**, *4*, 382-386
47. Balasenthil, S.; Nagini, S. *J. Biochem. Mol. Biol. Biophys.* **2000**, *4*, 35-39
48. Mostafa, M.G.; Mima, T.; Ohnishi, S.T.; Mori, K. *Planta Med* **2000**, *66*(2), 148-151

49. Numagami, Y.; Sato, S.; Ohnishi, S. T. *Neurochem. Int.* **1996**, *29*(2), 135-143
50. Numagami, Y.; Ohnishi, S.T. *J. Nutr.* **2001**, *131*, 1100S-1105S
51. Oi, Y.; Kawada, T.; Shishido, C.; Wada, K.; Kominato, Y.; Nishimura, S.; Ariga, T.; Iwai, K. *J. Nutr.* **1999**, *129*, 336-342.
52. Hageman, G.J.; van Herwijnen, M.H.M., Schilderman, P.A.E.L.; Rhijnsburger, E.H.; Moonen, E.J.C.; Kleinjans, J.C.S. *Nutr. Cancer* **1997**, *27*(2), 177-185.
53. Ide, N.; Lau, B.H.S. *Drug Dev. Ind. Pharm.* **1999**, *25*(5), 619-624.
54. Ide, N.; Lau, B.H.S. *J. Nutr.* **2001**, *131*, 1020S-1026S.
55. Ho, S.E.; Ide, N.; Lau, B.H.S. *Phytomedicine* **2001**, *8*(1), 39-46.
56. Liu, L.; Yeh, Y.-Y. *Lipids* **2000**, *35*(2), 197-203.
57. Gupta, N.; Porter, T.D. *J. Nutr.* **2001**, *131*, 1662-1667.
58. Liu, L.; Yeh, Y.-Y. *J. Nutr.* **2002**, *132*, 1129-1134.
59. Ohnishi, S.T.; Ohnishi, T.; Ogunmola, G.B. *Nutrition* **2000**, *16*, 330-338.
60. Ohnishi, S.T.; Ohnishi, T. *J. Nutr.* **2001**, *131*, 1085S-1092S.
61. Moriguchi, T.; Matsuura, H.; Kodera, Y.; Itakura, Y.; Katsuki, H.; Saito, H.; Nishiyama N. *Neurochem. Res.* **1997**, *22*(12), 1449-1452.
62. Ryu, K.; Ide, N.; Ichikawa, M.; Ogasawara, K.; Rosen, R.T. In *Food Factors in Health Promotion and Disease Prevention*; Shahidi, F.; Ho, C.-T., Eds.; *ACS Symposium Series*; American Chemical Society: Washington, D.C., *in press*.
63. Baumgartner, S.; Dax, T.G.; Praznik, W.; Falk, H. *Carbohydr. Res.* **2000**, *328*, 177-183.
64. Yaylayan, V.A.; Huyghues-Despointes, A. *Crit. Rev. Food Sci. Nutr.* **1994**, *34*(4), 321-369.
65. Matsuura, Y.; Zheng, Y.; Takaku, T.; Kameda, K.; Okuda, H. *J. Trad. Med.* **1994**, *11*: 256-263.
66. Takaku, T.; Han, L.K.; Kameda, K.; Ninomiya, H.; Okuda, H. *J. Trad. Med.* **1996**, *13*: 118-123.
67. Ide N.; Lau B.H.S.; Ryu, K.; Matsuura, H.; Itakura, Y. *J. Nutr. Biochem.* **1999**, *10*, 372-376.
68. Kitao, T.; Kon, K.; Nojima, K.; Takaku, T.; Maeda, N.; Okuda, H. *J. Trad. Med.* **1995** *12*: 294-295.
69. Yoo, B.C.; Park, G.H.; Okuda, H.; Takaku, T.; Kim, S.; Hwang, W.I. *Amino Acids* **1999**, *17*, 391-400.
70. Gutsche, B.; Herderich, M. *J. Agric. Food Chem.* **1997**, *45*, 2458-2462.
71. Peng, J.; Qiao, Y.; Yao, X. *Chin. J. Med. Chem.* **1995**, *5*(2),134-139.
72. Adachi, J.; Mizoi, Y.; Naito, T.; Ogawa, Y.; Uetani, Y.; Ninomiya I. *J. Nutr.* **1991**, *121*, 646-652.

73. Herraiz, T. *J. Agric. Food Chem.* **1999**, *47*, 4883-4887.
74. Herraiz, T. *J. Agric. Food Chem.* **2000**, *48*, 4900-4904.
75. Arutselvan, N.; Gopalan, S.; Kulkarni, V.G.; Balakrishna, K. *Arzneim.-Forsch./Drug Res.* **1999**, *49*(II), 729-731.
76. Pari, K.; Sundari, C.S.; Chandani, S.; Balasubramanian, D. *J. Biol. Chem.* **2000**, *275*(4), 2455-2462.

Chapter 20

Three New Sesquiterpene Lactones from *Inula britannica*

Naisheng Bai[1,2], Bing-Nan Zhou[2], Li Zhang[1], Shengmin Sang[1], Kan He[3], and Qun Yi Zheng[3]

[1]Department of Food Science, Rutgers, The State University of New Jersey, 65 Dudley Road, New Brunswick, NJ 08901-8520
[2]State Key Laboratory of Drug Research, Shanghai Institute of Materia Medica, Academia Sinica, Shanghai, People's Republic of China
[3]PureWorld Botanicals, Inc., 375 Huyler Street, South Hackensack, NJ 07606

Three new sesquiterpene lactones were identified from the flowers of *Inula britannica var. chinensis*: britannilide, oxobritannilactone and eremobritanilin. Their structures were established by spectroscopic methods. Britannilide and eremobritanilin inhibited the growth of human P_{388} cell lines.

Inula, from Compositae, has more than 100 species in the world, mainly found in Europe, Africa and Asia. There are more than 20 species in China. *Inula britannica* is a wild plant found in Eastern Asia, including China, Korea, and Japan. In traditional Chinese medicine, *Inula britannica* and *Inula japonica* are called "Xuanfuhua" and the flowers have been used for the treatment of digestive disorders, bronchitis and inflammation. Its extracts are reported to have anti-inflammatory, anti-bacteria, anti-hepatitis and anti-tumorigenic activities (*1*).

In our previous studies on Chinese *Inula* species, we isolated sesquiterpene lactones from *Inula britannica* (*2*), *I. salsoloides* (*3*), *I. hupehensis* and *I.*

helianthus-aquatica (*4*). Other compounds, such as kaurane glycosides, flavonoids and steroids have also been isolated from *I. britannica* (*5*). In these reports, the cytotoxic activities of nine sesquiterpene lactones were observed. In addition to our work, Park and Kim (*6*) reported four cytotoxic sesquiterpene lactones from *I. britannica*. The present paper describes the isolation and the structural determination of three new sesquiterpene lactones (**1, 2** and **3**) named britannilide, oxobritannilactone and eremobritanilin, respectively. Their cytotoxicity is also discussed.

Experimental

General Instrumentation

Mp: uncorr; $[\alpha]_D$ JASCO DIP-300 spectrophotometer; UV: Shimadzu UV-300; IR: Perkin-Elmer 599 B instrument; EIMS: MAT-711 and Finnigan MAT 4021 GC/MS; ^{1}H NMR, COSY and NOE: Bruker AM-400; ^{13}C NMR and DEPT: AC-100.

Plant Material

The flowers of *Inula britannica var. chinensis* were collected in July 1990 during the flowering stage, at the suburb of Yanan City, Shanxi Province, China. It was identified by Professor Tian-Lang Pan, Yanan Institute of Medicine Inspection, Shanxi Province, China. Voucher specimens are deposited in the Laboratory of Phytochemistry, Shanghai Institute of Materia Medica, Academia Sinica, Shanghai, China.

Extraction and Isolation of Compounds

The flowers (60 Kg) of *Inula britannica var. chinensis* were extracted three times with 95% EtOH at room temperature. The CHCl$_3$-soluble part of the EtOH extract was chromatographed on a silica gel column and packed in CHCl$_3$ using an CHCl$_3$-Me$_2$CO gradient solvent system. The fractions from chloroform-acetone (20:1 to 10:1) were evaporated under vacuum and repeatedly chromatographed on silica gel column, to give **1** (15 mg), **2** (3 mg) and **3** (21 mg).

Britannilide (1)

Obtained as needles, $C_{15}H_{22}O_3$, mp 76-78° (CHCl$_3$), [α]$_D$ +103.8 (CHCl$_3$; 0.249); UV λ$_{max}$ (MeOH, nm) (log ε): 214 (2.33); IR ν$_{max}$ (KBr, cm^{-1}): 2960, 2920, 2864, 1770, 1750, 1666, 1458, 1380, 1370, 1330, 1302, 1225, 1190, 1135, 1095, 1070, 1030, 1000 and 810; EIMS m/z (rel. int. %): 248(4) [M]$^+$, 233(100), 215(8), 204(7), 192(27), 187(18), 169(12), 152(33), 145(21), 137(18), 121(19), 107(40), 91(62), 77(37), 65(29) and 55(32). ^{1}H and ^{13}C NMR: Tables I and II.

Oxobritannilactone (2)

Obtained as needles, $C_{17}H_{22}O_5$, mp 65-67° (CHCl$_3$), UV λ$_{max}$ (MeOH, nm) (log ε): 204 (4.27). IR ν$_{max}$ (KBr, cm^{-1}): 2920, 2844, 1765, 1730, 1670, 1640, 1594, 1450, 1360, 1328, 1240, 1168, 1095, 1030, 940 and 890; EIMS m/z (rel. int. %): 307(8) [M+H]$^+$, 306(3) [M]$^+$, 288(2), 263(10), 247(65), 231(100), 217(72), 203(57), 189(21), 175(38), 161(37), 149(35), 133(15), 105(22), 91(26), 78(23), 67(49) and 55(30); ^{1}H NMR: Table I

Eremobritanilin (3)

Obtained as needles, $C_{15}H_{20}O_4$, mp 115-117° (CHCl$_3$), [α]$_D$ +27.86 (MeOH, c 0.047), UV λ$_{max}$ (MeOH, nm) (log ε): 216 (3.57); IR ν$_{max}$ (KBr, cm^{-1}): 3536, 3400, 3096, 2940, 1740, 1660, 1640, 1450, 1410, 1380, 1350, 1270, 1170, 1125, 1060, 1030, 970 and 925; EIMS m/z (rel. int. %): 264(2) [M]$^+$, 262(3), 247(12), 229(46), 211(11), 183(40), 174(14), 157(23), 145(20), 131(21), 119(20), 105(33), 91(53), 80(41), 67(46) and 55(100); ^{1}H and ^{13}C NMR: Tables I and II.

Cytotoxicity

Britannilide (1) and eremobritannilin (3) were evaluated for cytotoxicity in the P-388 system. The cultured cells were treated in duplicate with three concentrations and incubated for a period of 48 hours. The resulting cell number was determined by counting. The data are expressed as a percentage relative to controls treated only with solvent (DMSO), after correcting for the cell number at the beginning of the experiment.

Table I. ¹H NMR Spectral Data of Compounds 1-3.

H	1 (CDCl₃)	2 (CDCl₃)	3 (CD₃COCD₃)
1 a	3.55 m	4.00 t (6.7)	3.74 m
1 b	3.16 t	4.00 t (6.7)	3.74 m
2 a	1.75 m	1.73 m (6.7)	1.57 m
2 b	1.52 m	1.73 m (6.7)	1.82 m
3 a	1.64 m	1.60 m	2.38 m
3 b	1.00 m	1.50 m	2.22 m
4	2.45 m	2.87 dd	
6	5.65 d (4.7)		2.44 d (6.7)
7	3.58 m		3.34 m
8	4.82 m	5.11 m	4.83 m
9 a	2.04 dd	3.01 dd (16.4, 6.7)	2.04 dd (13.9, 6.3)
9 b	1.67 dd	2.46 dd (16.4, 10.1)	1.48 dd (13.9, 8.5)
13 a	6.27 d (3.6)	2.14 d (2.2)*	6.12 d (3.1)
13 b	5.58 d (3.0)		5.81 d (2.6)
14	1.24 s	2.00 s	0.86 s
15	1.11 d (7.1)	1.19 d (7.2)	5.29 d (1.6), 5.05 s
MeCOO-		2.07 s	

Run at 400 MHz.
J values (parentheses) in Hz, δ in ppm.
*This is a Me group.

Table II. ¹³C NMR Spectral Data of Compounds 1 and 3.

¹³C	1 (CDCl₃)	DEPT	3 (CD₃COCD₃)	DEPT
1	63.2	CH₂	71.4	CH
2	30.3	CH₂	27.7	CH₂
3	38.4	CH₂	29.8	CH₂
4	33.4	CH	146.4	C
5	152	C	44.6	C
6	116.7	CH	32.1	CH₂
7	39.2	CH	37.8	CH
8	75.3	CH	75.8	CH
9	38.4	CH₂	35	CH₂
10	76.6	C	86.4	C
11	138.2	C	139.9	C
12	missed	C	missed	C
13	121.6	CH₂	121.3	CH₂
14	27.1	CH₃	16.3	CH₃
15	21.7	CH₃	113.3	CH₂

Run at 100 M Hz, δ in ppm.

Results and Discussion

Repeated fractionation of the chloroform-soluble portion of the ethanol extract from the flowers of *I. britannica var. chinensis* gave three new sesquiterpene lactones (**1, 2** and **3**) (Figure 1).

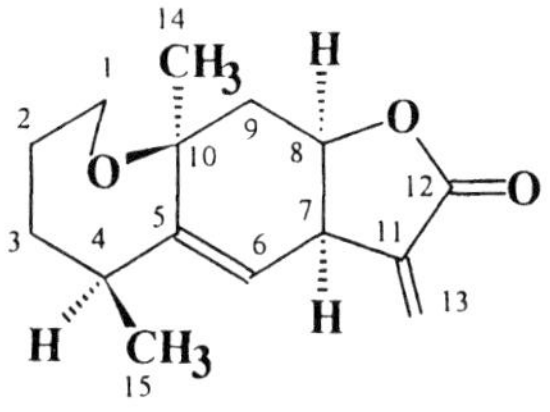

Figure 1. Sesquiterpene lactones identified in Inula Britannica.

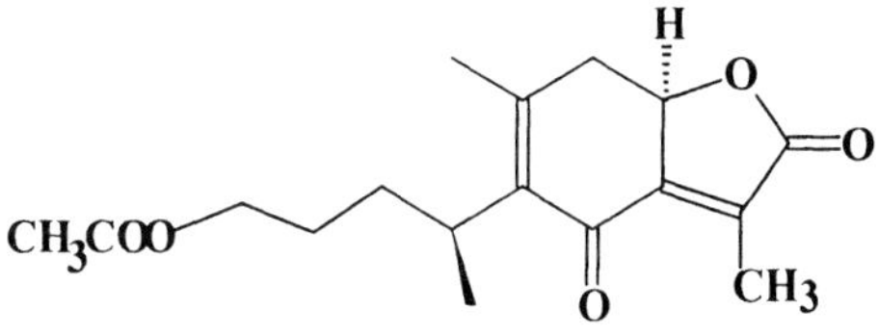

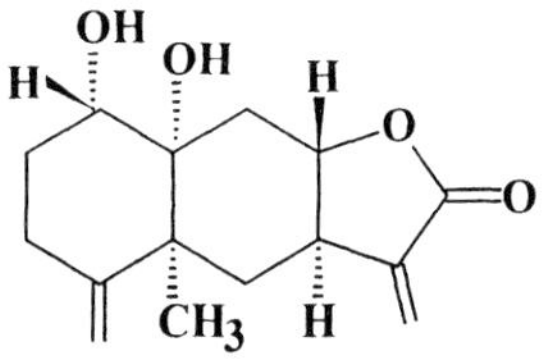

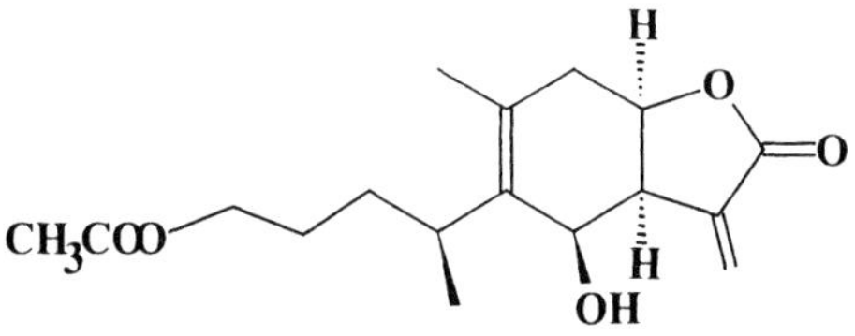

Britannilide (**1**) was obtained as needles. The ^{1}H NMR spectrum exhibited two doublets at δ 6.27 (1H, J = 3.6 Hz) and 5.58 ppm (1H, J = 3.0 Hz), which are characteristic of exocyclic α-methylene-γ-lactone. The IR absorptions at 1770 and 1750 cm^{-1} supported the presence of α,β-unsaturated-γ-lactone. However, it was different from common band because a Fermi effect occurred here, two peaks appeared. Its ^{13}C NMR and DEPT spectra revealed the occurrence of a tri-substituted double bond (δ 152.0 s, 116.7 d), an exomethylene (δ 138.2 s, 121.6 t) and three carbons (δ 75.3 d for C-8; δ 63.6 t for C-1 and δ 76.6 s for C-10) bearing oxygen atoms. The absence of hydroxyl group was evident by its IR spectra. The 2-D COSY spectrum of **1** showed correlation contours between protons. Thus, the structure of britannilide was deduced to be **1** and the ^{1}H and ^{13}C NMR signal parameters were assigned as in Tables I and II.

The relative stereochemistry of compound **1** was investigated on the basis of the coupling constants and NOE experiments. The coupling constants involving H-7 with H-8 (J = 8.1 Hz) suggested H-7 and H-8 in the α-configuration. In NOE experiments, irradiation of the H-8 (δ 4.82) increased the intensity of the H-7 (δ 3.58), allowing for the dispositions of protons at C-7 and C-8 to be assigned. The methyl group at C-10 and H-4 were assigned an α-orientation on the basis of the NOE irradiation of the H-14 at δ 1.24 which enhanced the H-4 and H-8 signals. Irradiation of H-8 had the same effect on both H-14 and H-4, and irradiation of H-4 had the same effect on H-14 and H-8. The results of all these spectral data and further comparison with britannilactone led us to assign the structure of **1**, named britannilide.

Oxobritannilactone (**2**) was obtained as needles. In the EIMS spectrum of **2**, a protonated molecular ion was observed at m/z 307 with the parent ion appearing at m/z 306. This mass spectral evidence, combined with the ^{1}H NMR data, disclosed that the molecular formula of **2** as $C_{17}H_{22}O_5$. The ^{1}H NMR spectra of **2** and 1-*O*-acetylbritannilactone (**4**) showed similarities. In the spectrum of **2**, a methyl doublet appeared instead of signals for the lactone exomethylene and H-7 disappeared indicating that **2** is a double bond moved derivative of 1-*O*-acetylbritannilactone (**4**). In view of the 1765 cm^{-1} band in the IR and the absorption at 204 nm in the UV spectra, **2** should contain a sesquiterpene γ-lactone. Its ^{1}H NMR spectra had no H-6 signal and IR spectra also confirmed the absence of a hydroxyl group, so that C-6 was identified as a carbonyl carbon when it was compared with the same position in 1-*O*-acetylbritannilactone (**4**), which matched the IR spectra of 1670 cm^{-1} (C=C-CO-C=C) unit. The UV absorptions at 204, 254 and 282 nm spectrum of **2** also matched with functional groups of (C=C-COO-), (C=C-CO-) and (-CO-C=C-COO-).

Spatial correlation and the relative stereochemistry were deduced by spin decoupling experiments. Irradiation of H-8 at δ 5.11 led to the decoupling of H-9a, H-9b and H-13 as shown by the collapse of the proton peaks at δ 3.01 (dd, H-9a), 2.46 (dd, H-9b) and 2.14 (d, H-13) into a doublet, a doublet and a singlet, respectively. During irradiation of H-1 at δ 4.00, H-4 at δ 2.87 and H-15 at δ 1.19, all the neighboring proton signals were decoupled. Based on all the spectral data and compared with 1-*O*-acetylbritannilactone (**4**), the structure of **2** was determined and named as oxobritannilactone.

Eremobritanilin (**3**) was also isolated as needles. Its structure was deduced from ^{1}H NMR (Table I), 2-D COSY, ^{13}C NMR and DEPT (Table II), EIMS and IR. The ^{1}H NMR spectrum of **3** showed four signals of exo-methylene protons: two doublets at δ 6.12 (1H, d, J = 3.1 Hz) and 5.81 (1H, d, J = 2.6 Hz) of exocyclic α-methylene-γ-lactone, which was confirmed by the IR spectrum at 1740 cm^{-1}, one doublet at δ 5.29 (1H, d, J = 1.6 Hz) and one singlet at δ 5.05 (1H, s). Spatial correlations and the relative stereochemistry were deduced by 2-D

COSY and then confirmed by NOE experiments. The coupling constant involving H-7 δ 3.34 (dddd, m, $J_{7,8}$ = 13.6) with H-8 δ 4.83 (ddd, q) suggested *trans*-diaxial configuration of the protons at C-7 (α) and C-8 (β). Its IR spectrum contained bands for two hydroxyls at 3540 cm^{-1} and 3400 cm^{-1}. The ^{13}C NMR spectrum showed one hydroxyl group at δ 71.4 (d, C-1) and the other at δ86.4 (s, C-10). In NOE experiments, irradiation of the H-1 increased the intensity of the H-8 (δ 4.83) and H-2a δ 1.57, and irradiation of H-8 had the same effect on H-1, so that the C-8 hydroxyl group must be α-oriented to meet H-1 and H-8 which are close in space. The methyl of C-5 was assigned an α-orientation on the basis of the NOE irradiation of the H-14 at δ 0.86 which enhanced the H-7 and H-2b at δ 1.82. Furthermore, the H-14 signal commonly at δ 1.26, such as in 1α-hydroxy-11α,β-dihydroalantolactone (7), was shifted upfield to δ 0.86 due to the shielding effect of the double bond, indicating that the methyl group is very close to the double bond at C-4. Thus, 3 was identified and named as eremobritanilin.

1 and 3 showed inhibitory effects against the P-388 (lymphocytic leukemia) cell line *in vitro* according to established protocol (8,9), with inhibition values of 1 at 100 μg/mL, 100% and 10.0 μg/mL, 60.1%, and 3 at 100 μg/mL, 100% and 10.0 μg/mL, 52.4%, respectively.

Acknowledgements

The authors wish to thank the Department of Pharmacology (Shanghai Institute of Materia Medica, Academia Sinica, Shanghai, China) for the biological test and Professor Tian-Lang Pan (Yanan Institute of Medicine Inspection, Shanxi Province, China) for the identification of plant materials.

References

1. Jiangsu, New Medical College. *Dictionary of traditional Chinese Materia Medica, Vol. 2*. Shanghai People's Press: Shanghai, China, 1977, pp. 2216-2219.
2. Zhou, B.N.; Bai, N.S.; Lin, L.Z.; Cordell, G.A. *Phytochemistry* **1993**, *34*, 249-252.
3. Zhou, B.N.; Bai, N.S.; Lin, L.Z.; Cordell, G.A. *Phytochemistry* **1994**, *36*, 721-724.
4. Wang, Q.; Zhou, B.N.; Zhang, R.W.; Lin, Y.Y.; Lin, L.Z.; Gil, R.R.; Cordell, G.A. *Planta Medica* **1996**, *62*, 166-169.
5. Shao, Y.; Bai, N.S.; Zhou, B.N. *Phytochemistry* **1996**, *42*, 783-786.
6. Park, E.J.; Kim, J. *Planta Medica.* **1998**, *64*, 752-754.

7. Oksuz, S.; Topcu, G.; Krawiec, M.; Watson, W.H. *Phytochemistry*, **1997**, 46, 1131-1134.
8. Kigodi, P.G.K.; Blasko, G.; Thebtaranonth, Y.; Pezzuto, J.M.; Cordell, G.A. *J. Nat. Prod.* **1989**, *52*, 1246-1251.
9. Jayasuriya, H.; McChesney, J.D.; Swanson, S.M.; Pezzuto, J.M. *J. Nat. Prod.* **1989**, *52*, 325-331.

Chapter 21

Chemistry and Bioactivity of the Seeds of *Vaccaria segetalis*

Shengmin Sang[1,4], Aina Lao[1], Zhongliang Chen[1], Jun Uzawa[2], and Yasuo Fujimoto[3]

[1]Institute of Materia Medica, Shanghai Institutes for Biological Sciences, Chinese Academy of Sciences, Shanghai 200031, People's Republic of China
[2]The Institute of Physical and Chemical Research (RIKEN), Wako, Saitama 351-01, Japan
[3]College of Pharmacy, Nihon University, 7-7-1 Narashinodai, Funabashi, Chiba 274, Japan
[4]Department of Food Science, Rutgers, The State University of New Jersey, 65 Dudley Road, New Brunswick, NJ 08901-8520

Vaccaria segetalis is an annual herb widely distributed in Asia, Europe and other parts of the world. The seeds of this plant, known as Wang-Bu-Liu-Xing in traditional Chinese medicine, have been used widely to promote diuresis and milk secretion, activate blood circulation and relieve carbuncle. Our research group recently isolated a wide range of chemical compounds, including triterpene saponins, alkaloids, cyclic peptide, phenolic acid, flavonoids and steroids from the seeds of this herb. Their structures were identified by a combination of FABMS, ESIMS, 1D and 2D-NMR (DQFCOSY, TOCSY, ROSEY, HMQC, HMBC and new techniques HMQC-TOCSY and HMBC-TOCSY). This paper summarizes the isolation, classification and biological activities of the components isolated from this species.

The plant *Vaccaria segetalis* (Neck.) Garcke (syn. V. pyramidata Medik) (Caryophyllaceae) is an annual herb widely distributed in Asia, Europe and other parts of the world. In Japan, it has been cultivated as a garden plant for several centuries. In China, it is distributed all over the country except the south. The seeds of this plant, known as Wang-Bu-Liu-Xing in traditional Chinese medicine have been widely used to promote diuresis and milk secretion, activate blood circulation and relieve carbuncle according to traditional Chinese medicine (*1*). Previous studies on the seeds of this species led to the isolation of several triterpenoid saponins, (*2-7*) seven cyclic peptides (*8-11*) and one flavonoid (Vacarin) (*12*). Recently, our research group re-investigated the components of *V. segetalis* (*13-24*). This paper mainly discusses the isolation, classification and biological activities of the components isolated from this species by our research group.

Materials and Methods

General Experimental Procedures. Optical rotation: JASCO-DIP-181 polarimeter. IR:Perkin-Elmer 599 infrared spectrometer. ^{1}H (600 MHz) and ^{13}C (150 MHz) NMR: JEOL α600 with NM-AFG type field gradient unit, TMS as internal standard. FAB-MS: MAT-95 Mass spectrometer. CC: silica gel 60H, TLC: HSGF254 (Qingdao Haiyang Chemical Group Co. of China).

Plant Material. The seeds of *Vaccaria segetalis* were purchased at Shijia Zhuang, Hebei Province (China) in 1995. The botanical identification was made by Professor Xuesheng Bao (Shanghai Institute of Drug Control). A voucher specimen has been deposited at the Herbarium of the Department of Phytochemistry, Shanghai Institute of Materia Medica, Chinese Academy of Sciences.

Extraction and Isolation. The powdered seeds of *V. segetalis* (50 Kg) were extracted successively with petroleum ether × 2 and 95% EtOH × 3. After evaporation of ethanol *in vacuo*, the residue was suspended in water and then extracted successively with CH_2Cl_2, EtOAc and *n*-BuOH. Using silica gel column, Sephadex LH-20 column and prepared TLC plate, we obtained six pure compounds from the methylene chloride fraction and 10 compounds from the ethyl acetate fraction (Figure 1). The *n*-BuOH fraction was subjected to Diaion HP-20 using a EtOH-H_2O gradient system (0%-100%). Four fractions were obtained: water fraction, 30% ethanol fraction, 70% ethanol fraction and ethanol fraction. By repeated column chromatographic purification over silica gel and RP C-18 gel, seven pure compounds were obtained from the 30% ethanol fraction and 20 compounds were obtained from the 70% ethanol fraction (Figure 1).

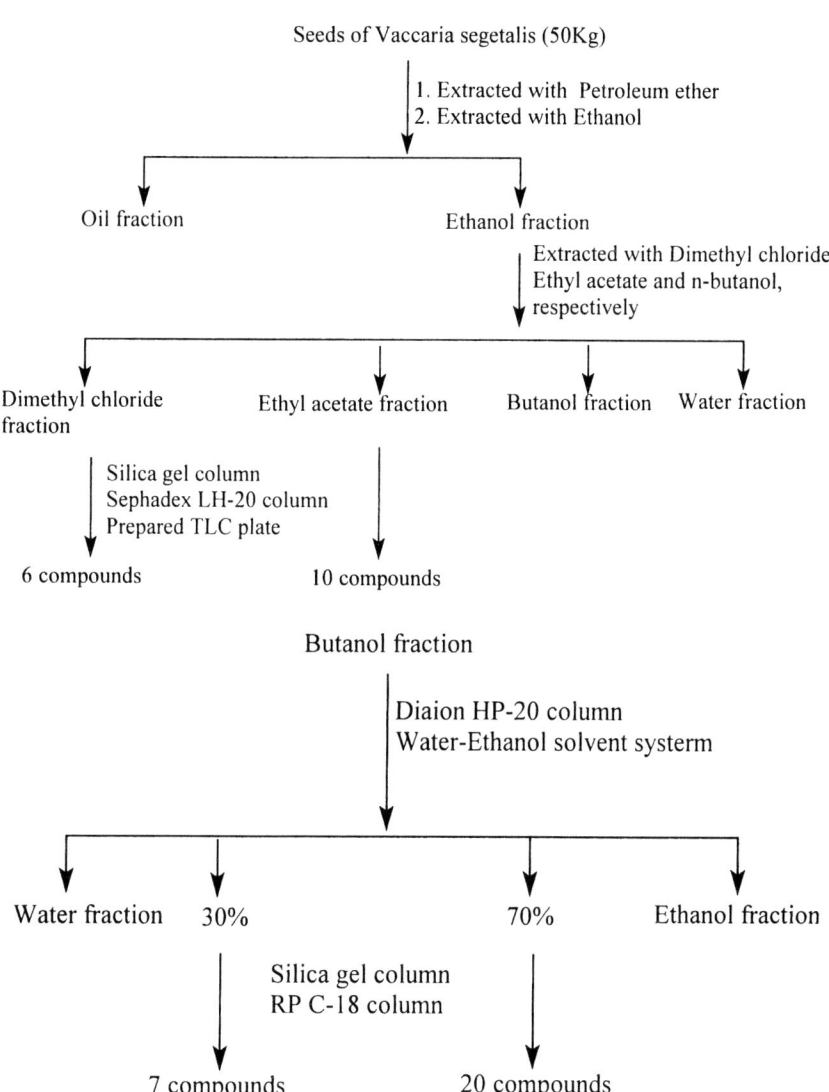

Figure 1. Extraction and isolation procedure.

Phytochemistry

The powdered seeds of *V. segetalis* (50 Kg) were extracted successively with petroleum ether × 2 and 95% EtOH × 3. After evaporation of ethanol *in vacuo*, the residue was suspended in water and then extracted successively with CH_2Cl_2, EtOAc and *n*-BuOH. 40 compounds were isolated and identified from these fractions. According to their structures, they can be classified as: triterpenoid saponins, alkaloids, cyclic peptides, phenolic acid, flavonoids, steroids and other compounds (Figures 2-10). 14 of them are new compounds. Among these new compounds, 12 are new triterpenoid saponins. Research has shown that saponins from a variety of sources, such as medicinal plants and foodstuffs, have considerable health benefits (*25,26*). Saponins occurring in *V. segetalis* seeds are glycosides of triterpenes. Triterpene aglycons can be substituted with several functional groups that result in a number of structurally different aglycons. In the triterpene saponins that we isolated from this seed, the aglycons received trivial names such as gypsogenin, quillaic acid, gypsogenic acid, 16-hydroxygypsogenic acid and 3,4-*seco*gypsogenic acid. The aglycons can be further substituted with 3-6 sugars, including glucopyranose, xylopyranose, arabinofuranose, fucopyranose, galactopyranose, glucurono-pyranosic acid and rhamnopyranose. Using repeated column chromatography on Diaion HP-20, silica gel and RP-18 silica gel, nineteen triterpene saponins, including six gypsogenin type saponins (**1-6**), five quillaic acid type saponins (**7-11**), seven gypsogenic acid type saponins (**12-18**) and one 3,4-*seco* derivative of gypsogenic acid type saponins (**19**) were isolated from the butanol fraction. It is notable that we isolated four cyclic peptides (Figure 6) from the ethyl acetate fraction of this seed. Cyclic peptides are natural products exhibiting a wide variety of biological functions. Large numbers of cyclic peptides with unique structures and various pharmacological activities are reported from marine organisms and microorganisms (*27*), whereas only few examples are known from higher plants (*28-32*). These compounds were first reported by Itokawa et al (*32*). Among them compounds **24** and **25** showed estrogen-like activity assayed by the increment of uterus against ovariectomized rats (*33*). In addition, four alkaloids (Figure 5), three phenolic acids (Figure 7), four flavonoids (Figure 8), two steroids (Figure 9) and four other compounds (Figure 10) were also isolated and identified from this seed. The structures of the isolated compounds were elucidated by spectral methods which include FABMS, ^{1}H-NMR, ^{13}C-NMR, DQFCOSY, TOCSY, ROESY, HMQC, HMBC, HMQC-TOCSY, and HMBC-TOCSY.

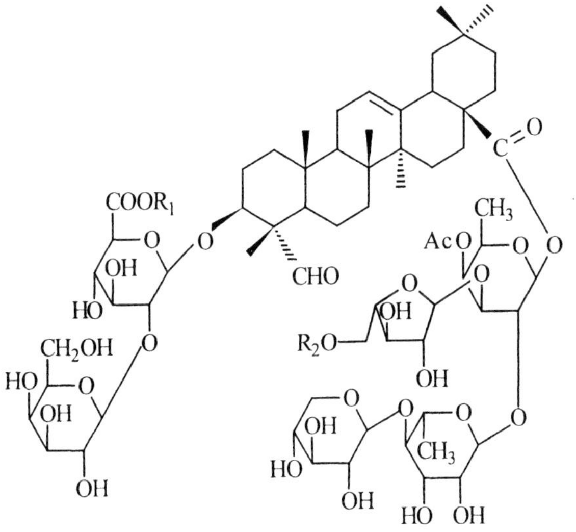

283

1 : R₁=R₂=H
2*: R₁=H, R₂=Ac
3*: R₁=CH₃, R₂=Ac
4*: R₁=CH₃, R₂=H
5*: R₁=CH₂CH₃, R₂=H
6*: R₁=CH₂CH₂CH₂CH₃
    R₂=H

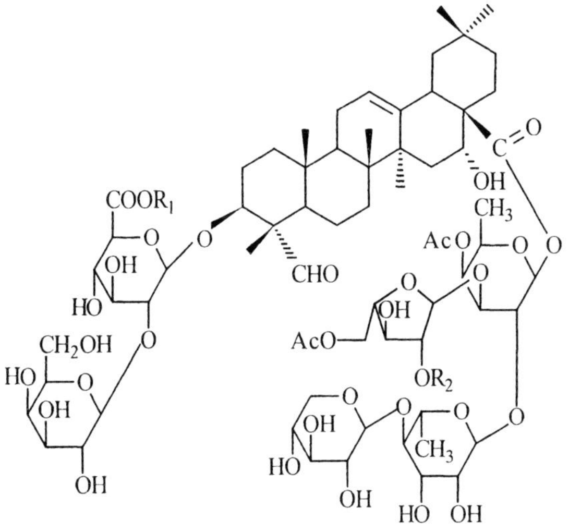

7*: R₁=R₂=H
8*: R₁=H, R₂=Ac
9*: R₁=CH₃, R₂=Ac
10*: R₁=CH₃, R₂=H
11*: R₁=CH₂CH₂CH₂CH₃
     R₂=H

Figure 2. Structures of triterpene saponins (compounds 1-11).

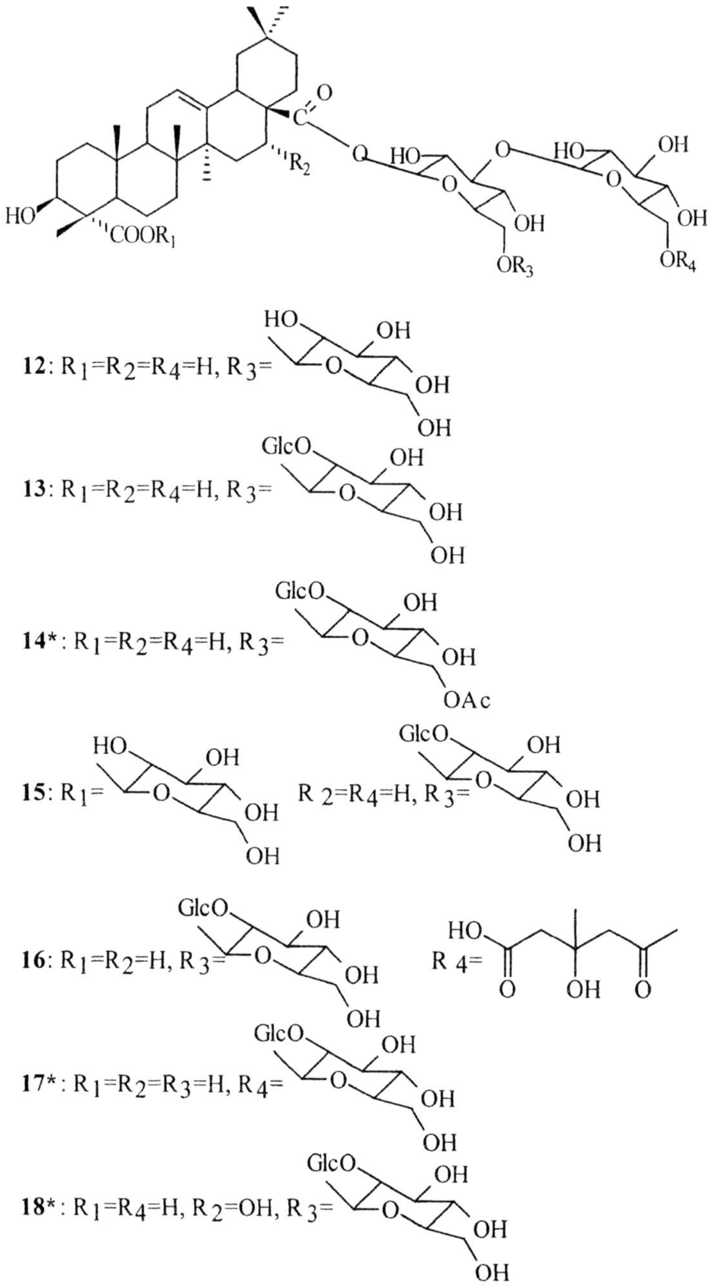

*Figure 3. Structures of triterpene saponins (compounds **12-18**).*

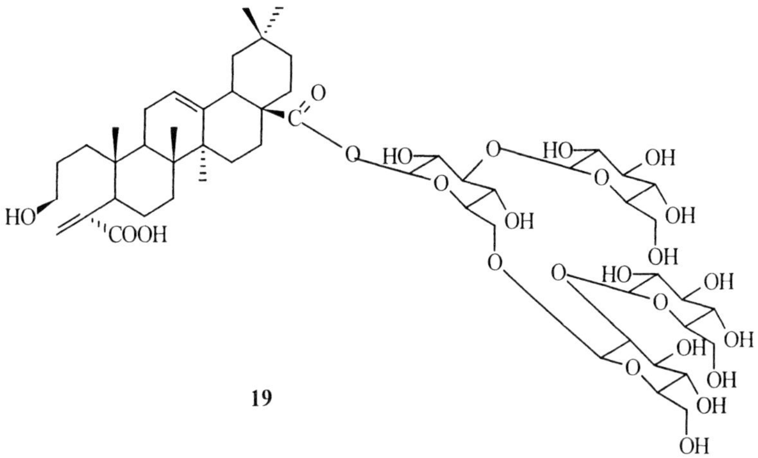

*Figure 4. Structures of triterpene saponin (compound **19**).*

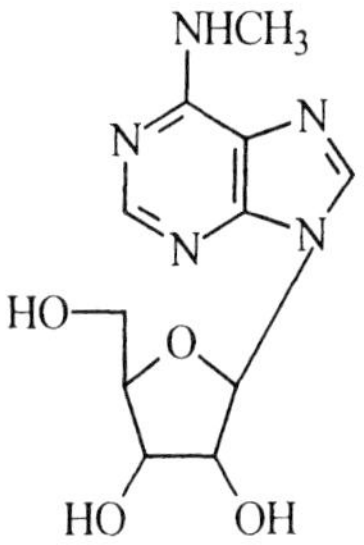

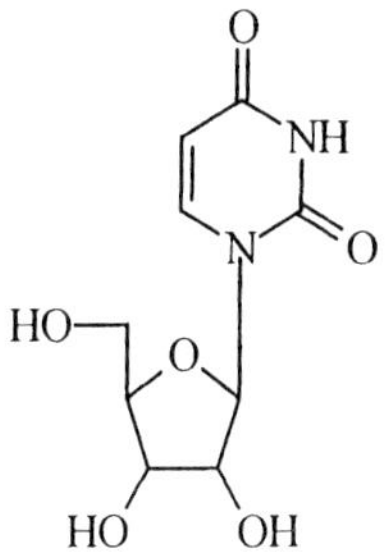

*Figure 5. Structures of alkaloids (compounds **20-23**).*

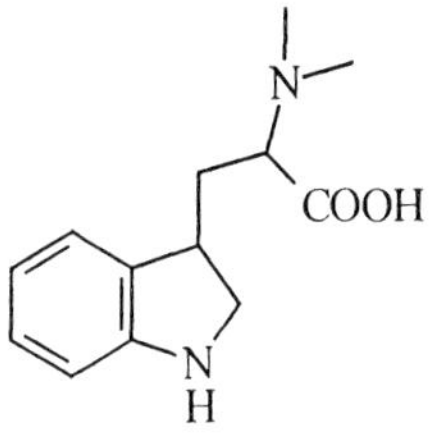

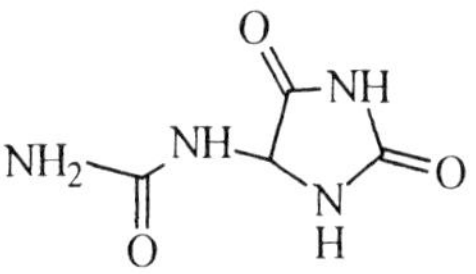

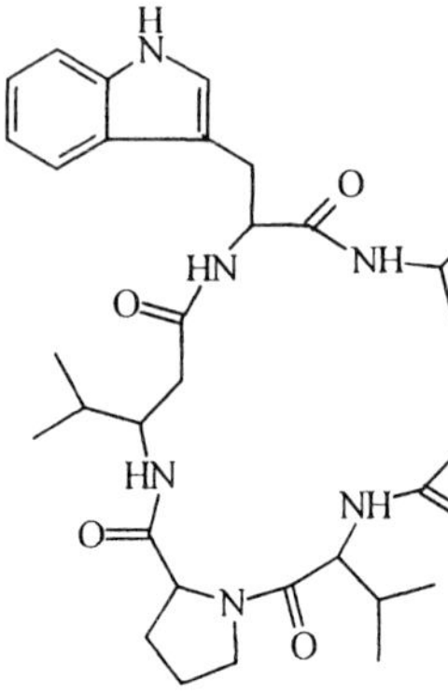

286

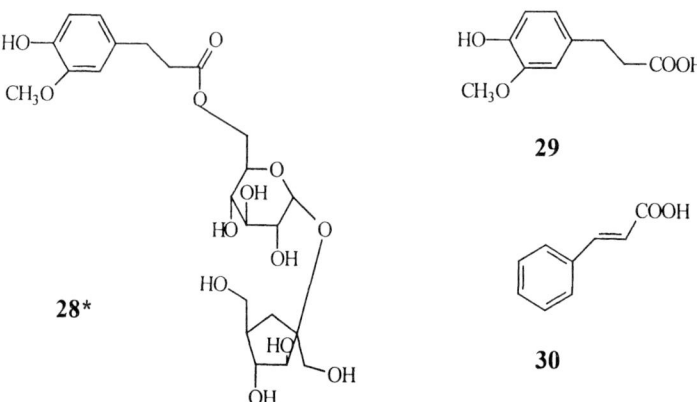

24 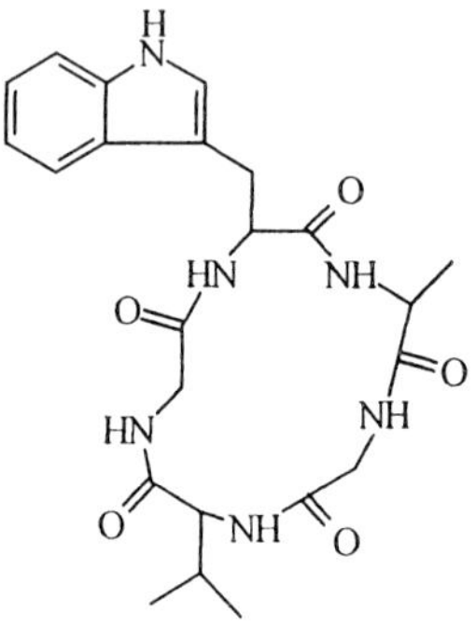 **25**

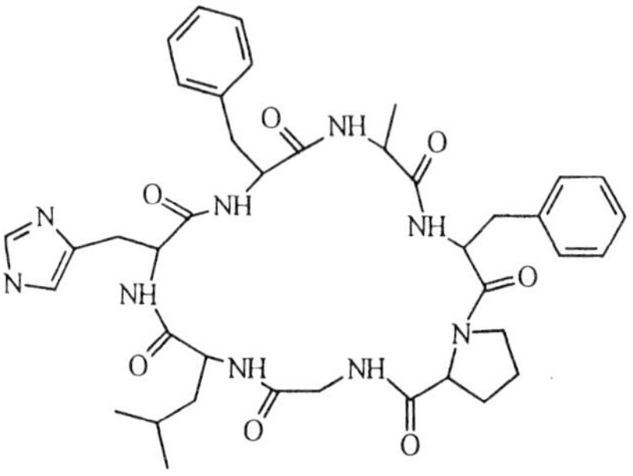

26 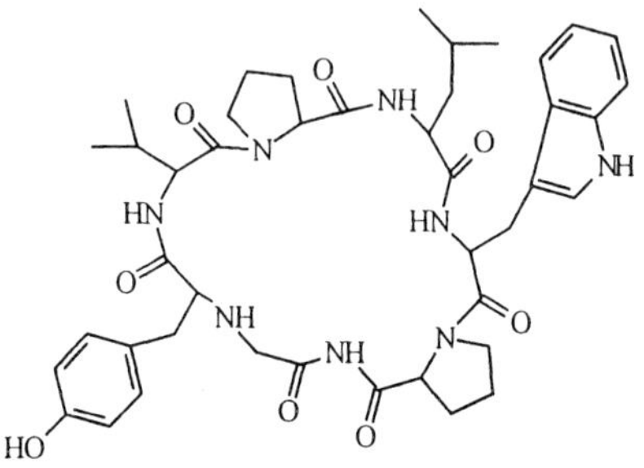 **27**

Figure 6. Structures of cyclic peptides (compounds **24-27**).

28* **29**

30

Figure 7. Structures of phenolic acids (compounds **28-30**).

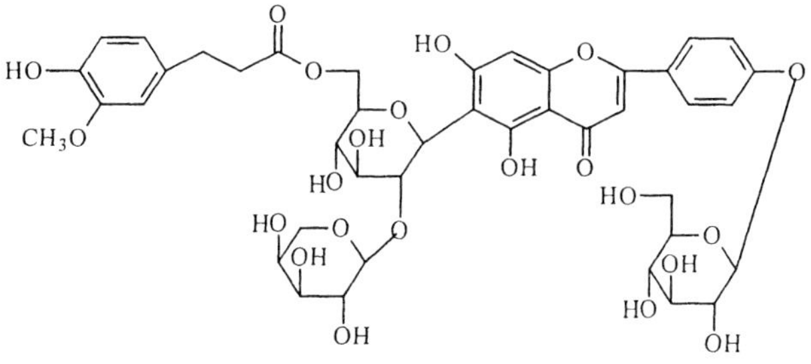

31*

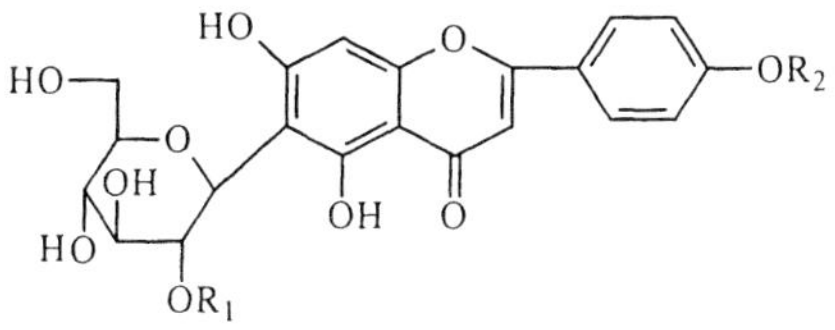

32. R₁=Ara, R₂=H
33. R₁=Glc, R₂=H
34. R₁=Ara, R₂=Glc

Figure 8. Structures of flavonoids (compounds 31-34).

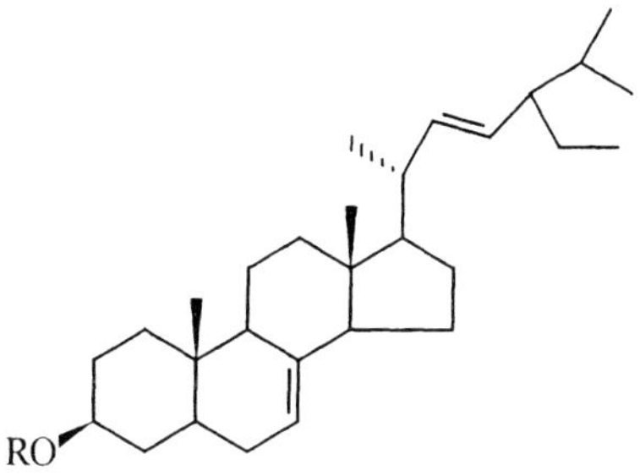

35. R=H
36. R=Glc

Figure 9. Structures of steroids (compounds 35 and 36).

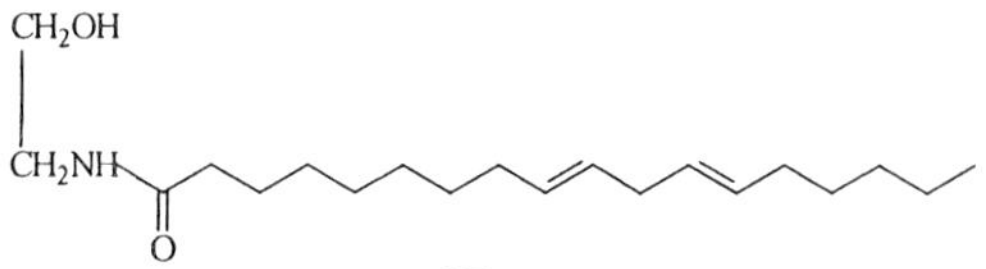

Figure 10. Structures of compounds 37-40.

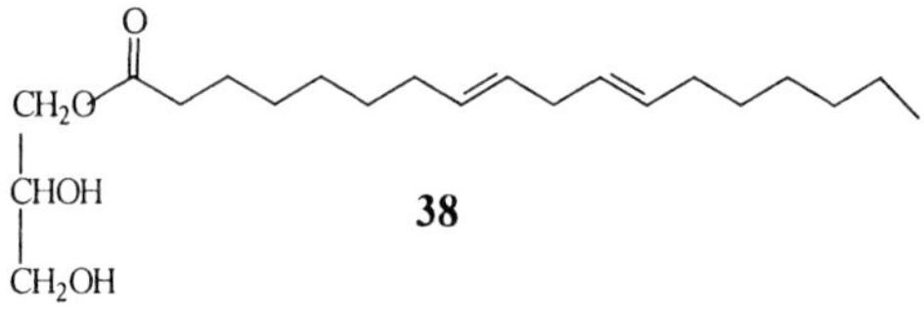

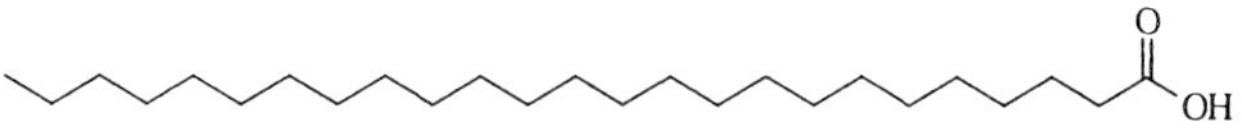

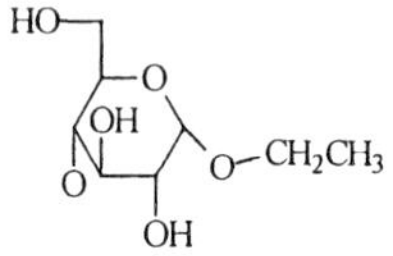

Bioactivities

In order to find the bioactive compounds responsible for the traditional use of this seed, two bioassays were conducted to investigated the inhibitory effect on the growth of both the human HL-60 cell line and the Luteal cells of rats. One new flavonoid (**31**) exerted inhibitory effect on the growth of the HL-60 cell line with IC$_{50}$ value of 10.9 µM. 18 compounds were tested for the inhibition of growth of Luteal cells of rats. Three new triterpenoids (compounds **4**, **10** and **11**) exhibited strong inhibitory effect on the growth of these cells resulting in 100% inhibition at a concentration of 20 µg/mL. It is known that steroid saponins have an inhibitory effect on the growth of luteal cells (*25*). However, these compounds are unique examples of triterpenoid saponins that show inhibitory activity on the growth of luteal cells.

Conclusion

In conclusion, 40 compounds were isolated and identified from the seeds of *V. segetalis*. 15 of them (**2-11, 14, 17, 18, 28, 31**) are new compounds, one of them (**37**) is a new natural product. Most of these new compounds have complex structures. New 2D techniques, HMQC-TOCSY and HMBC-TOCSY, were applied to determine their structures. 18 compounds were tested for the inhibition of Luteal cells in rats. Three new compounds showed very strong inhibition resulting in 100% at a concentration of 20 μg/mL. One new flavonoid showed inhibitory effect on the growth of the HL-60 cell line. Thus, the seeds of *V. segetalis* are a rich source of bioactive triterpene saponins, alkaloids, cyclic peptides, phenolic compounds and steroids.

Acknowledgement

This work was supported by the National Natural Science Foundation of China (No. 29632050).

References

1. Jiangsu New Medical College, *Zhong-yao-da-ci-dian*, Shanghai Science and Technology Publisher, 1986; p 311.
2. Litvinenko, V.I.; Amanmuradov, K.; Abubakirov, N.K. *Khim. Prir. Soed.* **1967**, *3*, 159-164.
3. Amanmuradov, K.; Abubakirov, N.K. *Khim. Geol. Nauk.* **1964**, *6*, 104-108.
4. Morita, H.; Yun, Y.S.; Takeya, K.; Itokawa, H.; Yamada, K.; Shirota, O. *Bioorg. Med. Chem. Letters* **1997**, *7*, 1095-1096.
5. Yun, Y.S.; Shimizu, K.; Morita, H.; Takeya, K.; Itokawa, H.; Shirota, O. *Phytochemistry* **1998**, *47*, 143-144.
6. Koike, K.; Jia, Z.H.; Nikaido, T. *Phytochemistry* **1998**, *47*, 1343-1349.
7. Jia, Z.H.; Koike, K.; Kudo, M.; Li, H.Y.; Nikaido, T. *Phytochemistry* **1998**, *48*, 529-536.
8. Morita, H.; Yun, Y.S.; Takeya, K.; Itokawa, H. *Tetrahedron* **1995**, *51*, 5987-6002.
9. Morita, H.; Yun, Y.S.; Takeya, K.; Itokawa, H. *Tetrahedron* 1995, **51**, 6003-6014.
10. Morita, H.; Yun, Y.S.; Takeya, K.; Itokawa, H. *Phytochemistry* 1996, **42**, 439-441.

11. Yun, Y.S.; Morita, H.; Takeya, K.; Itokawa, H. *J. Nat. Prod.* 1997, **60**, 216-218.
12. Baeva, R. T.; Karryev, M.O.; Litvinenko, V.I.; Abubakirov, N.K. *Khim. Prir. Soedin.* 1974, **10**, 171-176.
13. Sang, S.M.; Lao, A.N.; Wang, H.C.; Chen, Z.L.; Uzawa, J.; Fujimoto, Y. *Nat. Prod. Sci.* 1998, *4*, 268-273.
14. Sang, S.M.; Lao, A.N.; Wang, H.C.; Chen, Z.L.; Uzawa, J.; Fujimoto, Y. *J. Asian. Nat. Prod. Res.* 1999, *1*, 199-201.
15. Sang, S.M.; Lao, A.N.; Wang, H.C.; Chen, Z.L.; Uzawa, J.; Fujimoto, Y. *J. Asian. Nat. Prod. Res.* 2000, *2*, 187-193.
16. Sang, S.M.; Lao, A.N.; Wang, H.C.; Chen, Z.L.; Uzawa, J.; Fujimoto, Y. *Chin. Chem. Lett.* 2000, *11*, 49-52.
17. Sang, S.M.; Lao, A.N.; Wang, H.C.; Chen, Z.L.; Uzawa, J.; Fujimoto, Y. *Tetrahedron Lett.* 2000, *41*, 9205-9207.
18. Sang, S.M.; Lao, A.N.; Wang, H.C.; Chen, Z.L.; Uzawa, J.; Fujimoto, Y. *J. Asian. Nat. Prod. Res.* (in press).
19. Sang, S.M.; Lao, A.N.; Wang, H.C.; Chen, Z.L.; Uzawa, J.; Fujimoto, Y. *J. Asian. Nat. Prod. Res.* (Submitted)
20. Sang, S.M.; Lao, A.N.; Wang, H.C.; Chen, Z.L.; Uzawa, J.; Fujimoto, Y. *Phytochemistry* 1998, *47*, 569-871.
21. Sang, S.M.; Lao, A.N.; Wang, H.C.; Chen, Z.L. *Zhongcaoyao* 2000, *31*, 169-171.
22. Sang, S.M.; Lao, A.N.; Wang, H.C.; Chen, Z.L. *Tianran Chanwu Yanjiu Yu Kaifa* 2000, *12*, 12-15.
23. Sang, S.M.; Lao, A.N.; Wang, H.C.; Chen, Z.L. *Tianran Chanwu Yanjiu Yu Kaifa* 1999, *10*, 1-4.
24. Sang, S.M.; Xia, Z.H.; Lao, A.N.; Wang, H.C.; Chen, Z.L. *Zhongguo Zhongyao Zazhi* 2000, *25*, 221-222.
25. Rao, A.V.; Gurfinkel, D.M. *Drug Metabol. Drug Interact.* 2000, *17*, 211-235.
26. Hostettmann, K.; Marston, A. *Saponins*. Cambridge University Press: Cambridge, England, 1995.
27. Matsubara, Y.; Yusa, T.; Sawabe, A.; Lizuka, Y.; Takekuma, S.; Yoshida, Y. *Agric. Biol. Chem.* 1991, *55*, 2923-2929.
28. Yahara, S.; Shigeyama, C.; Ura, T.; Wakamatsu, K.; Yasuhara, T.; Nohara, T. *Chem. Pharm, Bull.* 1993, *41*, 703-709.
29. Witherup, K.M.; Bogusky, M.J.; Amderson, P.S.; Ramjit, H.; Ransom, R.W.; Wood, T.; Sardama, M. *J. Nat. Prod.* 1994, *57*, 1619-1625.
30. Van den Berg, A.J.J.; Horsten, S.F.A.J.; Kettenes-van den Bosch, J.J.; Kroes, B.H.; Beukelman, C.J.; Leeflang, B.R.; Labadie, R.P. *FEBS Lett.* 1995, *358*, 215-218.

31. Van den Berg, A.J.J.; Horsten, S.F.A.J.; Kettenes-van den Bosch, J.J.; Kroes, B.H.; Beukelman, C.J.; Leeflang, B.R.; Labadie, R.P. *Phytochemistry* **1996**, *42*, 129-133.
32. Itokawa, H.; Yun, Y. Morita, H.; Takeya K.;Yamada, K.. *Planta Med.* **1995**, *61*, 561-562.
33. Rae, M.T.; Menzies, G.S.; McNeilly, A.S.; Woad, K.; Webb, R. and Bramley, T.A. *Biology of Reproduction,* **1998**, *59*, 1016.

Chapter 22

Studies on the Chemical Constituents of Loquat Leaves (*Eriobotrya japonica*)

Qing-Li Wu[1], Mingfu Wang[1], James E. Simon[1], Shi-Chun Yu[2], Pei-Gen Xiao[2], and Chi-Tang Ho[3]

[1]New Use Agriculture and Natural Plant Products Program, Department of Plant Biology and Pathology, Cook College, Rutgers, The State University of New Jersey, 65 Dudley Road, New Brunswick, NJ 08901
[2]Institute of Medicinal Plant Development, Chinese Academy of Medical Sciences and Peking Union Medical College, Beijing 100094, People's Republic of China
[3]Department of Food Science, Cook College, Rutgers, The State University of New Jersey, 65 Dudley Road, New Brunswick, NJ 08901

Chemical study of *Eriobotrya japonica* leaves led to the isolation of three new compounds, linguersinol 9'-O-β-D-xylopyranoside, eriobotrin and isoeriobotrin, and together with seventeen known compounds, including three lignans, linguersinol, 2,6-dimethoxy-4-(2-propenyl)phenol and 2,6-dimethoxy-4-(2-propenyl)phenol 1-O-β-D-glucopyranoside, eight megastigmane derivatives, and (6R,7E,9R)-9-hydroxy-4,7-megastigmadien-3-one 9-O-β-D-apiofuranosyl-(1→6)-β-D-glucopyranoside, (6R,7E,9R)-9-hydroxy-4,7-megastigmadien-3-one 9-O-β-D-xylopyranosyl-(1→6)-β-D-glucopyranoside, (6R,7E,9R)-9-hydroxy-4,7-megastigmadien-3-one 9-O-α-L-arabinopyranosy-(1→6)-β-D-glucopyranoside, (6R,7E,9R)-9-hydroxy-4,7-megastigmadien-3-one, (6R,7E,9R)-9-hydroxy-4,7-megastigmadien-3-one 9-O-β-D-glucopyranoside, (6R,7E,9S)-9-hydroxy-4,7-megastigmadien-3-one 9-O-β-D-glucopyranoside, (6S,7E,9R)-6,9-dihydroxy-4,7-megastigmadien-3-one and (6S,7E,9R)-6,9-dihydroxy-4,7-megastigmadien-3-one

9-*O*-β-D-glucopyranoside, two flavonol glycosides, quercetin-3-*O*-β-D-glucoside and quercetin-7-α-L-rhamnoside and four triterpenes, 2α,3α,19α-trihydroxy-12-oleanen-28-oic acid, euscaphic acid, 2α-hydroxyoleanolic acid and oleanolic acid. Their structures were elucidated on the basis of spectral and chemical evidences.

Loquat (Chinese name: Pi Pa), *Eriobotrya japonica* (Thunb.) Lindle is a subtropical tree belonging to the rose family. It is indigenous to southeastern China and the southern end of Japan. Now it is widely cultivated all over the world. The tree is evergreen, with distinctly ribbed leaves, and grows to 5 to 10 meters high. The fruit is small (3 to 4 cm long), pale tangerine-colored, and pear-shaped, sometimes with a single almost spherical stone in the center. The taste is quite delicate, but distinctive, with a pleasant tartness. The leaves are glossy, dark green above and whitish or rusty-hairy beneath. These characteristics of the tree have made the loquat an excellent specimen in the home landscape.

The fruits, kernel and leaves of loquat are all used for medicinal purpose. The leaf of loquat is a well known Traditional Chinese Medicine used as anti-tussive and anti-inflammatory agent for acute and chronic bronchitis. Phytochemical studies (*1-5*) with this species have led to the isolation of many natural products including flavonoids, triterpenes and sesquiterpene glycoside. Some of the components have been found to have antioxidant (*2*), antiviral (*4*), cytotoxic (*1*) and hypoglycemic (*5*) properties. In the course of ongoing search for anti-tussive and anti-inflammatory components from this plant, an ethanolic extract of the leaves of *E. japonica* was examined. We report here the isolation and structural determination of three new components as well as seventeen known compounds.

Materials and Methods

Plant and Material

Leaves of *Eriobotrya japonica* (Thunb.) Lindle. (Rosaceace) were collected from Anhui province, China, in Sept. 1994. Voucher specimens are deposited in

Institute of Medicinal Plant Development, Chinese Academy of Medical Sciences, Beijing, P. R. China.

Equipment

All mps are uncorr. NMR spectra were recorded on a Bruker AM 500 or a Bruker DMX-600 spectrometer with TMS as int. standard. IR spectra were measured on Perkin-Elmer 599 IR instrument and MS were recorded on a MAT-711 mass spectrometer. Column chromatography was carried out on silica gel and Sephadex LH-20 and TLC was performed on silica gel G.

Extraction and Isolation

Air-dried leaves (9.5 Kg) were extracted with 95% EtOH for three times. The ethanolic extract (1.33 Kg) was dissolved in water, and then extracted with petrol, CH_2Cl_2, EtOAc and n-BuOH. The n-BuOH soluble fraction was concentrated under reduced pressure to get 47 g extract. The n-BuOH extract was chromatographed on silica gel column using a step-gradient $CHCl_3$-MeOH-H_2O and 1000 mL fraction was collected. Those fractions containing similar components as checked by TLC were combined and total 11 fractions were combined. Fr. 1 was further purified on Sephadex LH-20 (MeOH) to give compound **5** (15 mg). Fr. 2 was repeatedly re-chromatographed on Sephadex LH-20 (MeOH) and silica H to get compounds **2** (40 mg) and **10** (15 mg). Fr. 3 was repeatedly chromatographed on silica H and Sephadex LH-20 (MeOH) to get compound **14** (20 mg). Fr. 4 was further purified on Sephadex LH-20 (MeOH) to give pure compound **13** (230 mg). Fr. 5 was rechromatographed on Silica Gel ($CHCl_3$-AcOEt-MeOH-H_2O) and followed by gel filtration column chromatography on Sephadex LH-20 (MeOH) to give compounds **6** (40 mg), mixture of **3** and **4** (60 mg), **11** (70 mg) and **12** (30 mg). Re-separation of fr. 6 on Silica Gel ($CHCl_3$-MeOH) and Sephadex LH-20 (MeOH) gave compound **1** (20 mg). Re-purification of fr. 9 on silica gel ($CHCl_3$-EtOAc-MeOH) and Sephadex LH-20 (MeOH) gave compounds **9** (15 mg), **15** (700 mg) and **16** (10 mg). Fr. 10 was rechromatographed on silica gel and Sephadex LH-20 (MeOH) to afford compounds **7** (330 mg) and of **8** (620 mg).

The EtOAc extract (155 g) was chromatographed on silica gel column using step-gradient ($CHCl_3$-methanol) and totally 13 fractions were obtained. Rechromatography of fr. 4 on silica gel (petrol-$CHCl_3$) gave compound **20** (30 mg). Fr. 6 was further purified on silica H (petrol: EtOAc) and Sephadex LH-20 (MeOH) to afford compounds **17** (35 mg) **18**, (240 mg) and **19** (360 mg). The structures of compounds 1-14 are shown in Figure 1.

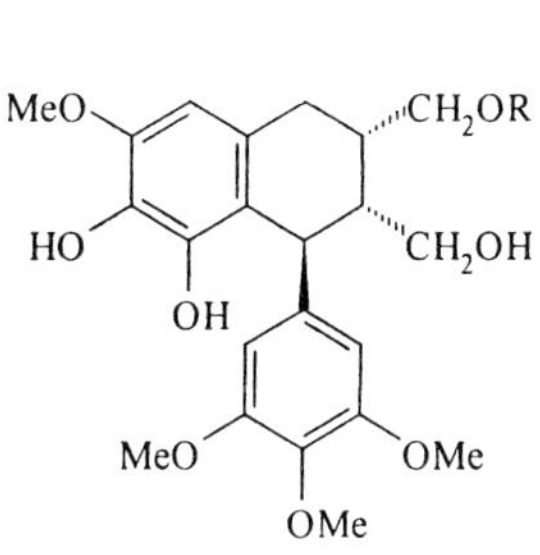

*1. R= Xyl, 2. R= H 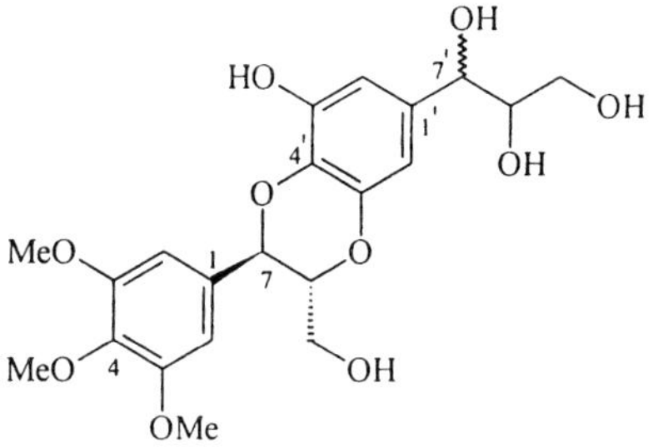   *3   *4. 7'-Epimer

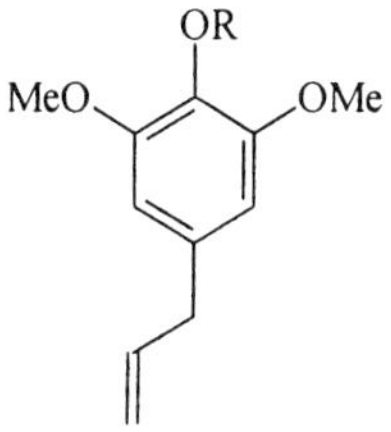

5. R=H   6. R=Glc

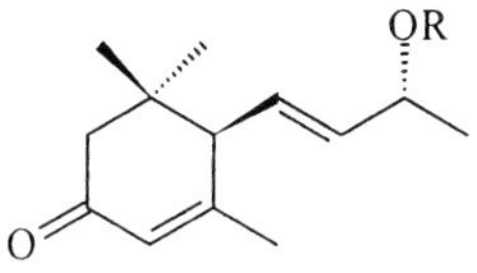

7. R=Glc 6-1 Api
8. R=Glc 6-1 Xyl
9. R=Glc 6-1 Ara
10. R=H, 11. R=Glc

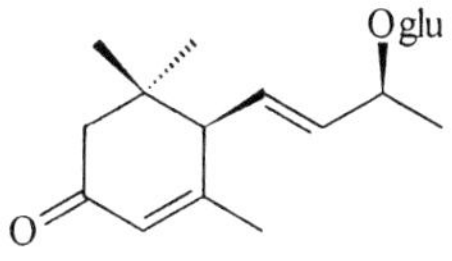

12

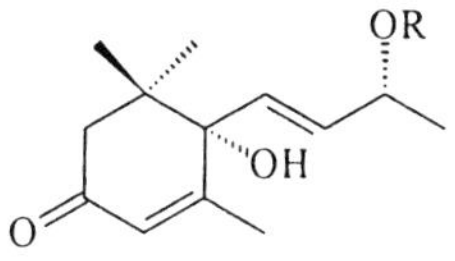

13. R=H, 14. R=Glc

*Figure 1. Structure of Compounds **1-14** identified in Loguat leaves.*

Acid Hydrolysis of 1

Compound **1** (2 mg) was dissolved in 5% HCl and heated at 100 °C for 3 h, cooled and filtered. The filtrate was evaporated to dryness. The residue was examined for sugar and aglycone by TLC on silica gel (n-BuOH-MeOH-H$_2$O and CHCl$_3$-MeOH).

Linguersinol 9'-*O*-β-D-xylopyranoside (1)

White powder, $[\alpha]_D^{25}$ +33.3° (c=0.900, MeOH); IRν_{max} (KBr, cm^{-1}): 3430, 2970, 2840, 1615, 1520, 1500, 1460, 1420, 1320, 1220, 1115, 1050; UVλ_{max} nm (MeOH): 238 (s), 276; EI-MS *m/z* (%): 552 [M]$^{+\cdot}$ (10.5), 420 for the aglycone (100), 402 [C$_{22}$H$_{26}$O$_7$]$^+$ (12.1), 371 [C$_{21}$H$_{23}$O$_6$]$^+$ (37.8), 248 [C$_{14}$H$_{16}$O$_4$]$^+$ (17.3), 217 [C$_{13}$H$_{13}$O$_3$]$^+$ (32.6), 205 [C$_{12}$H$_{13}$O$_3$]$^+$ (40.7), 183 [C$_9$H$_{11}$O$_4$]$^+$ (46.0), 173 [C$_{11}$H$_9$O$_2$]$^+$ (32.9), 167 [C$_9$H$_{11}$O$_3$]$^+$ (84.2); Positive SI-HRMS *m/z*: 553.2288 [M+H]$^+$, calcd. C$_{27}$H$_{37}$O$_{12}$, 553.2279. ^{1}H and ^{13}C NMR: Table I.

Linguersinol (2)

White powder, $[\alpha]_D^{25}$ +3.9° (c=0.770, MeOH), IRν_{max} (KBr, cm^{-1}): 3430, 2970, 2840, 1610, 1520, 1500, 1450, 1420, 1320, 1220, 1115, 1060; UVλ_{max} nm (MeOH): 240 (sh), 276; EI-MS *m/z* (%): 420 [M]$^{+\cdot}$ (100), 402 [C$_{22}$H$_{26}$O$_7$]$^+$ (6.2), 371 [C$_{21}$H$_{23}$O$_6$]$^+$ (25.5), 248 [C$_{14}$H$_{16}$O$_4$]$^{+\cdot}$ (15.1), 217 [C$_{13}$H$_{13}$O$_3$]$^+$ (42.0), 205 [C$_{12}$H$_{13}$O$_3$]$^+$ (55.5), 183 [C$_9$H$_{11}$O$_4$]$^+$ (64.0), 173 [C$_{11}$H$_9$O$_2$]$^+$ (28.5), 167 [C$_9$H$_{11}$O$_3$]$^+$ (59.1); SI-HRMS *m/z*: 420.1779 [M]$^{+\cdot}$, calcd. C$_{22}$H$_{28}$O$_8$, 420.1785. ^{1}H and ^{13}C NMR: Table I.

Eriobotrin and Isoeriobotrin (3 and 4)

White powder, $[\alpha]_D^{25}$ +24.0° (c=0.125, MeOH); IRν_{max} (KBr, cm^{-1}): 3430, 2920, 1600, 1515, 1460, 1340, 1220, 1120, 1030; UVλ_{max} nm (MeOH): 240 (sh), 268; ^{1}H NMR (500 MHz, CD$_3$OD) δ ppm: 6.92 (2H, brs, H-3, 5), [6.88 (1H, d, J=1.7 Hz, H-2'), 6.82 (1H, d, J=1.7 Hz, H-6') for **3**], [6.87(1H, brs, H-2'), 6.81 (1H, brs, H-6') for **4**], 5.07 (1H, d, J=8.0 Hz, H-7), 4.71 (1H, d, J=5.9 Hz, H-7'), 4.22 (1H, m, H-8), 4.07 (3H, s, 4-OMe), 4.05 (6H, s, 3, 5-OMe's), 3.92 (1H, dd, 12.4, 2.4 Hz, H-9a), 3.86 (1H, m, H-8'), 3.71 (2H, m, H-9b, 9'a), 3.58 (1H, m, H-9'b); ^{13}C NMR (125 MHz, CD$_3$OD) δ ppm: 149.9 (C-3'), 149.4 (C-3, 5), 145.4 (C-5'), 137.4 (C-4), 135.7 (C-1'), 133.8 (C-4'), 128.6 (C-1), [109.4 (C-

Table I. ^{1}H and ^{13}C NMR Spectral Data of 1 and 2 (δ in ppm, J in Hz)

	1		2	
	^{13}C	^1H	^{13}C	^1H
1	134.9		134.6	
2	107.5	6.61 brs	106.9	6.57 brs
3	149.1		149.1	
4	147.8		147.7	
5	149.1		149.0	
6	107.5	6.61 brs	106.9	6.57 brs
7	42.9	4.57 d (6.5)	42.3	4.50 d (5.7)
8	46.8	2.25 m	48.0	2.15 m
9	66.9	4.00 dd (9.8, 5.3)	66.8	3.97 m
		3.62 dd (9.8, 4.0)		3.78 m
1'	130.3		130.2	
2'	108.1	6.76 s	107.8	6.78 s
3'	148.8		148.7	
4'	139.4		139.3	
5'	139.0		138.9	
6'	126.4		126.3	
7'	33.8	2.92 dd (15.3, 4.7)	33.6	2.91 dd (15.1, 5.3)
		2.82 brd (15.3)		2.78 brd (15.1)
8'	40.9	1.90 m	40.9	1.82 m
9'	71.3	3.83 dd (10.9, 4.4)	64.2	3.69 m
		3.74 dd (10.9, 6.6)		
3,5-OMe	57.1	3.93 s	56.8	3.93 s
4-OMe	56.8	3.52 s	56.6	3.56 s
3'-OMe	60.2	4.04 s	60.2	4.05 s
1''	105.1	4.41 d (7.5)		
2''	74.9			
3''	78.0			
4''	71.7			
5''	66.4			

Measured in CD_3OD

6'), 104.7 (C-2') for **3**], [109.3 (C-6'), 104.6 (C-2') for **4**], 1.6.2 (C-2, 6), 79.9 (C-8), 77.8 (C-7), 77.4 (C-8'), [77.3 (C-7') for **3**], [77.2 (C-7') for **4**], 64.2 (C-9'), 62.1 (C-9), 56.9 (3, 5-Me's), 56.7 (C-9); EI-MS m/z (%) 438 [M]$^{+\cdot}$ (9.5), 420 (6.6), 402 (4.5), 377 (36.0), 347 (16.8), 210 (100), 192 (25.5), 182 (57.5), 167 (93.1), 153 (35.3), 137 (30.6), 109 (34.3).

2,6-Dimethoxy-4-(2-propenyl)phenol (5)

White powder, ^{1}H NMR (500 MHz, CD$_3$OD) δ ppm: 6.24 (2H, brs, H-2, 6), 6.14 (1H, m, H-8), 5.26 (1H, d, J=15.5 Hz, H-9r), 5.21 (1H, d, J=10.0 Hz, H-9s), 4.00 (6H, s, 3,5-OMe's), 3.49 (2H, d, J=8.3 Hz, H=7); ^{13}C NMR (125 MHz, CD$_3$OD) δ ppm: 149.3 (C-3, 5), 139.3 (C-5), 135.3 (C-8), 132.2 (C-1), 115.6 (C-9), 107.0 (C-6), 56.8 (3,5-OMe's), 41.2 (C-7).

2,6-Dimethoxy-4-(2-propenyl)phenol 1-*O*-β-D-glucopyranoside (6)

White powder, IRν$_{max}$ (KBr, cm^{-1}): 3420, 2940, 1595, 1500, 1460, 1420, 1335, 1240, 1130, 1070; UVλ $_{max}$ nm (MeOH): 236 (sh), 268; ^{1}H NMR (500 MHz, CD$_3$OD) δ ppm: 6.72 (2H, brs, H-2, 6), 6.15 (1H, m, H-8), 5.29 (1H, dd, J=17.0, 1.4 Hz, H-9r), 5.24 (1H, d, J=10.0 Hz, H-9s), 4.99 (1H, d, J=8.5 Hz), 4.02 (6H, s, 3,5-OMe's), 3.51 (2H, d, J=8.3 Hz, H=7); ^{13}C NMR (125 MHz, CD$_3$OD) δ ppm: 154.3 (C-3, 5), 138.6 (C-5), 138.5 (C-1), 135.1 (C-8), 116.1 (C-9), 108.0 (C-6), 105.8 (C-1'), 78.3 (C-3'), 78.0 (C-5'), 75.9 (C-2'), 71.6 (C-4'), 52.8 (C-6'), 57.2 (3,5-OMe's), 41.2 (C-7); EI-MS m/z (%) 194 (100, [M]$^{+\cdot}$ of aglycone), 179 (5.2), 163.2 (4.2), 131 (3.0), 105 (6.9), 91 (3.3).

(6R,7E,9R)-9-Hydroxy-4,7-megastigmadien-3-one 9-*O*-β-D-apiofuranosyl-(1→6)-β-D-glucopyranoside (7)

White powder, [α]$_D^{25}$ +55.5° (c=1.100, MeOH); IRν$_{max}$ (KBr, cm^{-1}): 3420, 2960, 2920, 2880, 1645, 1375, 1060; UVλ $_{max}$ nm (MeOH): 234; Positive SI-HRFABMS m/z: 525.2308 [M+Na]$^+$; ^{1}H NMR (500 MHz, CD$_3$OD) δ ppm: 6.07 (1H, s, H-4), 5.95 (H, dd, J=15.4, 6.6 Hz, H-8), 5.85 (H, dd, J=15.4, 9.1 Hz, H-7), 5.19 (1H, d, J=2.3 Hz, H-1''), 4.54 (1H, m, H-9), 4.53 (1H, d, J=7.9 Hz, H-1'), 2.85 (1H, d, J=9.1 Hz, H-6), 2.63 (1H, d, J=16.8 Hz, H-2α), 2.25 (1H, d, J=16.8 Hz, H-2β), 2.13 (3H, s, H-13), 1.48 (3H, d, J=6.4 Hz, H-10), 1.23, 1.20 (each 3H, s, H-11, 12); ^{13}C NMR (125 MHz, CD$_3$OD) δ ppm: 201.1 (C-3), 165.8 (C-5), 138.1 (C-7), 129.0 (C-8), 126.1 (C-4), 110.9 (C-1''), 102.6 (C-1'), 80.4 (C-3''), 78.1 (C-5'), 78.0 (C-3'), 77.1 (C-9), 76.9 (C-2''), 75.2 (C-2'), 74.9 (C-4''), 71.6 (C-4'), 68.5 (C-6'), 65.7 (C-5''), 56.7 (C-6), 48.4 (C-2), 37.1 (C-1), 28.0 (C-12), 27.5 (C-11), 23.7 (C-10), 21.0 (C-13).

(6R,7E,9R)-9-Hydroxy-4,7-megastigmadien-3-one 9-*O*-β-D-xylopyranosyl-(1→6)-*β*-D-glucopyranoside (8)

White powder, $[\alpha]_D^{25}$ +54.9° (c=0.820, MeOH); IRν_{max} (KBr, cm^{-1}): 3420, 2960, 2920, 2880, 1650, 1375, 1165, 1080, 1040; UVλ $_{max}$ nm (MeOH): 234; ^{1}H NMR (500 MHz, CD$_3$OD) δ ppm: 6.08 (1H, s, H-4), 5.96 (H, dd, J=15.4, 6.9 Hz, H-8), 5.85 (H, dd, J=15.4, 9.8 Hz, H-7), 4.55 (1H, m, H-9), 4.54 (1H, d, J=7.7 Hz, H-1'), 4.48 (1H, d, J=7.5 Hz, H-1''), 2.88 (1H, d, J=9.8 Hz, H-6), 2.62 (1H, d, J=17.0 Hz, H-2α), 2.24 (1H, d, J=17.0 Hz, H-2β), 2.13 (3H, s, H-13), 1.48 (3H, d, J=6.5 Hz, H-10), 1.22, 1.19 (each 3H, s, H-11, 12); ^{13}C NMR (125 MHz, CD$_3$OD) δ ppm: 201.0 (C-3), 165.8 (C-5), 138.1 (C-7), 128.9 (C-8), 126.1 (C-4), 105.4 (C-1''), 102.6 (C-1'), 77.9 (C-5'), 77.7 (C-3'), 77.1 (C-9), 76.8 (C-3''), 75.1 (C-2'), 74.8 (C-2''), 71.3 (C-4'), 71.1 (C-4''), 69.6 (C-6'), 66.8 (C-5''), 56.7 (C-6), 48.4 (C-2), 37.1 (C-1), 28.1 (C-12), 27.6 (C-11), 23.8 (C-10), 21.1 (13); Positive FAB-MS *m/z*: 503 [M+1]$^+$, 371[M-xyl+1]$^+$, 209 [M+1]$^+$ of aglycone.

(6R,7E,9R)-9-Hydroxy-4,7-megastigmadien-3-one 9-*O*-α-L-arabinopyranosy-(1→6)-β-D-glucopyranoside (9)

White powder, IRν_{max} (KBr, cm^{-1}): 3420, 2950, 2930, 2880, 1650, 1380, 1060; UVλ $_{max}$ nm (MeOH): 234; ^{1}H NMR (500 MHz, CD$_3$OD) δ ppm: 6.07 (1H, s, H-4), 5.93 (H, dd, J=15.6, 9.0 Hz, H-8), 5.79 (H, dd, J=15.4, 7.6 Hz, H-7), 4.57 (1H, m, H-9), 4.53 (1H, d, J=7.8 Hz, H-1'), 4.47 (1H, d, J=6.9 Hz, H-1''), 2.86 (1H, d, J=8.9 Hz, H-6), 2.65 (1H, d, J=16.4 Hz, H-2α), 2.27 (1H, d, J=16.4 Hz, H-2β), 2.17 (3H, s, H-13), 1.49 (3H, d, J=6.2 Hz, H-10), 1.25, 1.21 (each 3H, s, H-11, 12). ^{13}C NMR (125 MHz, CD$_3$OD) δ ppm: 202.1 (C-3), 165.9 (C-5), 138.0 (C-7), 129.0 (C-8), 126.2 (C-4), 105.4 (C-1''), 102.3 (C-1'), 78.2 (C-5'), 77.9 (C-3'), 77.2 (C-9), 75.3 (C-2'), 74.9 (C-2''), 72.4 (C-3''), 71.8 (C-4'), 69.9 (C-4''), 69.1 (C-6'), 66.7 (C-5''), 56.9 (C-6), 48.3 (C-2), 37.1 (C-1), 28.1 (C-12), 27.5 (C-11), 23.8 (C-10), 21.1 (C-13); EI-MS *m/z* (%): 208 [M]$^{+\cdot}$ (6.7), 192 (100), 177 (59.2), 137 (73.4), 135 (93.1), 123 (42.6), 108 (37.6), 91 (40.4), 83 (47.5), 55 (37.3), 43 (36.3).

(6R,7E,9R)-9-Hydroxy-4,7-megastigmadien-3-one (10)

White needles, IRν_{max} (KBr, cm^{-1}): 3420, 2950, 2920, 2885, 1660, 1380, 1150, 1075, 1040; UVλ $_{max}$ nm (MeOH): 234; EI-MS *m/z* (%): 208 [M]$^{+\cdot}$ (5.6), 192 (100), 177 (52.8), 137 (75.3), 135 (92.6), 123 (44.7), 108 (47.8), 91 (46.7), 83 (53.4), 55 (37.4), 43 (37.2); ^{1}H NMR (500 MHz, CD$_3$OD) δ ppm: 6.07 (1H, s, H-4), 5.89 (H, dd, J=15.2, 6.7 Hz, H-8), 5.81 (H, dd, J=15.2, 8.3 Hz, H-7), 4.41 (1H, m, H-9), 2.81 (1H, d, J=8.3 Hz, H-6), 2.55 (1H, d, J=16.9 Hz, H-2α), 2.19 (1H, d, J=16.9 Hz, H-2β), 2.08 (3H, s, H-13), 1.38 (3H, d, J=6.6 Hz, H-10), 1.17, 1.14 (each 3H, s, H-11, 12); ^{13}C NMR (125 MHz, CD$_3$OD) δ ppm: 202.1 (C-3), 165.9 (C-5), 140.3 (C-7), 127.1 (C-8), 126.1 (C-4), 68.9 (C-9), 56.8 (C-6), 48.5 (C-2), 37.2 (C-1), 28.1 (C-12), 27.6 (C-11), 23.7 (C-10), 22.4 (C-13); EI-MS m/z (%): 208 [M]$^{+•}$ (5.6), 192 (100), 177 (52.8), 137 (75.3), 135 (92.6), 123 (44.7), 108 (47.8), 91 (46.7), 83 (53.4), 55 (37.4), 43 (37.2).

(6R,7E,9R)-9-Hydroxy-4,7-megastigmadien-3-one 9-O-β-D-glucopyranoside (11)

White powder, [α]$_D^{25}$ +120.0° (c=0.450, MeOH), IRν$_{max}$ (KBr, cm^{-1}): 3420, 2960, 2925, 2880, 1650, 1375, 1160, 1080, 1040; UVλ$_{max}$ nm (MeOH): 234; ^{1}H NMR (500 MHz, CD$_3$OD) δ ppm: 6.08 (1H, s, H-4), 5.97 (H, dd, J=15.4, 6.4 Hz, H-8), 5.85 (H, dd, J=15.4, 8.6 Hz, H-7), 4.59 (1H, m, H-9), 4.55 (1H, d, J=7.9 Hz, H-1'), 2.87 (1H, d, J=8.6 Hz, H-6), 2.62 (1H, d, J=17.8 Hz, H-2α), 2.23 (1H, d, J=17.8 Hz, H-2β), 2.13 (3H, s, H-13), 1.48 (3H, d, J=6.4 Hz, H-10), 1.22, 1.19 (each 3H, s, H-11, 12); ^{13}C NMR (125 MHz, CD$_3$OD) δ ppm: 201.9 (C-3), 165.7 (C-5), 138.2 (C-7), 128.8 (C-8), 126.2 (C-4),102.5 (C-1'), 78.1 (C-5'), 77.9 (C-3'), 76.9 (C-9), 75.2 (C-2'), 71.6 (C-4'), 62.7 (C-6'), 56.8 (C-6), 48.3 (C-2), 37.1 (C-1), 28.1 (C-12), 27.5 (C-11), 23.7 (C-10), 21.0 (13); EI-MS m/z (%): 208 [M]$^{+•}$ (5.8), 192 (100), 177 (56.3), 137 (89.1), 135 (98.4), 123 (44.1), 108 (42.4), 91 (40.1), 83 (41.3), 55 (35.5), 43 (37.7).

(6R,7E,9S)-9-Hydroxy-4,7-megastigmadien-3-one 9-O-β-D-glucopyranoside (12)

White powder, [α]$_D^{25}$ +77.5° (c=1.110, MeOH); IRν$_{max}$ (KBr, cm^{-1}): 3420, 2960, 2920, 2870, 1655, 1375, 1160, 1080, 1040; UVλ$_{max}$ nm (MeOH): 234; ^{1}H NMR (500 MHz, CD$_3$OD) δ ppm: 6.08 (1H, s, H-4), 5.95 (1H, dd, J=15.3, 9.4 Hz, H-8), 5.78 (1H, dd, J=15.4, 7.5 Hz, H-7), 4.67 (1H, m, H-9), 4.58 (1H, d, J=7.9 Hz, H-1'), 2.88 (1H, d, J=7.5 Hz, H-6), 2.67 (1H, d, J=16.8 Hz, H-2α), 2.25 (1H, d, J=16.8 Hz, H-2β), 2.17 (3H, s, H-13), 1.48 (3H, d, J=6.6 Hz, H-10), 1.22, 1.18 (each 3H, s, H-11, 12); ^{13}C NMR (125 MHz, CD$_3$OD) δ ppm: 201.9 (C-3), 165.5 (C-5), 137.0 (C-7), 131.0 (C-8), 126.2 (C-4),101.2 (C-1'), 78.4 (C-5'), 78.1 (C-3'), 74.9 (C-2'), 74.8 (C-9), 71.7 (C-4'), 62.8 (C-6'), 56.9 (C-6), 48.3 (C-2), 37.1 (C-1), 28.0 (C-12), 27.3 (C-11), 23.8 (C-10), 22.2 (13); Positive SI-HRMS m/z: 393.1883 [M+Na]$^+$.

(6*S*,7*E*,9*R*) 6,9-Dihydroxy-4,7-megastigmadien-3-one (13)

White needles, $[\alpha]_D^{25}$ +190.2° (c=0.820, MeOH); IRν_{max} (KBr, cm^{-1}): 3370, 2980, 2920, 2880, 1660, 1430, 1380, 1120, 1070, 1020; UVλ_{max} nm (MeOH): 234; ^{1}H NMR (500 MHz,CD$_3$OD) δ ppm: 6.07 (1H, s, H-4), 5.99 (2H, m, H-7, 8), 4.51 (1H, m, H-9), 2.67 (1H, d, J=16.9 Hz, H-2α), 2.35 (1H, d, J=16.9 Hz, H-2β), 2.11 (3H, s, H-13), 1.43 (3H, d, J=6.5 Hz, H-10), 1.23, 1.21 (each 3H, s, H-11, 12); ^{13}C NMR (125 MHz, CD$_3$OD) δ ppm: 201.1 (C-3), 167.3 (C-5), 137.0 (C-7), 130.0 (C-8), 127.1 (C-4), 79.9 (C-6), 68.6 (C-9), 50.7 (C-2), 42.4 (C-1), 24.5, 23.4 (C-11, 12), 23.8 (C-10), 19.5 (13); EI-MS *m/z* (%): 206 [M-H$_2$O]$^+$ (1.7), 168 (6.6), 150 (5.9), 135 (5.3), 124 (100); Positive SI-HRFABMS: 225.1484 [M+H]$^+$.

(6*S*,7*E*,9*R*) 6,9-Dihydroxy-4,7-megastigmadien-3-one 9-*O*-β-D-glucopyranoside (14)

White powder, $[\alpha]_D^{25}$ +69.7° (c=1.320, MeOH); IRν_{max} (KBr, cm^{-1}): 3410, 2960, 2920, 2880, 1650, 1375, 1160, 1075, 1040; UVλ_{max} nm (MeOH): 234; ^{1}H NMR (500 MHz, CD$_3$OD) δ ppm: 6.06 (2H, m, H-7, 8), 6.05 (1H, s, H-4), 4.61 (1H, m, H-9), 4,53 (1H, d, J=7.9 Hz, H-1'), 2.70 (1H, d, J=17.0 Hz, H-2α), 2.36 (1H, d, J=17.0 Hz, H-2β), 2.11 (3H, s, H-13), 1.48 (3H, d, J=6.8 Hz, H-10), 1.23, 1.22 (each 3H, s, H-11, 12); ^{13}C NMR (125 MHz, CD$_3$OD) δ ppm: 201.1 (C-3), 167.0 (C-5), 135.3.0 (C-7), 131.6 (C-8), 127.2 (C-4), 80.0 (C-6), 78.2 (C-5'), 78.0 (C-3'), 77.2 (C-9), 75.3 (C-2'),71.8 (C-4'), 63.0 (C-6'), 50.8 (C-2), 42.4 (C-1), 24.7, 23.4 (C-11, 12), 21.2 (C-10), 19.4 (13); EI-MS *m/z* (%): 206 [M$^+$-H$_2$O] (1.7), 168 (6.6), 150 (5.9), 135 (5.3), 124 (100); Positive SI-HRFABMS: 225.1484[M+H]$^+$.

Quercetin-3-*O*-β-D-glucoside (15)

Yellow powder, IRν_{max} (KBr, cm^{-1}) 3410 (OH), 2940, 1660 (α,β-unsaturated C=O), 1615, 1510, 1450, 1370, 1310, 1215, 1090; UVλ_{max} nm 254, 300 (sh), 356 (MeOH); ^{1}H NMR (500 MHz, DMSO-d$_6$) δ ppm: 12.60 (5-OH), 7.65 (1H, dd, J=8.8, 1.7, H-6'), 7.52 (1H, d, J=1.7 Hz, H-2'), 6.81 (1H, d, J=8.8 Hz, H-5'), 6.39 (1H, d, J=1.8 Hz, H-8), 6.19 (1H, d, J=1.8 Hz, H-6), 5.35 (1H, d, J=7.7Hz, H-1''); ^{13}C NMR (125 Hz, DMSO-d$_6$) δ ppm 177.4 (C-4), 164.0 (C-7), 161.2 (C-9), 156.3 (C-2), 156.2 (C-5), 148.4 (C-4'), 144.7 (C-3'), 133.5 (C-3), 121.9 (C-6'), 121.1 (C-1'), 115.9 (C-5'), 115.1 (C-2'), 103.9 (C-10), 101.9 (C-

1″), 98.6 (C-6), 93.4 (C-8), 75.8 (C-5c), 75.2 (C-3″), 71.2 (C-2″), 67.9 (C-4″), 60.1 (C-6″); EI-MS m/z (%) 302 [M-glc]$^+$ (100).

Quercetin-7-α-L-rhamnoside (16)

Yellow powder, IRν$_{max}$ (KBr, cm^{-1}) 3400 (OH), 1665 (α,β-unsaturated C=O), 1620, 1600, 1510, 1220, 1130, 1035; UVλ $_{max}$ nm 254, 370 (MeOH); ^{1}H NMR (500 MHz, DMSO-d$_6$) δ ppm 12.50 (1H, s, chelated HO-5), 7.74 (1H, d, J=1.8, H-2′), 7.58 (1H, dd, J=8.5, 1.8 Hz, H-6′), 6.88 (1H, d, J=8.5 Hz, H-5′), 6.75 (1H, d, J=1.6 Hz, H-8), 6.41 (1H, d, J=1.6 Hz, H-6), 5.55 (1H, brs, H-1‴), 1.13 (3H, d, J=6.7, H-6″); ^{13}C NMR (125 MHz, DMSO-d$_6$) δ ppm 175.7 (C-4), 161.2 (C-7), 160.3 (C-9), 155.6 (C-5), 147.6 (C-4′), 147.3 (C-2), 145.0 (C-3′), 136.0 (C-3), 121.4 (C-1′), 120.3 (C-6′), 115.5 (C-5′), 115.2 (C-2′), 104.5 (C-10), 98.6 (C-1″), 98.5 (C-6), 94.0 (C-8), 71.4 (C-4″), 70.2 (C-3″), 70.1 (C-2″), 69.9 (C-5″), 17.6 (C-6″); EI-MS m/z (%) 302 [M-rha]$^+$ (100).

2α,3α,19α-Trihydroxy-12-oleanen-28-oic Acid (17)

White powder, IRν$_{max}$ (KBr, cm^{-1}): 3200, 2970, 1695, 1455, 1390, 1040, 995, 940; ^{1}H NMR (500 MHz, DMSO-d$_6$) δ ppm: 5.14 (1H, brs, H-12), 4.87 (1H, brs, OH), 4.30 (1H, brd, J=11.0 Hz, H-2), 3.74 (1H, brs, H-3), 3.60 (1H, brs, H-18), 3.58 (1H, m, H-16α), 2.78 (1H, brt, J=11.1 Hz, H-15β), 1.54 (3H, s, H-27), 1.25 (3H, s, H-23), 1.16 (3H, s, H-26), 1.09 (3H, s, H-29), 1.06 (3H, s, H-30), 0.97 (3H, s, H-25), 0.89 (3H, s, H-24); ^{13}C NMR (125 MHz, Pyridine-d$_5$) δ ppm: 180.9 (C-28), 144.9 (C-13), 123.1 (C-12), 81. 4 (C-19), 79.5 (C-3), 66.1 (C-2), 54.8 (C-18), 49.0 (C-1), 48.3 (C-9), 46.1 (C-17), 44.9 (C-5), 42.7 (C-14), 40.3 (C-4), 38.9 (C-8), 38.8 (C-10), 35.7 (C-20), 33.7 (C-22), 33.4 (C-7), 29.4 (C-23), 29.3 (C-15), 29.1 (C-21), 28.8 (C-29), 28.5 (C-27), 24.9 (C-23), 24.8 (C-30), 24.3 (C-16), 22.3 (C-11), 18.7 (C-6, 25), 17.7 (C-26), 16.6 (C-24); EI-MS m/z (%): 488 [M]$^{+.}$ (2.1), 264 (a, 83.7), 246 (a-H$_2$O, 92.3), 231 (246-Me, 41.6), 223 (b-1, 12.4), 219 (a-COOH, 14.3), 201 (219- H$_2$O, 100), 146 (18.8).

Euscaphic acid (18)

White powder, IRν$_{max}$ (KBr, cm^{-1}): 3200, 2970, 1695, 1455, 1390, 1160, 1040, 940; ^{1}H NMR (500 MHz, Pyridine-d$_5$) δ ppm: 5.57 (1H, brs. H-12), 4.87 (s, OH), 4.28 (1H, brd, J=11.3 Hz, H-2), 3.74 (1H, br, H-3), 3.07 (1H, dt, J=13.1, 4.6 Hz, H-16α), 3.02 (1H, s, H-18), 2.32 (1H, dt, J=13.7, 4.2 Hz, H-15β), 1.65 (3H, s, H-27), 1.49 (3H, s, H-29) 1.32 (3H, s, H-23) 1.16 (3H, s, H-

26) 1.16 (3H, d, J=6.5 Hz, H-30) 0.99 (3H, s, H-25) 0.89 (3H, s, H-24); ^{13}C NMR (125 MHz, Pyridine-d$_5$) δ ppm: 180.6 (C-28), 140.0 (C-13), 128.1 (C-12), 79.4 (C-3), 72.8 (C-19), 66.1 (C-2), 54.7 (C-18), 48.9 (C-5), 48.3 (C-17), 47.7 (C-9), 42.9 (C-14), 42.4 (C-1), 42.3 (C-20), 40.7 (C-8), 38.8 (C-4), 38.7 (C-10), 38.5 (C-22), 33.6 (C-7), 29.4 (C-23), 29.3 (C-15), 27.2 (C-29), 27.0 (C-21), 26.5 (C-16), 24.7 (C-11), 24.1 (C-27), 22.3 (C-24), 18.7 (C-6), 17.3 (C-26), 16.8 (C-25), 16.7 (C-30); EI-MS m/z (%): 488 [M]$^{+\cdot}$ (5.3), 264 (a, 29.0), 246 (a-H$_2$O, 47.2), 231 (246-Me, 15.4), 223 (b-1, 21.5), 218 (a-HCOOH, 27.3), 201 (219-H$_2$O, 47.5), 146 (100).

2α-Hydroxyoleanolic acid (19)

White needles, IRv$_{max}$ (KBr, cm^{-1}): 3200, 2970, 1690, 1450, 1380, 1170; ^{13}C NMR (125 MHz, Pyridine-d$_5$) δ ppm: 180.1 (C-28), 143.1 (C-13), 123.0 (C-12), 83.9 (C-3), 68.7 (C-2), 56.1 (C-18), 47.8 (C-9), 46.9 (C-17), 46.2 (C-19), 42.6 (C-14), 42.3 (C-18), 39.2 (C-4), 38.7 (C-1), 37.3 (C-10), 35.2 (C-21), 34.6 (C-22), 33.4 (C-7), 32.8(C-29), 31.1 (C-20), 28.8 (C-23), 28.6 (C-15), 26.7 (C-27), 24.1 (C-11, 16), 23.9 (C-29, 30), 18.7 (C-6), 17.8 (C-26), 16.3 (C-24), 15.4 (C-25); EI-MS m/z (%) 472 [M]$^{+\cdot}$ (20.6), 248 (a, 45.6), 223 (b-1, 25.1).

Oleanolic acid (20)

White powder, ^{1}H NMR (500 MHz, Pyridine-d$_5$) δ ppm: 5.49 (1H, brs, H-12), 4.89 (brs, OH), 3.44 (1H, m, H-3), 3.31 (1H, brd, J=12.4 Hz, H-16α), 1.28 (3H, s, H-27), 1.24 (3H, s, H-29), 1.02 (6H, s, H-29, 30), 1.01 (3H, s, H-26), 0.94 (3H, s, H-25), 0.90 (3H, s, H-24); ^{13}C NMR (125 MHz, Pyridine-d$_5$) δ ppm: 180.2 (C-28), 145.0 (C-13), 122.6 (C-12), 78.2 (C-3), 55.9 (C-18), 48.2 (C-9), 46.8 (C-17), 46.7 (C-19), 42.3 (C-14), 42.1 (C-18), 39.4 (C-4), 39.0 (C-1), 37.5 (C-10), 34.4 (C-21), 34.4 (C-22), 33.3 (C-7, 29), 31.2 (C-20), 28.8 (C-23), 28.4 (C-15), 28.2 (C-2), 26.2 (C-27), 23.9, 23.8 (C-11, 16, 29, 30), 18.9 (C-6), 17.5 (C-26), 16.5 (C-24), 15.6 (C-25).

Results and Discussion

The ethanolic extracts of the dried leaves of *E. japonica* were fractionated by solvent-partitioning process into four fractions soluble in petrol, CH$_2$Cl$_2$, EtOAc and *n*-BuOH. The pharmacological study showed that the *n*-BuOH fraction had strong anti-tussive and anti-inflammatory effects. Further purification of this fraction and Ethyl acetate fraction resulted in the isolation of three new lignans, linguersinol 9′-O-β-D-xylopyranoside 1, eriobotrin 3 and isoeriobotrin 4, as well as seventeen known compounds, three lignans, linguersinol 2 (6), 2,6-dimethoxy-4-(2-propenyl)phenol 5 and 2,6-dimethoxy-4-(2-propenyl)phenol 1-O-β-D-glucopyranoside 6, eight known megastigmane derivatives, (6R,7E,9R)-9-hydroxy-4,7-megastigmadien-3-one 9-O-β-D-apiofuranosyl-(1-6)-β-D-glucopyranoside 7 (3), (6R,7E,9R)-9-hydroxy-4,7-*megastigmadien-3-one 9-O-β-D-xylopyranosyl-(1-6)-β-D-glucopyranoside 8* (7), (6R,7E,9R)-9-Hydroxy-4,7-megastigmadien-3-one 9-O-α-L-arabinopyranosy-(1-6)-β-D-glucopyranoside 9 (8), (6R, 7E, 9R)-9-hydroxy-4, 7-megastigmadien-3-one 10 (9-10), (6R,7E,9R)-9-hydroxy-4,7-megastigmadien-3-one 9-O-β-D-glucopyranoside 11 (9), (6R,7E,9S)-9-hydroxy-4,7-megastigmadien-3-one 9-O-β-D-glucopyranoside 12 (9), (6S,7E,9R)-6,9-dihydroxy-4, 7-megastigmadien-3-one 13 (11) and (6S,7E,9R)-6,9-dihydroxy-4,7-megastigmadien-3-one 9-O-β-D-glucopyranoside 14 (11-12), two known flavonol glycosides quercetin-3-O-β-D-glucoside (15) and quercetin-7-α-L-rhamnoside (16) and four known triterpenes, 2α,3α,19α-trihydroxy-12-oleanen-28-oic acid 17 (13-14), euscaphic acid 18 (13), 2α-hydroxyoleanolic acid 19 and oleanolic acid 20.

Compound 1, was isolated as white crystals. Its molecular formula was deduced as $C_{27}H_{36}O_{12}$ from SI-FAB-HRMS m/z 553.2288 [M+H]$^+$ (Calcd. $C_{27}H_{37}O_{12}$, 553.2297). Its UV (MeOH) absorption maxima λ_{max} nm: 238 (sh), 276 was typical of 4-phenyl-tetrahydronaphthalene lignan nucleus (6). The ^{1}H NMR (500 MHz, CD$_3$OD) of 1 showed two same aromatic proton signals at δ 6.61 (2H, brs, H-2, 6), two methoxyl singlets at δ 3.93 (6H, s, 3,5-OMe's) and another methoxyl singlet at δ 3.52 (3H, s, 4-OMe), indicating the presence of a 3,4,5-trimethoxyphenyl group. This is proven by EI-MS which gave a trimethoxybenzyl rearrangement ion peak at m/z 167. In addition, its ^{1}H NMR spectrum also gave several upfield proton signals at δ 4.57 (1H, d, J=6.5 Hz, H-7), 4.00 (1H, dd, J=9.8, 5.3 Hz, H-9a), 3.83 (1H, dd, J=10.9, 4.4 Hz, H-9′a), 3.74 (1H, dd, J=10.9, 6.6 Hz, H-9′b), 3.62 (1H, dd, J=9.8, 4.0 Hz, H-9b), 2.92 (1H, dd, J=15.3, 4.7 Hz, H-7′a), 2.82 (1H, brd, J=15.3 Hz, H-7′b), 2.25 (1H, m, H-8) and 1.90 (1H, m, H-8′). The above spectral data are similar to those of the known compound lingueresinol (6). Acidic hydrolysis of 1 yields D-xylose and the aglycone, lingueresinol 2. The anomaric proton signal at δ 4.41 (1H, d, J=7.5 Hz, H-1″) established the β-configuration of xylose. In its ^{13}C NMR spectrum, the characteristic glycosilation shift (+7.1 ppm) was observed for C-9′, indicating the location of the xylose moiety to be the C-9′. Thus 1 was established as linguersinol 9′-O-β-D-xylopyranoside.

Compound 3 and 4 are mixture of epimers. The UV, IR, NMR (including 2D NMR) and MS spectra suggested that 3 and 4 are benzyl-dioxyhexane lignan (6,15) 3-[3-(3,4,5-trimethoxyphenyl)-2,3-dihydro-2-hydroxymethyl-hydroxy-

(1,4-benzodioxin-6-yl)]-2,3-dihydroxypropanol. Compound **3** and **4** have the molecular formula $C_{21}H_{26}O_9$ based on their EI-MS m/z 438 $[M]^{+}$, ^1H and ^{13}C NMR spectra. Their ^1H NMR and ^{13}C NMR spectral data are very similar, except for two aromatic proton and carbon signals linked to C-7'. The proton and carbon signals at δ 6.88 (1H, d, J=1.7Hz), 6.82 (1H, d, J=1.7Hz), 104.7 and 109 2 were due to H-2', H-6', C-2'and C-6' of **3**, and 6.87 (1H, brs), 6.81 (1H, brs), 104.6 and 109.3 to H-2', H-6', C-2'and C-6' of **4**, indicating **3** and **4** are couple of 7' position epimers. The ^1H NMR spectrum of **3** and **4** gave two aromatic protons with the same chemical environment at δ 6.92 (2H, s, 2, 6-H) and three methoxyl group at δ 4.07 (3H, s, 4-Me) and 4.05 (6H, s, 3, 5-OMe's), indicating the presence of a 3,4,5-trimethoxylphenyl substitution. Its EI-MS also gave a trimethoxylbenzyl rearrangement ion peak at m/z 167. In their HMBC spectrum, the proton signal at δ 4.71 (1H, d, J=5.9Hz, H-7') corresponded to C-2' and C-6'. The *trans*-configuration of C-7 and C-8 was inferred from their coupling constant (J=8.0 Hz). The $^3J_{CH}$-2D correlation spectrum (HMBC) was used at the beginning to differentiate the structures **3**, **4** from **3a**, **4a** (*16*), but no long distance coupling could be detected between H-7 and H-8 and the ring carbons C-4' and C-5', since the coupling constants in such cases are of the order of 1.0 Hz. Fortunately the structures of **3** and **4** could be assigned on the basis of lower power hetero-coupling experiment (600 MHz, CD_3OD). Irradiation of H-7 signal resulted in considerable sharpening of C-4', while irradiation of H-8 signal resulted in considerable sharpening of C-5'. Hence, the structures were assigned as **3** and **4**, named as eriobotrin and isoeriobotrin. All the carbon and proton carbon signals were assigned .based on their ^1H-^1H COSY ^{13}C-^1H COSY, HMBC and NOESY spectra.

References

1. Ito, H.; Kobayashi, E.; Takamatsu, Y.; Li, S.H.; Hatano, T.; Sakagami, H.; Kusama, K.; Satoh, K.; Sugita, D.; Shimura, S.; Itoh, Y.; Yoshida, T. *Chem. Pharm. Bull.* **2000**, *48*, 687-93.
2. Jung, H.A.; Park, J.C.; Chung, H.Y.; Kim, J.; Choi, J.S. *Arch. Pharm. Res.* **1999**, *22*, 213-218.
3. De Tommasi, N.; Aquino, R.; Simone, F.D.; Pizza, C. *J. Nat. Prod.* **1992** *55*, 1025-1032.
4. De Tommasi, N.; De Simone, F.; Pizza, C.; Mahmood, N.; Moore, P.S.; Conti, C.; Orsi, N.; Stein, M.L. *J. Nat. Prod.* **1992**, *55*, 1067-73.
5. De Tommasi, N.; De Simone, F.; Cirino, G.; Cicala, C.; Pizza, C. *Planta Med.* **1991**, *57*, 414-416.
6. Sepulveda-Boza, S.; Delhvi, S.; Cassels, B.K. *Phytochemistry* **1990**, *29*, 2357-2358.

7. Ito, H.; Kobayashi, E; Li, S.H.; Hatano, T.; Sugita, D;, Kubo, N.; Shimura, S.; Itoh, Y.; Yoshida, T. *J. Nat. Prod.* **2001**, *64*, 737-740.
8. Matsuda, N.; Isawa, K.; Kikuchi, M. *Phytochem.* **1997**, *45*, 777-779.
9. Pabst, A.; Barron, D.; Semon E.; Shreler, P. *Phytochem.* **1992**, *31*, 1649-1652.
10. Macías, F.A.; Varela, R.M.; Torres, A.; Oliva, R.M.; Molinillo, J.M.G. *Phytochem.* **1998**, *48*, 631-636.
11. Schwab,W.; Schreier, P. *J. Nat. Prod.* **1990**, *29*, 161-164.
12. Herderich, M.; Feser W.; Shreier, P. *Phytochem.* **1992**, *31*, 895-897.
13. Liang, G.Y.; Gray, A.I.; Waterman, P.G. *J. Nat. Prod.* **1989**, *52*, 162-166
14. Lee, T.H.; Lee, S.S.; Kuo, Y.C.; Chou, C.H. *J. Nat. Prod.* **2001**, *64*, 865-869
15. Kumar, S.; Ray, A.B.; Konno, C.; Oshima, Y; Hikino, H. *Phytochem.* **1988**, *27*, 636-638.
16. Shamsuddin, T.; Rahman, W.; Khan, S.A.; Shamsuddin, K.M.; Kintzinger, J. *Phytochem.* **1988**, *27*, 1908-1909.

Chapter 23

Changes in Some Components of Tea Fungus Fermented Black Tea

Hui-Yin Fu and Den-En Shieh

Department of Food Science and Technology, Tajen Institute of Technology, 907, Pingtung, Taiwan, Republic of China

Starter cultures of *Acetobacter xylium* or *Acetobacter aceti* were added to a mixture of green tea, ethanol (3%) and glucose (3%), and the resulting mixture allowed to ferment for 25 days. Significant amounts of acetic acid, gluconic acid and glucuronic acid were formed during this 25 day fermentation period, as well as increase in DPPH free radical scavenging capability. Reducing power throughout the fermentation period resulted in the accumulation of reduced type ascorbic acid.

Introduction

The tea fungus first appeared in 220 B.C. in Manchuria and spread over Russia to Central Europe during World War II. It is now accepted in the United States (*1*). According to the different types of consumption, tea fungus can be divided into two categories. When the main interest is the fermented broth, tea fungus is called "Kocha Kinoko' in Japan (*2*), 'Hongo' in Germany (*3*) and commonly designated as 'Kombucha' (*1*).

Kombucha tea is a fermented beverage produced by a symbiosis of acetic acid bacteria and yeasts. Although *Acetobacter xylinum* dominated in Kombucha beverage, other researchers isolated and characterized pure cultures of *A. intermedius* sp. from the same source (*4*). For yeasts, the genera *Brettanomyces, Zygosaccharomyces, Saccharomyces, Pichia, Schizosaccharomyces, Saccharo-*

mycodes and *Candida* were identified (*5,6*). Kombucha beverage is mainly cultivated in sugared black tea by inoculating a previously grown culture and incubating statically under aerobic conditions for 7-10 days (*7*). Sucrose was hydrolyzed into glucose and fructose, and part of the glucose was fermented by yeast to produce ethanol. *Acetobacter xylinum* synthesizes a floating cellulose network accompanyed by organic acids. A cellulose network floating at the surface of various fruit juices (namely coconut and pineapple), fermented by a symbiotic culture composed of *Acetobacter xylinum and yeasts*, and was named "nata" in Philippines (*8*).

A resent report (*9*) indicated that kombucha processes potent antioxidant and immunopotentiating activities. It helps in excretion of chromium from body tissues. The strong antioxidant activity decreases peroxidation, enhances antibody titers and delays hypersensitivity response in control (chromium treated) rat. However, there has been a lot of attention regarding the possible toxicity of kombucha tea. The presence of *Bacillus anthrax* in kombucha tea fermented in unhygienic conditions has been reported (*10*). Gastrointestinal toxicity of kombucha has also occurred in four patients (*11*). Recent FDA studies found no evidence of contamination in kombucha products fermented under sterile conditions. FDA and State of California inspections of the facilities of a major Kombucha tea supplier also found that its product was being manufactured under sanitary conditions.

The Kombucha culture produces gluconic and glucuronic acids, carbonic acid and acetic acid (anti-streptococci, anti-diplococci, etc), plus a range of B vitamins (*1-3,7,13*) folic acid, usinic acid (antibacterial and antiviral) and many enzymes.

In this study, to develop a pleasantly sour and sparkling beverage, without the risk of contamination, a pure culture of *Acetobacter xylinum* (CCRC 10589) (Figure 1), instead of the traditional fermented broth, was inoculated into fully fermented tea infusions. Tea leaves were purchased from a local tea farm (Taitung, Taiwan) and processed into Black Tea following the standard procedure [Black Tea (Fully fermented Tea): Fresh leaves → Withering → Rolling → Fermentation → Drying → Tea]. Ten grams of black tea was soaked in 1 liter of water for 1 hour and filtered through a cheese cloth. 30 grams of glucose and 30 g of ethanol were added to the tea infusion (1 liter) and autoclaved at 120 °C for 20 min. The *A. aceti* (CCRC10382) and *A. xylinum* was purchased from the Food Industry Research Development Institute. The fermentation was initiated by adding 10 % of pure starter culture and the incubation was carried out at 30 $\pm$ 1 °C for 25 days. The fermented medium (4, 9, 13 17 and 25 days) was centrifuged at 7000 rpm for 30 min and stored in glass vials at – 20 °C until further use.

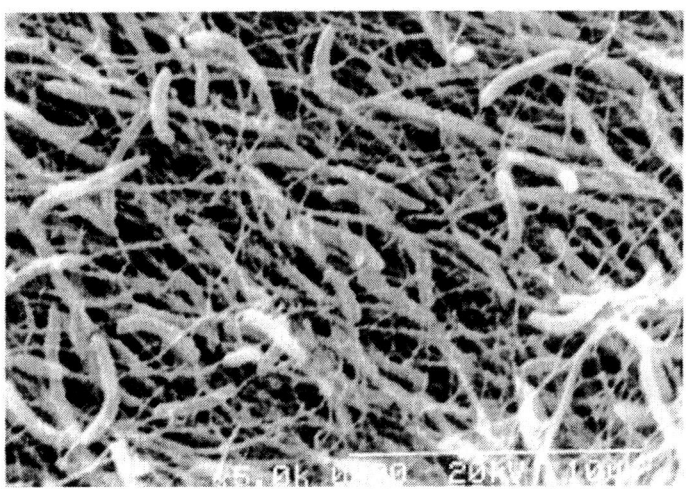

Figure 1. Scanning electron micrographs of A. xylinum and cellulose fibril.

Cell Mass Development

A. xylinum developed faster in black tea infusion than did *A. aceti*. Cell mass synthesized by *A. aceti* and *A. xylinum* increased with incubation time, up to 5 mg % and 22 mg %, respectively, after 25 days as shown in Figure 2. As shown by scanning electron microscope, Figure 2 shows that some rod-shape *A. xylinum* cells adhered on the surface of the bacterial cellulose fibrils. Caffeine and theophylline were identified as potent stimulators for bacterial cellulose synthesis in *A. xylinum*. Methylxanthine probably blocks the action of the specific diguanyl cyclic phosphodiesterase and then avoids or postpones the normal "switch off" of active cellulose synthase (*12*). The level of caffeine in the black infusion used in this experiment was determined to be in the range of 2.4 to 3.4 %. This may account for the accumulation of more cell mass in the medium inoculated with *A. xylinum* rather than in the one with *A. aceti*.

Organic Acid Production

The pH value of the fermented black tea infusion decreased during fermentation from 6.75 to 3.11 and 2.43 for *A. aceti* and *A. xylinum*, respectively, as shown in Table I. After 25 days incubation, in addition to the acetic acid (18.2 g/L) produced in the infusion inoculated with *A. aceti* and *A. xylinum* in the fully fermented tea sample, a significant amount of gluconic acid

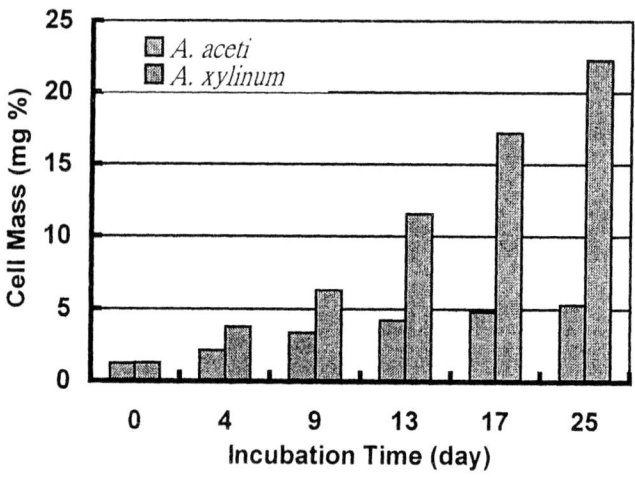

Figure 2. Cell mass synthesized by Acetobacter xylinum and Acetobacter aceti.

(8.4 g/L) and glucuronic acid (0.4 g/L) was synthesized. Our results were in good accordance with that of Sievers, et al. (5). Ethanol concentration rose to a maximum and subsequently declined. The maximal concentration (1.34 g/L) was obtained after 5 days of incubation in the flask with 100 g/L initial concentration of sucrose. Acetic acid converted from ethanol rose to a maximum (4.5-5.6 g/L) until the 15th day of incubation and subsequently declined (1). Acetic acid (22.3 g/L) dominated in the infusion inoculated with *A. aceti,* while only minute amounts of gluconic acid (1.1 g/L) existed after 25 days incubation.

Table 1. Concentration of Organic Acids in Medium Inoculated with *A. xylinum* and *A. aceti* After 25 days.

	A. xylinum	A. aceti
pH	2.43	3.11
Acetic acid	18.2 g/L	22.3 g/L
Gluconic acid	8.4 g/L	1.1 g/L
Glucuronic acid	0.4 g/L	trace
Ketogluconic acid	trace	trace

The major metabolites of tea fungus were subjects of intensive study. A symbiotic culture of *A. xylinum* and two yeasts, *Zygosaccaromyces rouxii* and *Candida sp.*, in sugared tea was analyzed (*1*). Several metabolites, ethanol, lactic acid, gluconic and glucuronic acids were identified and quantified. Similar major components have been reported (*6,7*). The pH value of the kombucha beverage decreased during fermentation from 3.75 to 2.42 in a mixed culture of *A. xylinum* and *Zygosaccaromyces* sp. Siever et al. (*6*) ascribed the decrease in pH value to the transformation of sucrose into glucose, fructose, ethanol, acetic acid and gluconic acid during a 60 day tea fungus fermentation. Our results confirmed that *A. xylinum* is responsible for the synthesis of organic acids. Glucuronic acid is able to combine with over two hundred known xenobiotics or endobiotics, which leads to the ultimate excretion of the substances into the urine or the bile (*13*). Glucuronic acid is known to be manufactured in the liver, but those who suffer from long-term illness do not produce it in sufficiently large quantities to assist the body in rapid cleaning that is often a vital part of the recovery process. The therapeutic effectiveness of the Kombucha-beverage for gout, rheum and arthritis, etc. is attributed to the elimination of toxins through a chemical combination of products that are easily excreted and eliminated from the organism. *A. xylinum* produces an exopolysaccharide, acetan, which has the following structure for its chemical repeat unit: a side-chain of α,L-rha-(1,6)-β ,D-glc-(1,6)- α,D-glc-(1,4)- β-D-glcA-(1,2)- α,D-man-(1,3) linked to a cellobiose unit in the backbone (*14*). UDP-Glucuronic acid is a precursor for sugar nucleotides, which are needed for the biosynthesis of many components of bacterial polysaccharides. Various biochemical studies have suggested that the production of UDP-Glucuronic acid may be the rate-limiting step in providing precursors for the expanding cell wall (*15*). Glucuronic acid, being an intermediate metabolite in *A. xylinum* cells, should under normal circumstances, not be present outside of bacterial cells. However, large quantities of intermediate compounds found in solutions actually represent an over-production caused by the availability of excessive quantities of carbohydrates and frequently by insufficient amounts of trace elements (*16*). Another possibility for the detection of glucuronic acid in the *A. xylinum* inoculated black tea infusion is that autolysis may occurrs after prolonged incubation leading to the leaching of glucuronic acid.

DPPH Free Radical Scavenging Activity

The 2,2-diphenyl-1-picrylhydrazyl radical (DPPH) method was used to investigate the scavenging activity of tea samples. The data for DPPH free radical scavenging activity, measured as scavenging percentage at room temperature after addition of various fermented tea infusions are presented in Figure 3. DPPH free radical scavenging activity of tea infusion inoculated with *A. aceti* and *A. xylinum* increased with incubation time up to 22.1% and 21.9%, respectively after 25 days. DPPH is a reproducible and practical way to measure antioxidative activity. It has been used to evaluate the potential of tea catechins and their epimerized, acylated and glucosylated derivatives as antioxidants (*17*). Tea catechins and their epimers were shown to have 50% radical scavenging ability in the concentration range of 1 to 3 µM. No significant differences were observed between the scavenging activities of tea catechins and their epimers. It is suggested that the galloyl moiety attached to flavan-3-ol at 3 position has a strong scavenging ability on the DHHP radical as well as the ortho-trihydroxyl group in the B ring. The antioxidative activity of phenolic acid increases with increased number of hydroxyl (OH) groups (*18*). Total amount of phenolic compounds in fermented black tea infusion was measured with a Folin-Ciocalteu reagent (*19*) and gallic acid was used as a standard substance for calibration curve. The results showed no significant increase in the total phenolics level in both mediums inoculated with *A. aceti* (from 83 µg/mL to 85 µg/mL) and *A. xylinum* (from 83 µg/mL to 89 µg/mL) for 25 days (data not shown). Based on the theory that the immunosuppressive effect of chromium is attributed to the production of oxygen free radicals, Sai Ram et al. (*9*) ascribed the capability of Kombucha tea to relieve the immunosuppressive activity of chromium to its antioxidant activity.

Reducing Power

The total reducing potential of the fermented tea samples was determined using the method developed by Langley-Evans (*20*). FRAP reagent was prepared from 300 mmol/L acetate buffer, pH 3.6, 20 mmol/L ferric chloride and 10 mmol/L 2,4,6-tripyridyl-s-triazine made up in 40 mmol/L hydrochloric acid. All three solutions were mixed together in the ratio 25:2.5:2.5 (v:v:v). The FRAP assay was performed using reagent preheated to 37 °C. To 0.5 mL of sample was added 7 mL of reagent and reactions incubated at 37 °C for 4 min. Absorption at 593 nm was determined relative to a reagent blank and is shown in Figure 4. The relative reducing power of the infusion with *A. xylinum* increased with increasing incubation time from 0.59 to 0.91 after 25 days incubation.

Asai (*21*) proposed that the primary Kombucha bacterium, *Acetobacter*, initially oxidizes ethanol to acetaldehyde and then to acetic acid. Ethanol is oxidized to acetaldehyde by alcohol-cytochrome-553 reductase and the resulting electrons are successively delivered to the heme iron of cytochrome-553. The acetaldehyde thus formed is oxidized further by coenzyme-independent aldehyde dehydrogenase or by NADP-dependent aldehyde dehydrogenase. Observations on alcohol dehydrogenase have been made by Ameyama and Adachi (*22*) who

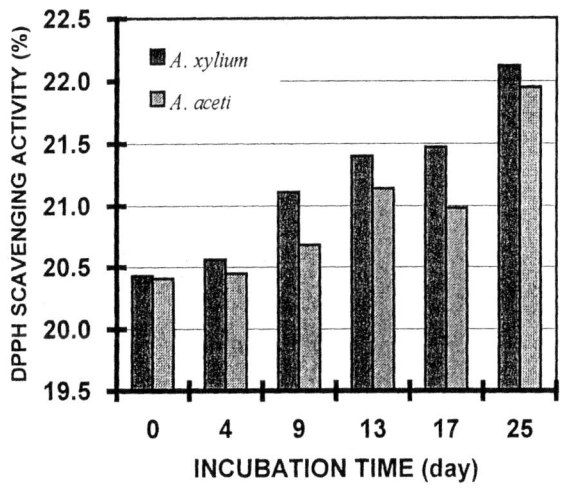

Figure 3. DPPH radical scavenging activity of fully fermented tea during incubation.

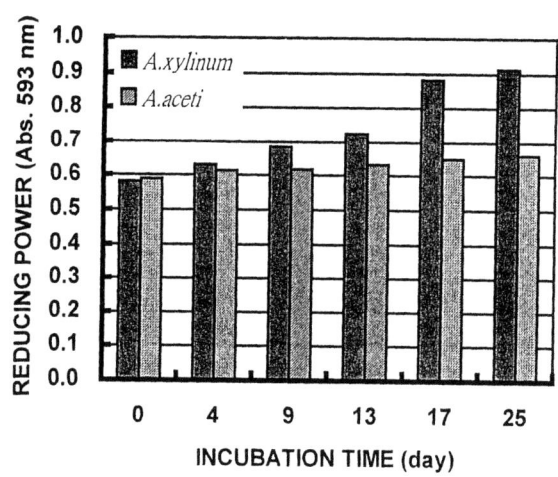

Figure 4. Reducing power of fully fermented tea during incubation.

reported that the enzyme of *A. suboxydans* was NAD-linked. The acetaldehyde dehydrogenase required NADP as a cofactor. Ameyama and Adachi also (23) reported that both *A. suboxydans* and *A. acetic* possess an NADP-linked acetaldehyde dehydrogenase, and that *A. acetic* possesses an NAD-linked enzyme in addition. A large amount of organic acids was found in the fermented broth in the present study.

Reduced Type Ascorbic Acid

The reduced type ascorbic acid in tea infusions of fully fermented tea was characterized by using a HPLC following the method of Rizzolo et al (24) and found that the concentration increased slightly with increased incubation time. The concentration of the reduced type ascorbic acid in black tea infusion with *A. xylinum* and *A. acetic* after 25 days incubation was 64.48 and 63.35 mg/mL respectively, as shown in Figure 5. During tea processing, part of the ascorbic acid was found to be oxidized due to exposure to atmospheric oxygen and heat treatment. The NADH or NADPH generated from the redox reaction during acetic acid, gluconic acid synthesis may account for the conversion of oxidized ascorbic acid to the reduced form in the fermentation condition. The increase in the reducing power and ascorbic acid in *A. xylinum* fermented tea infusion could be attributed to the generation of NADH or NADPH during acetic acid and

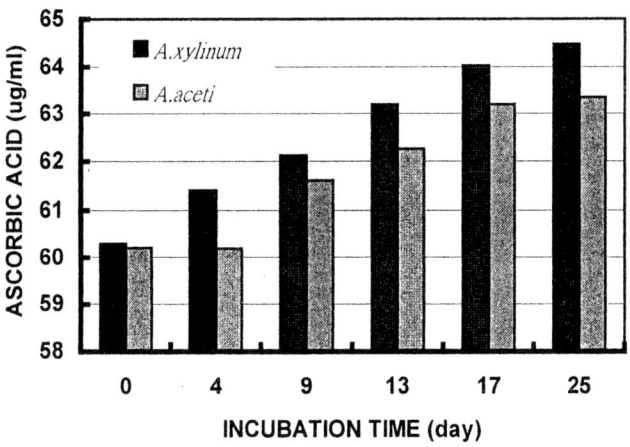

Figure 5. Ascorbic acid content of fully fermented tea during incubation.

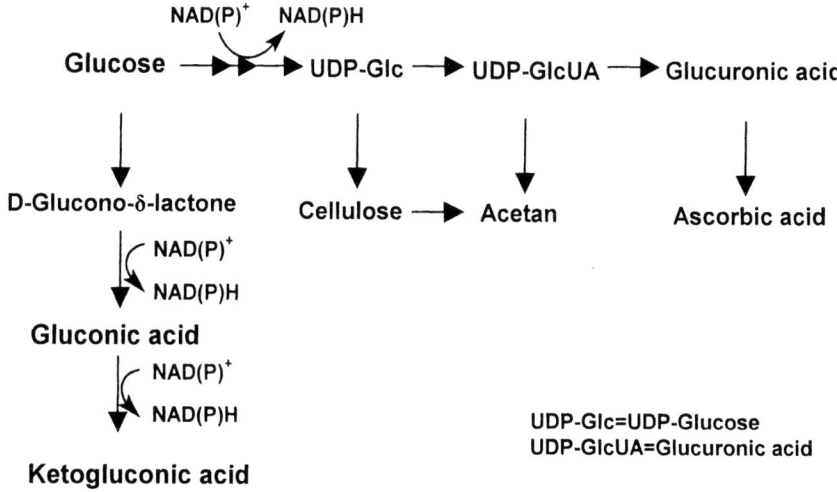

Figure 6. Scheme of the metabolic activities and Glucuronic Acid Pathway of Acetobacter xylinum

gluconic acid fermentation and acetan synthesis as shown in Figure 6. In addition to its reducing capability, ascorbic acid may also be synthesized; following a "side arm" of the glucuronic acid pathway branching off from glucuronic acid, followed by several additional steps, and leading to the formation of ascorbic acid (25).

References

1. Blanc, P.J. *Biotech. Letters.* **1996**, *18*, 139-142.
2. Terada, S.; Nishimura, A. *Ichimura Gakuen Tanki Daigaku Shizen Kagaku Kenkyukai Kaishi.* **1976**, *10*, 15-18.
3. Benk, E. *Verbraucherdienst.* **1988**, *33*, 213-214.
4. Boesch,C.; Traek, J.; Sievers, M.; Teuber, M. *System. Appl. Microbiol.* **1998**, *21*, 220-229.
5. Mayser, P.; Fromme, S.; Leitzmann, C.; Grunder, K. *Mycoses.* **1995**, *38*, 289-295.
6. Sievers, M.; Lanini, C.; Weber, A.; Schuler-Schmid, U. *Appl. Microbiol.* **1995**, *18*, 590-594.
7. Chen, C.; Liu, B.Y. *J. Appl. Microbiol.* **2000**, *89*, 834-839.
8. Yoshinaga, F.; Tonouchi, N.; Watanabe, K. *Biosci. Biotech. Biochem.* **1997**, 61, 219-224.

9. Sai Ram, M.; Anju, B., Pauline, T.; Prasad, D.; Kain, A.K.; Mongia, S.S.; Sharma, S.K.; Singh, B.; Singh, R.; Ilavazhagan, G.; Kumar, D. *J. Ethnophamacology* **2000**, *71*, 235-240.
10. Sadjiadi, J. J. Amer. Medic. Assoc. **1980,** *280*, 1567-1568.
11. Srinivasan, R.; Smolinske, S.; Greenbaum, D. *J. General Internal Medicine* **1997**, *12*, 643-644.
12. Fontana, J.D.; Franco, V.C.; De Souza, S.J.; Lyra, I.N.; De Souza, A.M. *Appl. Biochem. Biotech.* **1991**, *28/29*, 341-351.
13. Tephly, T.R.; Burchell, B. *TiPS* **1990**, *11*, 276-279.
14. Griffin, A.M.; Edwards, K.J.; Morris,V.J.; Gasson, M.J. *Biotech. Lett.* **1997**, *19*, 469-474.
15. Tenhaken, R.; Thuke, O. *Plant Physiol.* **1996**, *112*, 1127-1134.
16. Schlegel, H.G. *Allgemeine Mikrobiologie*, Thieme Verlag, Stuttgart. 1985 pp. 124
17. Nanjo, F.; Goto, K.; Seto, R.; Suzuki, M.; Sakai, M.; Hara, Y. *Free Radic. Biol. Med.* **1996**, *21*, 895-902.
18. Dziedric, S.Z.; Hudson, B.J.F. *Food Chem.* **1984**, *14*, 45-51.
19. Folin-Ciocalteu Index. *Off. J. Eur. Communities* **1992**, 178-179.
20. Langley-Evans S.C. *Inter. J. Food Sci. Nutrit.* **2000**, 51, 181-188.
21. Asai, T. *Aceti Acid Bacteria: Classification and Biochemical Activities.* University of Tokyo Press: Tokyo, **1968**, pp. 316.
22. Ameyama, M.; Adachi, O. *Methods Ezymol.* **1982**, *89*, 450-457.
23. Ameyama, M.; Adachi, O. *Methods Ezymol.* **1982**, *89*, 491-497.
24. Rizzolo, A.; Brambilla, A.; Valsecchi, S.; Eccher-Zerbini, P. *Food Chem.* **2002**, *77*, 257-262.
25. Kirk, R.E.; Othmer, D.E. *Encyclopedia of Chemical Technology.* John Wiley: New York. 1978.

Chapter 24
Chemistry and Bioactivity of *Allium tubersom* Seeds

Shengmin Sang[1,2], Aina Lao[1], and Zhongliang Chen[1]

[1]Institute of Materia Medica, Shanghai Institutes for Biological Sciences, Chinese Academy of Sciences, Shanghai 200031, People's Republic of China
[2]Department of Food Science, Rutgers, The State University of New Jersey, 65 Dudley Road, New Brunswick, NJ 08901-8520

The scientific name of Chinese chives is *Allium tuberosum* Rottl. (Liliaceae). It is known as "Jiucai" in China and "Nira" in Japan. It is a perennial plant and both the leaves and the inflorescences are edible. It has also been used as an herbal medicine for many diseases. According to the dictionary of Chinese medicines, the leaves have been used for the treatment of abdominal pain, diarrhea, hematemesis, snakebite and asthma; while the seeds are used as a tonic and aphrodisiac. In the present study, 39 compounds were isolated and identified from the ethanol extract of the seeds of *Allium tuberosum*. Among them, 23 are new compounds and include spirostanol saponins, furostanol saponins, cholesterol saponins and alkaloids. Their structures were identified by a combination of ESIMS, 1D, and 2D-NMR (COSY, TOCSY, ROSEY, HMQC, and HMBC). The antitumor activities of some of these compounds will be discussed.

The scientific name of Chinese chives is *Allium tuberosum* Rottl. (Liliaceae). It is known as "Jiucai" in China and "Nira" in Japan. It is believed to have originated in China. It grows naturally in central and northern parts of Asia, and is cultured in China, Japan, Korea, India, Nepal, Thailand and the Philippines (*1*). It is a perennial plant and both the leaves and the inflorescences are eaten. It has also been used as an herbal medicine for many diseases. According to the dictionary of Chinese medicines (*2*), the leaves have been used for the treatment of abdominal pain, diarrhea, hematemesis, snakebite and asthma; while the seeds are used as a tonic and aphrodisiac. In earlier studies, various volatile and nonvolatile sulfur-containing compounds (*3-5*), N-*p*-coumaroyl tyramine and bis-(*p*-hydroxyphenyl) ether (*6*), purine nucleoside, adenosine, and major free amino acids, alanine, glutamic acid, aspartic acid and valine (*7*), 3-*O*-rhamnogalactosyl-7-*O*-rhamnosylkaempferol (*8*), and acylated flavonol glucosides (*9*) have been isolated from the leaves of *A. tuberosum*. However, there are no reports on phytochemicals in the seeds of this plant, it was thought to be desirable to carry out systematic chemical investigations on the seeds of this plant. This paper mainly discusses the isolation, classification and biological activities of the components isolated from this species (*10-17*).

Materials and Methods

General Procedures

Optical rotations were obtained on a JASCO DIP-181 polarimeter (Jasco Inc.., Norwalk). IR spectra were recorded on a Perkin-Elmer model 599 Infrared spectrometer (PerkinElmer Co., Shelton, CT). ^{1}H (400 Hz), ^{13}C (100 Hz) and all 2D NMR spectra were run on a Bruker AM-400 NMR spectrometer (Bruker Co., Fallanden), with TMS as internal standard. FABMS were recorded on a MAT-95 mass spectrometer (Finnegan Co., Bremen). GLC analysis: Shimadzu GC-9A (Shimadzu Co., Kyoto), glass column (300 × 0.32 cm) packed with OV 225, carrier gas, N_2, flow rate, 30 mL/min. Silica gel 60H and HSGF$_{254}$ (Qingdao Haiyang Chemical Group Co., Qingdao, People's Republic of China) were used for column chromatography and TLC respectively.

Plant Material

The seeds of *Allium tuberosum* were purchased from Shanghai Traditional Chinese Medicine Inc. in September 1997, and were identified by Professor Xuesheng Bao (Shanghai Institute of Drug Control). A voucher specimen (No. 334) has been deposited at the Herbarium of the Department of Phytochemistry, Institute of Materia Medica, Shanghai Institutes for Biological Sciences, Chinese Academy of Sciences.

Extraction and Isolation Procedures

The powdered seeds of *A. tuberosum* (50 Kg) were extracted successively with petroleum ether × 2 and 95% EtOH × 3. After evaporation of ethanol *in vacuo*, the residue was suspended in water and then extracted successively with petroleum ether, EtOAc and *n*-BuOH. Using silica gel column, sephadex LH-20 column and prepared TLC plate, we obtained 16 pure compounds from the ethyl acetate fraction. Two of them are new compounds. The water fraction was subjected to Diaion HP-20 column eluted with water first and then methanol. The methanol fraction was subjected to sephadex LH-20 column eluted with 95% ethanol to give two compounds (Figure 1). The *n*-BuOH fraction was subjected to Diaion HP-20 using a EtOH-H$_2$O gradient system (0%-100%). Four fractions were obtained: water fraction, 30 percent ethanol fraction, 70 percent ethanol fraction and ethanol fraction. By repeated column chromatographic purification over silica gel and RP C-18 gel, 23 pure compounds were obtained from the combination of 30% fraction and 70% fraction (Figure 1). All of them are steroid saponins. Twenty-one of them are new compounds.

Phytochemistry

The powdered seeds of *V. segetalis* (50 Kg) were extracted successively with petroleum ether × 2 and 95% EtOH × 3. After evaporation of ethanol *in vacuo*, the residue was suspended in water and then extracted successively with petroleum ether, EtOAc and n-BuOH. Thirty nine compounds were isolated and identified from these fractions. According to their structures they can be classified as: spirostanol glycosides, furostanol glycosides, cholesterol glycosides, alkaloids, lignins, fatty acid, stigmasterol glucoside, and phenolic comounds (Figures 2-12). Twenty-three of them are steroidal saponins. Steroidal saponins are naturally occurring glycosides that possess properties such as froth formation, hemolytic activity, toxicity to fish and complex formation with cholestern (*18*). During recent years, steroidal glycosides have attracted a growing interest owing to the wide range of their biological action on living organisms, including antidiabetic (*19*), antitumor (*20*), antitussive (*21*) and platelet aggregation inhibition (*22*). It is well known that the *Allium* genus, with about 500 species, has a wide distribution in the northern hemisphere and is a rich source of steroidal saponins (*23*). The steroidal saponins that we isolated from the seeds of this plant contained three different aglycons: spirostanol, furaostanol and cholesterol, and only two different sugars: glucose and rhamnose. These compounds contained 1-4 sugar units. According to the configuration at postion 5 and 25, these spirostanol and furostanol saponins can be divided as 25S, 5β-type spirostanol saponins (compounds 1-5), 25S, 5α-type

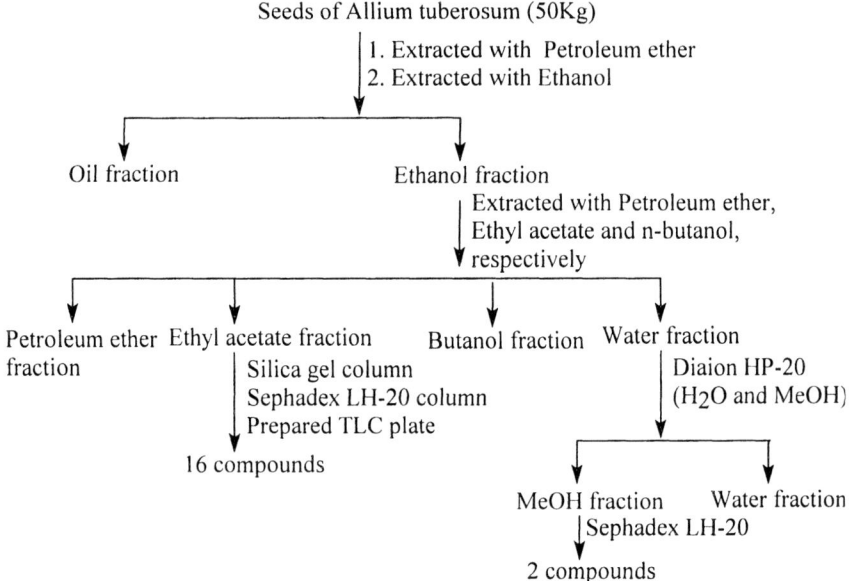
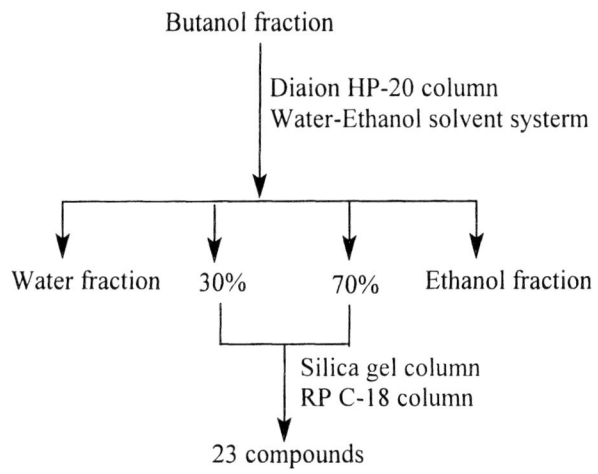

Figure 1. Extraction and isolation procedure.

spirostanol saponin (compounds **6-8**), 25R, 5β-type furostanol saponin (compounds **9-11**) and 25R, 5β-type furostanol saponins (compounds **12, 13**), 25S, 5α-type spirostanol saponins (compounds **14-21**). Among them, compounds **18-21** are all unusual 22(23)-ene-furostanol saponins. In addition, it is novel in spirostanols and furostanols that the C-20 of compounds **18** and **19** possess all R-configurations. Compounds **22** and **23** are cholesterol saponins. Compounds **24-31** are all alkaloids. Compounds **32** and **33** are lignins. Compound **34** is an oxygenated unsaturated fatty acid, which was isolated from the rice plant and shown to be active against fungus (*24*). Compound **35** is a stigmasterol glucoside, while compounds **36-39** are all phenolic compounds. Among these compounds, compounds **1-7, 9-22, 24** and **27** are new compounds. The structures of the isolated compounds were elucidated by spectral methods which include FABMS, ^{1}H-NMR, ^{13}C-NMR, COSY, TOCSY, ROESY, HMQC, and HMBC.

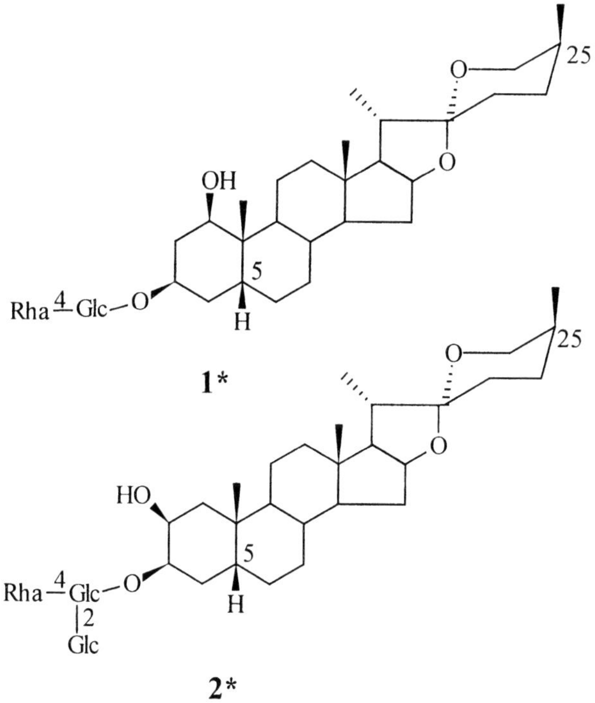

Figure 2. Structures of spirostanol glycosides (Compounds 1 and 2).

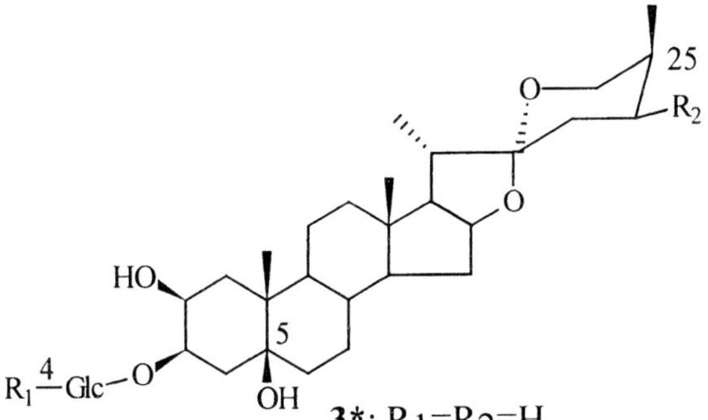

3*: R₁=R₂=H
4*: R₁=Rha, R₂=H
5*: R₁=Rha, R₂=OH

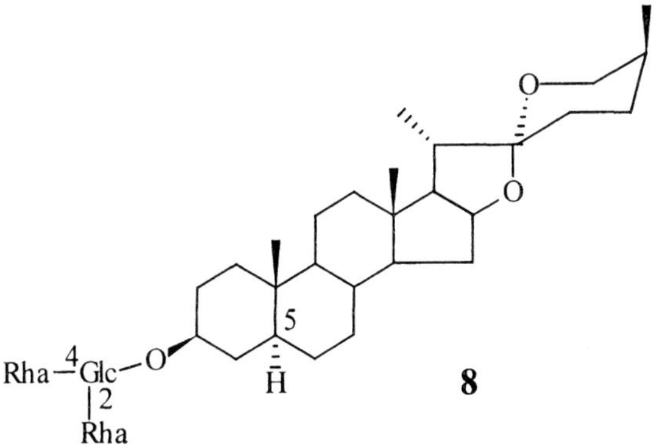

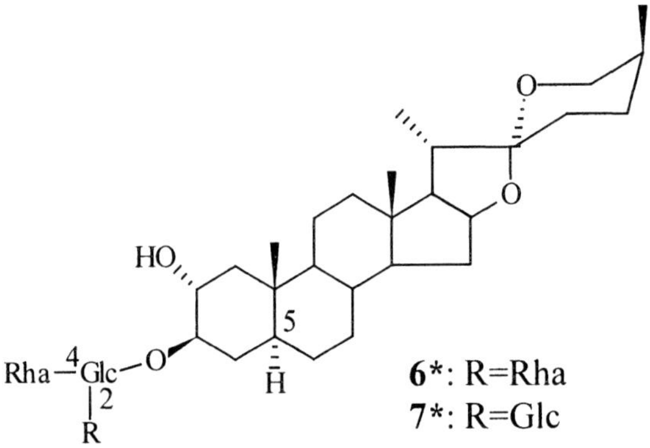

6*: R=Rha
7*: R=Glc

Figure 3. Structures of spirostanol glycosides (Compounds 3-7)

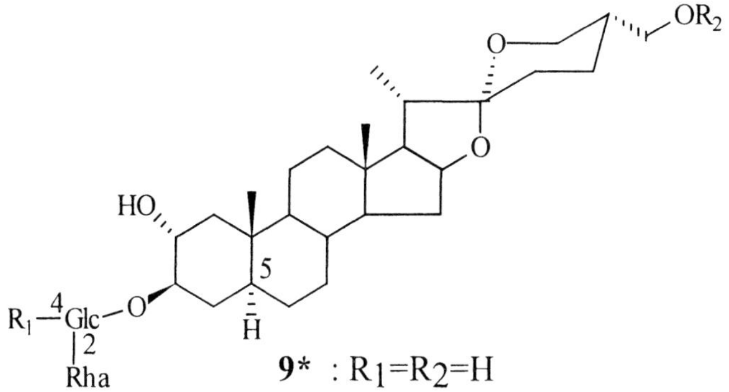

323

9* : R₁=R₂=H
10*: R₁=Rha, R₂=H
11*: R₁=Rha, R₂=Glc

*Figure 4. Structures of spirostanol glycosides (Compounds **9-11**).*

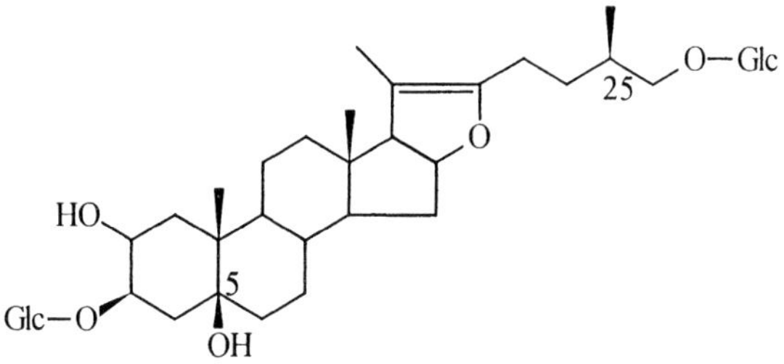

12*

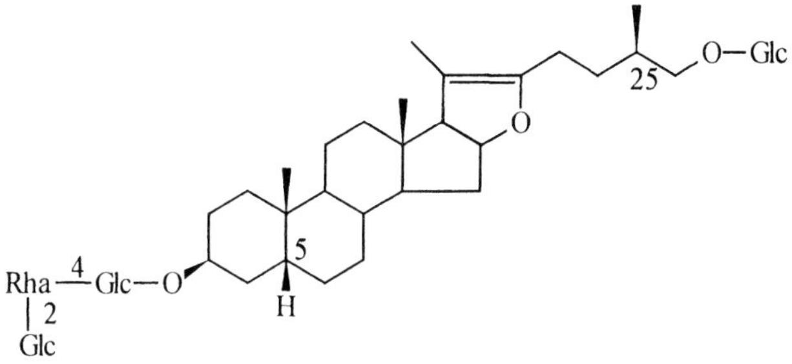

13*

*Figure 5. Structures of furostanol glycosides (Compounds **12** and **13**).*

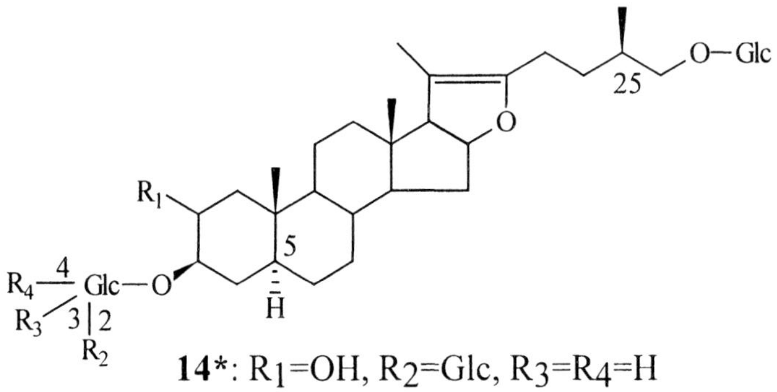

14*: R$_1$=OH, R$_2$=Glc, R$_3$=R$_4$=H
15*: R$_1$=OH, R$_2$=R$_4$=Rha, R$_3$=H
16*: R$_1$=OH, R$_2$=Rha, R$_3$=Glc, R$_4$=H
17*: R$_1$=R$_3$=H, R$_2$=R$_4$=Rha

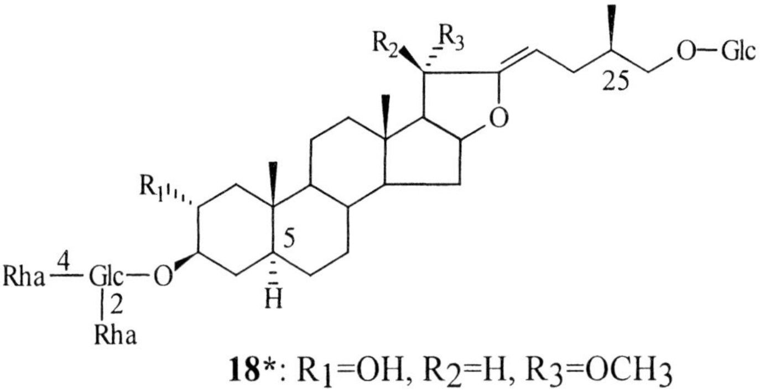

18*: R$_1$=OH, R$_2$=H, R$_3$=OCH$_3$
19*: R$_1$=OH, R$_2$=H, R$_3$=OH
20*: R$_1$=OH, R$_2$=OH, R$_3$=H
21*: R$_1$=H, R$_2$=OH, R$_3$=H

Figure 6. Structures of furostanol glycosides (Compounds 14-21)

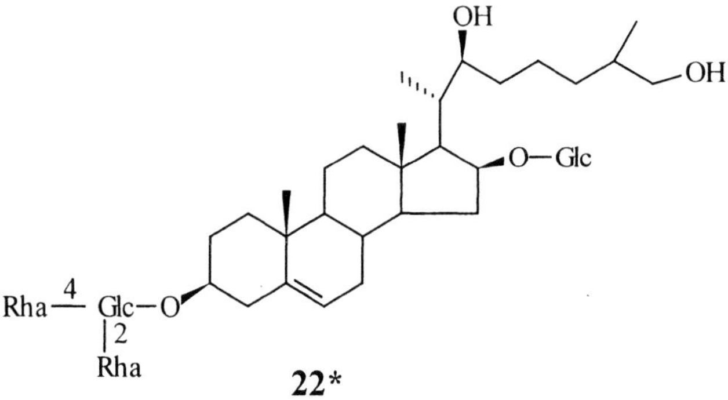

22*

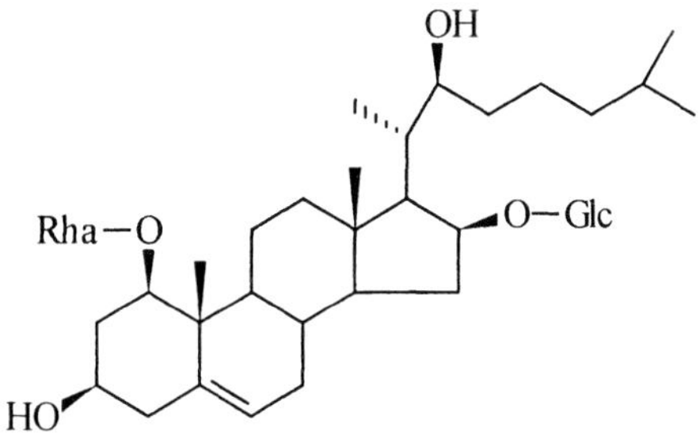

23

*Figure 7. Structures of cholesterol glycoside Compounds (**22** and **23**).*

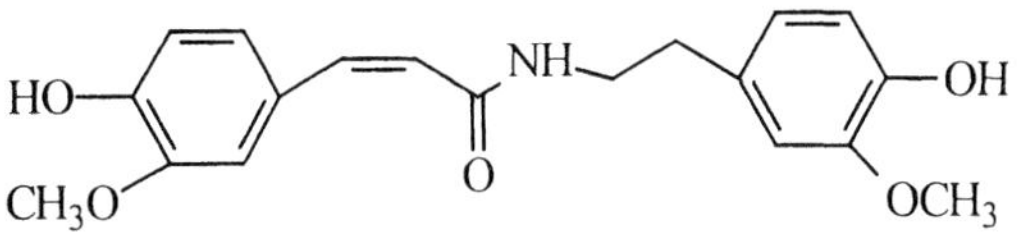

24*

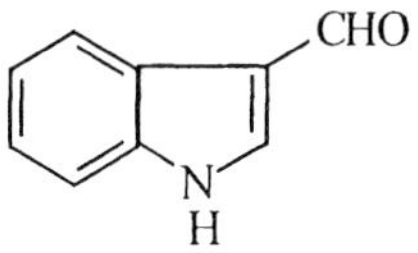

25

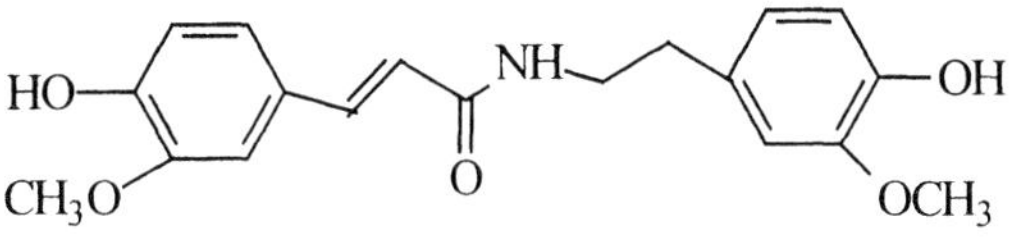

26

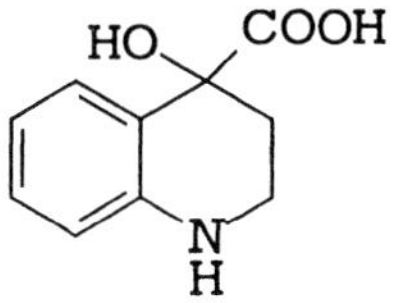

27*

*Figure 8. Structures of alkanoids (Compounds **24-27**).*

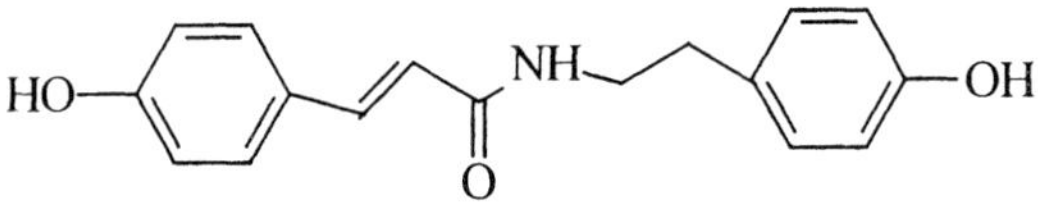

Figure 9. Structures of alkaloids (Compounds 28-31).

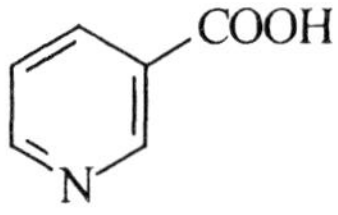

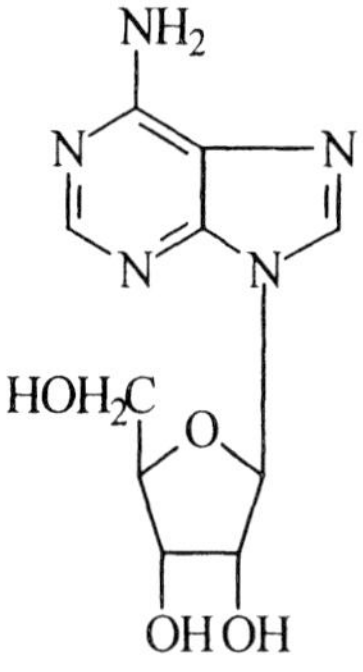

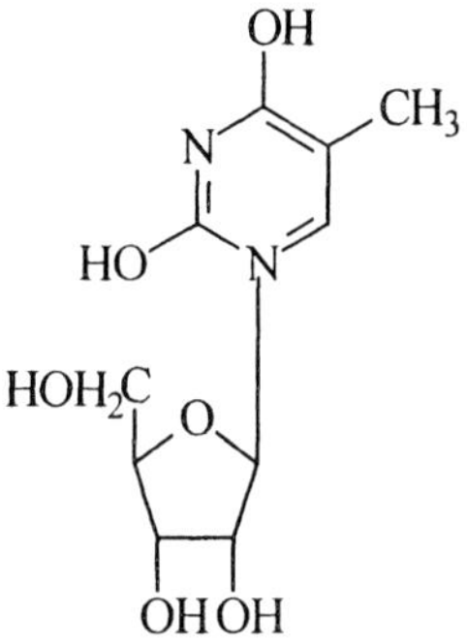

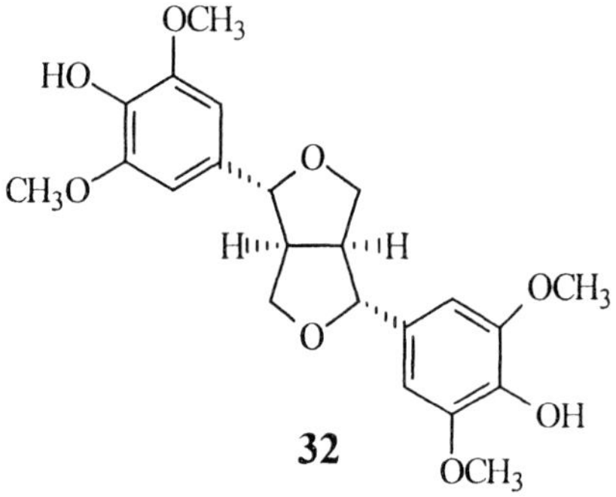

Figure 10. Structures of lignins (32 and 33).

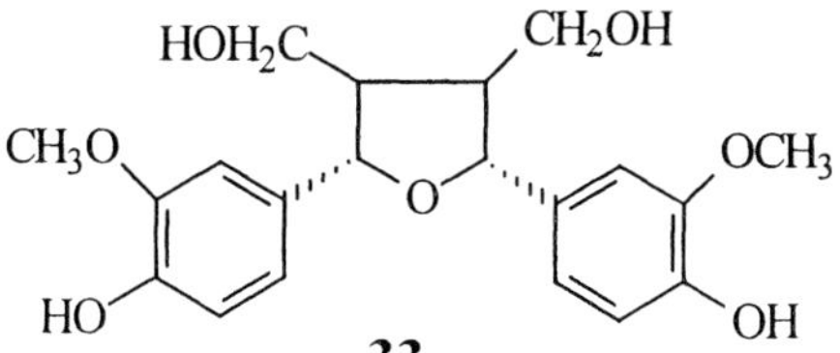

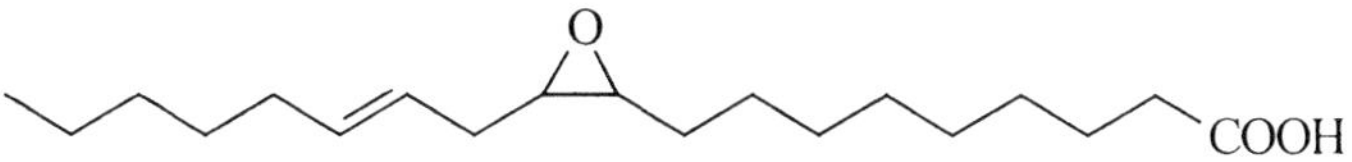

34: vernolic acid

Figure 11. Structure of oxygenated unsaturated fatty acid (Compound 34).

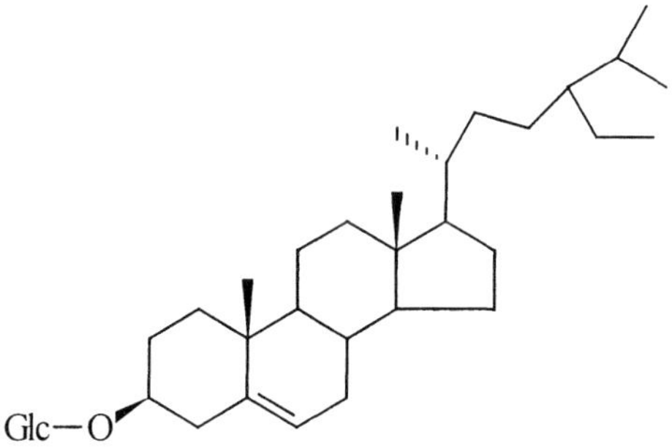

35

Figure 12. Structure of stigmasterol glucoside (Compound 35)..

Conclusion

In this research, we systematically studied the chemical components of the seeds of *Allium tuberosum*. A total of 39 compounds were isolated. Twenty-three of them are new compounds (marked in *). The structures of the isolated compounds were eludidated by spectral methods which include APCI-MS, ^{1}H-NMR, ^{13}C-NMR, ^{1}H-^{1}H COSY, TOCSY, ROESY, HMQC, and HMBC. In addition, 12 of them have been tested for the inhibition of the HL-60 cell line. One new spirostanol saponin (**1**) showed very strong inhibition with an IC$_{50}$ values of 6.8 μg/mL. Further investigations on the prevention of disease and improvement of human health by the seeds of *Allium tuberosum* and/or its bioactive components are required.

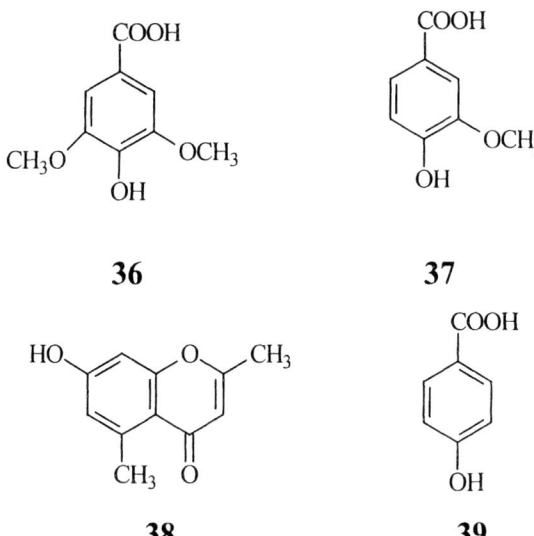

Figure 13. Structures of phenolic compounds (Compounds 36–39).

References

1. Brewster, J.L.; Rabinowitch, H.D. *Onions and Allied crops; Chinese chives Allium tuberosum rottl.*; Saito, S.; CRC Press, Inc.: Boca Raton, Florida, **1990**; 220.
2. Jiangsu New Medical College. *Zhong-yao-da-ci-dian*; Shanghai Science and Technology Publisher. Shanghai, **1986**: 1646.
3. Block, E.; Putmand, D.; Zhao, S.-H. *J. Agirc. Food Chem.* **1992**, *40*, 2431-2438.
4. Mackenzie, A.; Ferns, D. *Phytochemisty* **1977**, *16*, 763-764.
5. Meng, Z.; Wang Y.; Cao, Y. *Zhongguo Yaoke Daxue Xuebao.* **1996**, *27*, 139.
6. Choi, J.S.; Go, C.H. *Archives of Pharmacal Research.* **1996**, *19*, 60-61.
7. Choi, J.S.; Kim, J.Y.; Lee, J.H.; Young, H.S.; and Lee, T.W. *J. Korean Soc. Food &Nutrition.* **1992**, *21*, 286-290.
8. Kaneta, M.; Hikichi, H.; Endo, S.; Sugiyama, N. *Agric. Biol. Chem.* **1980**, 44, 1405.
9. Yoshida, T.; Saito, T.; Kadoya S. *Chem. Pharm. Bull.* **1987**, *35*, 97-107.
10. Sang, S.M.; Zhang, X.W.; Lao, A.N.; Chen, Z.L. **2002**, *3* (in press).

11. Sang, S.M.; Zou, M.L.; Lao, A.N.; Chen, Z.L.; Ho, C.T. *J. Agric. Food Chem.* **2001**, *49*, 4780-4783.
12. Sang, S.M.; Mao, S.L.; Lao, A.N.; Chen, Z.L.; Ho, C.T. *J. Agric. Food Chem.* **2001**, *49*, 1475-1478.
13. Sang, S.M.; Mao, S.L.; Lao, A.N.; Chen, Z.L. *Tianran Chanwu Yanjiu Yu Kaifa,* **2000,** *12,* 1-3.
14. Sang, S.M.; Mao, S.L.; Lao, A.N.; Chen, Z.L. *Zhongcaoyao,* **2000**, *31,* 244-245.
15. Sang, S.M.; Mao, S.L.; Lao, A.N.; Chen, Z.L. *Zhongguo Zhongyao Zazhi*, **2000**, *25,* 286-288.
16. Sang, S.M.; Lao, A.N.; Wang, H.C.; Chen, Z.L. *J. Nat. Prod.* **1999**, *62*, 1028-1029.
17. Sang, S.M.; Lao, A.N.; Wang, H.C.; Chen, Z.L. *Phytochemistry* **1999**, *52*, 1611-1615.
18. Agrawal, P. K.; Jain, D. C.; Guptq, R.K.; Thakur, R.S. *Phytochemistry.* **1985**, *24*, 2479-2496.
19. Nakashima, N.; Kimura, I.; Kimura, M.; Matsuura, H. *J. Nat. Prod.* **1993**, *56*, 345-350.
20. Wu, R.T.; Chiang, H.C.; Fu, W.C. *Int. J. Immunopharmac.* **1990**, *12*, 777-781.
21. Miyata, T. *J. Trad. Sino-Jpn. Med.* **1992**, *13*, 276-279.
22. Niwa, A.; Takeda, O.; Ishimaru, M. *Yakugaku Zasshi* **1988**, *108*, 555-558.
23. Hostettmann, K.; Marston, A. *Saponins*, Cambridge University press, Cambridge, **1995**; 76.
24. Mimaki, Y.; Kawashima, K.; *J. Nat. Prod.* **1998**, *61*, 1279-1282.

Indexes

Author Index

Bai, Naisheng, 271
Chen, Chieh-Fu, 32, 104
Chen, Nien-Tsu, 87
Chen, Yen-Chou, 113
Chen, Yen-Ju, 202
Chen, Yong, 224
Chen, Zhongliang, 279, 318
Chiou, Wen-Fei, 32
Chou, Cheng-Jen, 32
Fu, Hui-Yin, 215, 307
Fujimoto, Yasuo, 279
He, Kan, 271
Ho, Chi-Tang, 66, 121, 224, 234, 292
Hsu, Foun-Lin, 113
Huang, Tzou-Chi, 215
Hung, Chien-Ya, 202
Hwang, Lucy Sun, 87
Inoue, Toshio, 142
Itokawa, Hideji, 2
Kikuzaki, Hiroe, 176
Kobayashi, Yoko, 142
Kozuka, Mutsuo, 2
Lao, Aina, 279, 318
Lee, Kuo-Hsiung, 2
Li, Guolin, 247
Lin, Jen-Kun, 66, 121
Lin, Lan-Chi, 87
Lin, Yun-Lian, 104
Lin-Shiau, Shoei-Yn, 66, 121
Liu, Tsung-Yun, 104
Liuchang, Huei-Chiuan, 87
Lo, Ai-Hsiang, 66, 121
Maeng, Il-Kyung, 48

Masuda, Hideki, 142
Nakatani, Nobuji, 166
Oda, Yoshimitsu, 104
Pandey, Ravindra K., 247
Park, Sang Shin, 104
Rafi, Mohamed M., 48
Rosen, Robert T., 258
Ryu, Kenjiro, 258
Sang, Shengmin, 234, 271, 279, 318
Shen, Shing-Chuan, 113
Shen, Yuh-Chiang, 32
Shi, John, 154
Shiao, Ming-Shi, 87
Shieh, Den-En, 307
Shyu, Chi-Chuo, 104
Simon, James E., 234, 292
Tadmor, Yaakov, 234
Ueng, Yune-Fang, 104
Uzawa, Jun, 279
Wang, Guei-Jane, 32
Wang, Mingfu, 234, 292
Wang, Shiow Y., 190
Wu, Qing-Li, 234, 292
Xiao, Pei-Gen, 292
Yadav, Prem N., 48
Yen, Gow-Chin, 202
Yu, Shi-Chun, 292
Zhang, Li, 271
Zheng, Qun Yi, 271
Zhou, Bing-Nan, 271
Zhu, Dayuan, 247
Zhu, Nanqun, 224

Subject Index

A

Acetobacter aceti, 308, 309, 310–311
Acetobacter intermedius sp., 307
Acetobacter xylinum
  cell mass development, 309, 310*f*
  cellulose fibril production, 308, 309*f*
  glucuronic acid pathway, 314–315
  glucuronic acid pathway of
    Acetobacter xylinum, 314–315
  glucuronic acid production, 308, 310–311
  stimulation by caffeine and theophylline, 309
  use in kombucha preparation, 307–308
  See also Tea fungus fermented black tea
Aflatoxin B$_1$ (AFB$_1$), 105, 106–108
Aged garlic extract (AGE)
  biological effects, 259, 264*t*
  compounds found in AGE, 260–261
  effects of processing conditions, 259
  fructosylarginine (Fru-Arg), 262–264
  preparation, 258, 259
  S-allyl-L-cysteine (SAC), 261, 262*t*, 264*t*
  1,2,3,4-tetrahydro-β-carboline-3-carboxylic acids, 265–266
AHH [benzo(a)pyrene hydroxylation]. *See* Benzo(a)pyrene
Allicin, 19, 259, 260*f*
Alliin, 19, 259
Allithiamine, 19
Allium tubersom Rottl. (Chinese chives)
  alkanoids, 326*f*–327*f*
  cholesterol glycoside, 326*f*
  extraction and isolation procedures, 320, 321*f*
  furostanol glycosides, 324*f*–325*f*
  general information, 319
  lignins, 327*f*
  phenolic compounds, 329*f*
  spirostanol glycosides, 322*f*–324*f*
  steroidal saponins, 320, 322, 322*f*–326*f*
  stigmasterol glucoside, 328*f*
  use in traditional Chinese medicine, 319
  vernolic acid, 328*f*
Alpinia flabellata Ridley (Iriomotekumatakeran), 168, 169
Alpinia specosa K. Schum (Getto), 168, 169, 171–172, 173*f,* 174*f*
Amomum sublatum (greater cardamom), 181–182, 183*f*
Anthocyanins, 191, 192*t,* 195–197
Apoptosis
  characteristics and function, 116, 118, 122
  cytochrome c increase during apoptosis, 123, 126, 136–137
  DNA fragmentation, 122, 128, 131*f*
  DNA fragmentation factor (DFF), 131, 133*f*
  induction by rosemary phytochemicals, 127–128
  induction by wogonin, 118
  inhibitor of caspace-activated DNase (ICAD), 123, 131, 133*f*
  lamins, 123
  mitochondrial transmembrane potential, 133–136, 137
  oligomerized Apaf-1 (apoptotic-protease-activating factor 1), 123

poly (ADP-ribose) polymerase (PARP), 125–156, 131, 133*f*
role of caspases, 122–123
sub-G1 cell population, 128, 130*t*
Aristolochic acid, 37, 38*f*–39*f*
Artemisia campestris L (Ryukyuyomgi), 168, 169, 170–171, 172*f*
Artemisinin (Qing Hao Su), 24
8-C-Ascorbyl(-)-epigallocatechin, 18
Aspergillus flavus, 105
Assistant herbs, 4
Astragalosides, 12*f*
Astragalus or Huang-qi (*Astragalus membranaceus*, Leguminosae), 11–12

B

Baicalein
  beneficial effects against liver damage, 106
  dietary effects on benzo(a)pyrene DNA adduct formation, 108–109
  dietary effects on CYP, UGT, and GST in mouse liver, 110–110
  inhibition of AHH and AFO activities, 107–110
  presence in Scutellariae Radix as glucuronide, 105
  reduction of genotoxicities and oxidations of benzo(a)pyrene and aflatoxin B_1, 106–108
  toxicity, 106
  tumor inhibition, 106
Benzo(a)pyrene
  activation by CYPs, 105
  environmental pollutant, 105
  genotoxicity, 106–107
  inhibition of benzo(a)pyrene hydroxylation (AHH) by baicalein and wogonin, 107–108
  inhibition of binding to epidermal DNA by rosemary extract, 122
  reduction of genotoxicity by baicalein and wogonin, 106–107
  tumorigenic effects, 105
Benzoxazinone, 13
Benzoxazolinone (coixol), 13
Berries, antioxidant capacity
  anthocyanins, 191, 192*t*, 195–197
  antioxidant activity, contributions of phenolic acids and flavones, 193, 194*t*
  antioxidant capacity in berries, 191–193
  caffeic acid in chokeberries, 193, 194*t*
  chlorogenic acid in blueberries, 193, 194*t*
  cinnamic acid, 193, 194*t*
  p-coumaric acid, 193, 194*t*
  ellagic acid, 193
  flavonoids, 193
  oxygen radical absorbance capacity (ORAC), 191, 192*t*, 195, 197
  peroxyl radical (ROO˙), antioxidant capacity against, 191, 192*t*, 195, 197
  phenolic acids, 193, 194*t*
  phenolic content, total, 191, 192*t*
  proanthocyanidins, 195
  scavenging capacity for radicals other than peroxyl, 191, 193, 195, 197
  vanillic acid in cranberries, 193, 194*t*
Bioactivity-directed fractionation and isolation (BDFI), 22
Black tea, 308
  See also Tea fungus fermented black tea
Britannilide
  cytotoxicity, 273, 277
  NMR spectra, 274*t*
  physical properties, 273, 275
  stereochemistry, 276
N-Butylidenephthalide, 14–15

C

Caffeic acid
  effect on lipid peroxidation, 208, 209f
  effect on peroxide-induced DNA damage, 210, 212f
  genotoxicity, 210, 211f
  in chokeberries, 193, 194t
Caries, 142, 149f
Carnosic acid
  antioxidative properties, 73, 75, 122
  cleavage of poly (ADP-ribose) polymerase and DFF45/ICAD, 131, 133f
  cytotoxicity, 127, 128f
  effect on cell viability, 73, 75f
  effect on LPS-induced NO production, 75–77, 78f
  effect on nitrite production, 76
  induction caspace activities, 128, 130, 132f
  induction of apoptosis, 127–128, 129f, 130t, 136–139
  induction of DNA fragmentation, 131f
  structure, 74f, 124f
Carnosol
  antioxidative properties, 73, 75, 122
  cleavage of poly (ADP-ribose) polymerase and DFF45/ICAD, 131, 133f
  cytotoxicity, 127, 128f
  effect on activation of IKK, 82, 83
  effect on cell viability, 73, 75f
  effect on LPS-induced NF-κB activation, 79, 83
  effect on LPS-induced phosphorylation and degradation of IκBα, 79–81
  effect on NF-κB/Rel family subunits, 78–79, 83
  effect on nitrite formation, 75, 76f
  from rosemary, 66, 67, 73
  induction caspace activities, 128, 130, 132f
  induction of apoptosis, 127–128, 129f, 130t, 136–139
  induction of DNA fragmentation, 131f
  inhibition of tumor initiation and promotion, 122
  structure, 74f, 124f
  suppression of iNOS protein and mRNA expression, 76–78, 83
Caspases
  effects of rosemary phytopolyphenols on caspace activities, 128, 130, 132f
  measurement of caspace activity, 125–126
  procaspase-3, 123
  procaspase-9, 123
  role in apoptosis, 122–123
Cell proliferation inhibition assay, 221–222
Cell viability assay, 124–125
Chinese Materia Medica books, 3–4
Chlorogenic acid in blueberries, 193, 194t
Chuan Chiung (*Cnidium officinale, Umbelliferae*), 15
Chung Hua Pen Tsao (Chinese Materia Medica), 4
Chung Yao (traditional drugs), 3
Cinnamic acid, 193, 194t
Coix seeds (*Coix lachryma-jobi* var. ma-yuen Stapf), 13
Comet assay (alkaline single-cell gel electrophoresis), 205
Coronary heart disease (CHD), 88
P-coumaric acid, 193, 194t
Curcumin, 50f, 51
Curry leaf *(Murraya koenigii)*, 185–187
Cyclanoline, 37, 38f–39f
Cyclooxygenase-2 (COX-2)
  effect of diarylheptanoids, 56f–57f
  effects of nutraceuticals, 49, 57
  inhibition of COX-2 gene expression, but not COX-2 activity by

wogonin, oroxylin A, and
  quercetin, 117f
inhibition of COX-2 gene expression
  by wogonin, oroxylin A, and
  quercetin, 115–116
Cytochrome c, 123, 126, 136, 137f
Cytochrome P450 (CYP)-dependent
  monooxygenase, 105, 110–111

D

Daidzein, 16
Daidzin, 16
Dang Gui Lu Hui Wan, 24
Dencichine, 9f
Dental caries, 142, 148, 149f
Diallyl sulfide, 19
Diarylheptanoids
  anti-inflammatory properties, 54
  effect on cyclooxygenase-2 (COX-
    2), 56f–57f, 59f
  effect on inducible nitric oxide
    synthase (iNOS), 56f–57f
  effect on nitric oxide, 55f
  effect on nuclear factor-κB (NF-κB),
    58f
  effects on cytokines, 56f
  HMP [7-(4′-hydroxy-3′-
    methoxyphenyl)-1-phenylhept-4-
    en-3-one], 50f, 54, 55f–57f, 59f
  nutraceutical use, 49, 51
  structure, 50f
  yakuchinone A, 54
  yakuchinone B, 54
4,5-Dihydroxy-butylidenephthalide,
  15
N,N′-dimethylindirubin, 23, 24–25
DNA fragmentation assay, 125
Dong Quai or Tang Kuei (*Angelica
  sirensis*, Umbelliferae), 14–15
DPPH (α,α-diphenyl-β-
  picrylhydrazyl) free radical
  scavenging
  carbazoles isolated from curry leaf,
    186f, 187t

diarylheptanoids isolated from
  ginger, 184f
Du-Zhong (*Eucommia ulmodies*
  Oliver), 227f, 228–230
Hsian-tsao (*Mesona procumbens*
  Hemsl.), 205, 207–208
Okinawan herbs, 169
onion (*Allium cepa* L.), 219–220
Papua mace lignans, 180f
phenolics isolated from greater
  cardamom, 183f
Pu-Erh tea, 91, 94, 100
rosemary phytochemicals, 69, 73, 75
tea fungus fermented black tea, 311–
  312, 313f
Du-Zhong (*Eucommia ulmodies*
  Oliver)
  antioxidant activities of leaves, 225
  DPPH scavenging activity of extracts
    and fractions, 227f, 228–230,
    228f
  extraction, separation, and
    identification procedures, 226
  medicinal uses, 225
  (+)-pinoresinol-*O*-β-D-glucose, 226,
    229–230
  structure determination of isolated
    compounds, 226
  syringic acid, 228, 229–230

E

EC [(-)-Epicatechin], 17
ECG [(-)-Epicatechin-3-gallate], 17,
  18f, 53
EGC [(-)-Epigallocatechin], 18f
EGCG. *See* (-)-Epigallocatechin-3-
  gallate
Electrophoretic mobility shift assay
  (EMSA), 72
Ellagic acid, 16, 17f, 193
Ephedra, 34
Ephedras (*Ephedra spp.*,
  Ephedraceae), 23–24
(-)-Ephedrine, 23–24

(-)-Epicatechin-3-gallate (ECG), 17, 18f, 53
(-)-Epicatechin (EC), 17, 18f, 53
(-)-Epigallocatechin-3-gallate (EGCG)
  cholesterol-lowering effect, 88, 99
  inhibition of Cu^{2+} induced LDL oxidation, 88
  inhibition of tumor promotion, 17
  nutraceutical properties, 17, 53
  structure, 18f, 50f, 89f
  suppression of NO production, 68
(-)-Epigallocatechin (EGC), 18f
Eremobritanilin
  cytotoxicity, 273, 277
  NMR spectra, 274t
  physical properties, 273, 276–277
  stereochemistry, 277
Eriobotrya japonica. See Loquat
Etoposide, 21–22
Evodia lepta (Spreng.) Merr.
  acetylation to determine hydroxyl location, 249–250
  alkaloids from *Evodia lepta*, 247, 248f
  anti-fungus activity tests, 255
  anti-HIV activity tests, 255
  anti-tumor activity tests, 255
  chemical correlation between chromenes, 249–250
  clovandiol (sesquiterpene), 253f
  dichromenes, 250–251
  2,2-dimethyl-chromenes, 248–250
  2,2-dimethylchromans, 251, 252f, 253f
  evodione, 249
  extraction and isolation of phytochemicals, 248
  flavones, 253f
  insecticide activity tests, 253–254
  isoevodionol, 249
  methylevodionol, 249
  use in Traditional Chinese Medicine, 247
Evodia rutaecapar (E.R.), 34, 35

F

Fang-Ji
  Aristolochia heterophylla, 37, 38f–39f
  Aristolochia weslandi, 37, 38f–39f
  chemical constituents, 37, 38f–39f
  Cocculus trilobus (Thunb.) DC., 37
  Sinomenium acutum, 37, 38f–39f
  See also Radix Stephania tetrandra (RST)
Fangchinoline, 37, 38f–39f
Ferulic acid, 15
Flow cytometry analysis of DNA content, 125, 129f
Folk drugs (Min Chien Yao), 3
Free radical scavenging. See DPPH (α,α-diphenyl-β-picrylhydrazyl) free radical scavenging
Fructosylarginine (Fru-Arg), 262–264
Functional foods (Yao Shan), 3, 4

G

Ganoderic acids, 5, 6
Ganolucidic acid, 5, 6
Garlic or Ta Suan (*Allium sativum,* Liliaceae), 18–19, 259
  See also Aged garlic extract (AGE)
Geniposidic acid, 10
Genistein, 16
Ginger (*Zingiber officinale,* Zingiberaceae), 16–17, 182, 184f
Gingerol, 16–17
Ginsengs
  American ginseng (*Panax quinquefolium,* Araliaceae), 7, 9
  Asian ginseng (*Panax ginseng,* Araliaceae), 7
  chemical composition and properties, 7–9
  ginsenosides (Rb_1 and Rb_2), 7, 8f–9f, 34–35

Sanchi ginseng (*Panax notoginseng,* Araliaceae), 9
variability, 34–35
GL-331, 22
Glucans, 5, 6, 142
Glucosyltransferases (GTases)
  effect of isothiocyanates on insoluble glucan synthesis by GTases, 148, 150*f*
  effect of oolong tea extract on insoluble glucan synthesis by GTases, 148, 151*f*
  glucan synthesis, 142, 144
  Streptococcus sobrinus, 143
Glutamic-oxaloacetic transaminase (GOT), 145, 148, 149*f*
Glutamic-pyruvic transaminase (GPT), 145, 148, 150*t*, 235
Glutathione S-transferase (GST), 105, 110–111
Glycans (β-D-glycans), 5
Gomisins, 11
GOT. *See* Glutamic-oxaloacetic transaminase
GPT. *See* Glutamic-pyruvic transaminase
Greater cardamom. *See Amomum sublatum*
Green tea polyphenols (GTPs), 17–18, 53
  See also (-)-Epicatechin-3-gallate (ECG); (-)-Epicatechin (EC); (-)-Epigallocatechin-3-gallate (EGCG); (-)-Epigallocatechin (EGC); Tea polyphenols
Green tea (*Thea sinensis* or *Camellia sinensis,* Theaceae), 17–18
  See also Green tea polyphenols (GTPs); Tea *(Camellia sinensis)*
GST (glutathione S-transferase), 105, 110–111
GTPs. *See* Green tea polyphenols
"Guidance for Industry Botanical Drug Products" (U.S. FDA), 35–36

H

Herbal medicines research, 22–25
Herbal properties, 4
Herbs, antioxidant capacity
  antioxidant activity compared to berries, fruits, and vegetables, 197–198
  anthocyanins, 192*t*
  oxygen radical absorbance capacity (ORAC), 192*t*
  phenolic content, total, 192*t*
  garden sage (*Salvia officinalis*), 198
  oregano *(Poliomintha longiflora, Origanum vulgare spp. hirtum* and *riganum x majoricum),* 197–198
  thyme *(Thymus vulgaris),* 198
  See also Rosemary (*Rosmarinus officinalis* Labiatae)
Herbs, categories, 4
  See also Lower Class herbs; Middle Class herbs; Upper Class herbs
Hsia Ku Tsao (*Prunella vulgaris* L., Labiatae), 20
Hsian Ku mushrooms [*Lentinus edodes* (Berk.) Sing] (shiitake mushrooms), 6
Hsian-tsao (*Mesona procumbens* Hemsl.)
  antioxidant activity, 202, 206
  cytotoxicity of WEHT, 208
  effects of WEHT and phenolic compounds on hydrogen peroxide-induced DNA damage, 210, 211*f,* 212*f*
  effects of WEHT and phenolic compounds on intracellular ROS level, 210, 212*f,* 213, 213*f*
  effects on lipid peroxidation in cells, 205, 208, 209*f,* 209*t*
  phenolic compounds in WEHT, 206
  scavenging effects of WEHT on DPPH radical, superoxide anion, hydrogen peroxide, and nitric oxide, 205, 207–208

use and composition, 203
water extracts from Hsian-tsao
  (WEHT), preparation, 204
Huang-qi or astragalus (*Astragalus membranaceus,* Leguminosae), 11–12
Huang Qui (*Scutellariae baicalensis* George), 113, 114
  See also Oroxylin A; Scutellariae Radix; Wogonin
Huang Ti Nei Ching (The Yellow Emperor's Classic on Internal Medicine), 4

I

IκB kinase (IKK), 71–72, 82
Immunoprecipitation, 71–72
Imperial herbs, 4
Indirubin (from Dang Gui Lu Hui Wan), 24–25
Inducible nitric oxide synthase (iNOS)
  effect of carnosol, 76–77, 78*f*
  effect of diarylheptanoids, 56*f*–57*f*
  effect of nutraceuticals, 49, 57–58, 59*f*
  inhibition of iNOS gene expression, but not iNOS activity by wogonin, oroxylin A, and quercetin, 117*f*
  inhibition of iNOS gene expression by wogonin, oroxylin A, and quercetin, 115–116
  Northern blot analysis, 71, 77, 78*f*
  Western blot analysis, 70
Inhibitor of caspace-activated DNase (ICAD), 123
Inula britannica, 271
  See also Sesquiterpene lactones
Isothiocyanates
  effect of aryl isothiocyanates on sucrose dependent adherence by growing cells of *S. mutans,* 147*f*
  effect of isothiocyanates on body weight of rats, 148, 149*f*
  effect of isothiocyanates on caries score of rats, 148, 149*f*
  effect of isothiocyanates on insoluble glucan synthesis by GTases, 148, 150*f*
  effect of isothiocyanates on sucrose dependent adherence by growing cells of *S. mutans,* 145–148
  effect of ω-alkenyl isothiocyanates on sucrose dependent adherence by growing cells of *S. mutans,* 146*f*
  effect of ω-methylsulfinylalkyl isothiocyanates on sucrose dependent adherence by growing cells of *S. mutans,* 147*f*
  effect of ω-methylthioalkyl isothiocyanates on sucrose dependent adherence by growing cells of *S. mutans,* 146*f*
  in wasabi, 142–143
  MIC values of isothiocyanates, 148, 151*f*

K

Kansui (*Euphorbia kansui,* Euphorbiaceae), 19–20
Kansuiphorins, 20
Karmen unit of rats fed with or without isothiocyanates, 150*t*
Ko Ken or kudzu vine (*Puerialia lobata,* Leguminosae), 15–16
Kombucha tea. *See* Tea fungus fermented black tea
Kou Chi Tzu (*Lycium barbarum,* Solanaceae), 12
Kuei Chiu (*Podophyllum emodi,* Berberidaceae), 21–22

L

Lentinan (β-D-glucan), 6
Ligustilide, 14–15

Ling Chih mushrooms [*Ganoderma lucidum* (Leyss.: Fr.) P. Karst, Polyporaceae], 5, 6
Lipid analysis, 92
Livingstona chinensis R. Br var. *subglobosa* Becc (Biro), 168, 169
Loquat [Pi Pa, *Eriobotrya japonica* (Thunb.) Lidle]
   2,6-dimethoxy-4-(2-propenyl)phenol, 295*f*, 298
   2,6-dimethoxy-4-(2-propenyl)phenol 1-*O*-β-D-glucopyanoside, 295*f*, 298
   eriobotrin, 295*f*, 296, 298, 304–305
   extraction and isolation of compounds, 294
   general description, 293
   isoeriobotrin, 295*f*, 296, 298
   linguersinol, 295*f*, 296, 297*t*
   linguersinol 9'-*O*-β-D-xylopyranoside, 295*f*, 296, 297*t*, 304
   megastigmane derivatives, 295*f*, 298–301
   pharmacological activity, 303
   quercetin-7-α-L-rhamnoside, 295*f*, 302
   quercetin-3-*O*-β-D-glucoside, 295*f*, 301–302
   structure of compounds identified in loquat leaves, 295*f*
   triterpenes, 295*f*, 302–303
   use in traditional Chinese medicine, 293
Lovastatin, 89*f*, 91, 94, 95*f*, 100
Low-density lipoprotein cholesterol (LDL-C), 88
Low-density lipoprotein (LDL), 88, 91, 93, 100
Lower Class herbs, 4, 19–22
Lycopene
   all-*trans*-isomer, structure and properties, 155, 156*f*, 162
   antioxidant activity, 155–156
   bioactivity potency, 156, 162–163
   biological properties, 155
   cancer prevention, 155
   chemical properties, 155
   cis-isomers, properties, 155, 162
   effects of heat treatment on lycopene oxidation and isomerization, 157, 158*f*, 159*f*, 162–163
   effects of light irradiation on lycopene degradation, 159–161, 162
   in tomatoes, 156–157
   losses during tomato processing, 157
   pathway of lycopene degradation, 157
   quenching constant, 155

M

Magnoflorine, 37, 38*f*–39*f*
Mahuang Apricot Seed Combination, 4
Mahuang Combination, 4
Menisine and menisidine, 37
N-methylindirubin, 24–25
Micellar electrokinetic capillary chromatography (MEKC), 241–242
Middle Class herbs
   Chuan Chiung (*Cnidium officinale*, Umbelliferae), 15
   definition, 4
   ginger (*Zingiber officinale*, Zingiberaceae), 16–17
   green tea (*Thea sinensis* or *Camellia sinensis*, Theaceae), 17–18
   Ko Ken or kudzu vine (*Pueralia lobata*, Leguminosae), 15–16
   Shi Liu Pi or pomegranate (*Punica granatum*, Punicaceae), 16
   Ta Suan or garlic (*Allium sativum*, Liliaceae), 18–19
   Tang Kuei or Dong Quai (*Angelica sirensis*, Umbelliferae), 14–15
Min Chien Yao (folk drugs), 3
Ministerial herbs, 4
Mitochondrial transmembrane potential, 133–136, 137

Murraya koenigii (curry leaf), 185–187
Musa balbisiana Colla (Ryukyubasho), 168, 169
Mushrooms, medicinal, 5–6
Mutans streptococci, 142, 143, 145–148, 151*f*
Myristica argentea Warb, 177, 179–181

N

National Cancer Institute (NCI), 23
New medicines from herbal products, 22–25
NF-κB. *See* Nuclear factor-κB
Nitric oxide (NO)
  biological functions, 67–68, 84
  effect of carnosic acid on LPS-induced NO production, 75–76
  effect of carnosol on LPS-induced NO production, 75–76, 82–83
  effect of diarylheptanoids, 55*f*
  effect of nutraceuticals, 49, 58
  effect of rosmarinic acid on LPS-induced NO production, 75–76
  effect of ursolic acid on LPS-induced NO production, 75–76
  inhibition of LPS-induced NO production by wogonin, quercetin, and oroxylin A, 114, 116*f*
  scavenging effect of water extract from sian-tsao, 203, 207–208
  suppression of NO production by EGCG, 68
  suppression of NO production by resveratrol, 68
Nitric oxide synthase, inducible. *See* Inducible nitric oxide synthase
NO synthase (NOS), 67
Northern blot analysis, 71, 77, 78*f*
Nuclear factor-κB (NF-κB), 49, 57–58, 59*f*, 78–79
Nutraceuticals
  antioxidative and anti-inflammatory activities of nutraceuticals, 48–49, 57–58
  effect on cytokines, 49, 57
  effect on inducible nitric oxide synthase (iNOS), 49, 57
  effect on nitric oxide, 49, 58
  effect on nuclear factor-κB (NF-κB), 49, 57–58
  effects on cyclooxygenase-2 (COX-2), 49, 57
  targets of nutraceuticals, 59*f*
  See also Curcumin; Diarylheptanoids; Green tea polyphenols (GTPs); Resveratrol

O

Oblongine, 37, 38*f*–39*f*
Oil stability index (antioxidant activity), 180*f*, 181
Okinawa, 166
Okinawan herbs
  antioxidant activities, 168
  DPPH radical scavenging activity, 169
  extraction procedure, 167
  inhibition of xanthine oxidase (XOD), 169–170
  isolation of antioxidants, 167, 168*f*
  scavenging activity on superoxide anion radical (O_2-), 169
  See also Alpinia flabellata Ridley; *Alpinia specosa* K. Schum; *Artemisia campestris* L; *Livingstona chinensis* R. Br var. *subglobosa* Becc; *Musa balbisiana* Colla; *Peucedanum japonicum* Thunb; *Smilax china* L var. *kuru* Sakaguchi ex Yamamoto; *Smilax nervo-marginate* Hayata
Onion (*Allium cepa* L.)
  antioxidant activity, 220
  cell proliferation inhibition assay, 221–222

change in quercetin glycosides by different heat treatments, 218
DPPH scavenging capacity, 219–220
flavonoid loss during cooking, 216
β-glucosidase activity, 218, 219f
moisture sorptions of onion powder, 216
nutraceutical uses and health effects, 216
Oxidative Stability Index, 220
quercetin loss during cooking, 216
Oolong tea, 148, 151f
Oroxylin A
in Huang Qui, 114
induction of vasorelaxation, 118, 119f
inhibition of iNOS and COX-2 gene expression, 115–116
inhibition of iNOS and COX-2 gene expression, but not iNOS and COX-2 activities, 117f
inhibition of nitric oxide production, 114, 116f
inhibition of PGE$_2$ production, 115, 116f
structure, 115f
Oxidative Stability Index, 220
Oxidized LDL (OxLDL), 88
Oxobritannilactone, 273, 274t, 276
Oxygen radical absorbance capacity (ORAC) in berries, 191, 192t, 195, 197
Oxygen radical absorbance capacity (ORAC) in herbs, 192t, 197–198

P

P53 tumor suppressor gene, 118
Paua mace. See *Myristica argentea Warb*
Pen Tsao Kan Mu (A General Catalog of Herbs), 4
Peroxyl radical (ROO·), 191
PET (dry, powdered Pu-Erh tea)
antioxidant activity, 94 (*See also* Pu-Erh tea)
DPPH free radical scavenging, 94, 100
effects on body, liver, and adipose tissue weights of hamsters, 96, 97t
effects on fecal cholesterol of hamsters, 97, 98t
effects on liver and serum lipids in hamsters, 97, 98t
inhibition of cholesterol synthesis, 90, 93–94, 96t, 100
inhibition of LDL oxidation *in vitro*, 94
lovastatin, 91
production, 90
Peucedanum japonicum Thunb (Botanbofu), 168, 169–170, 171f
Pinoresinol, 226, 229–230
Poly (ADP-ribose) polymerase (PARP), 123
Pomegranate or Shi Liu Pi (*Punica granatum*, Punicaceae), 16
Precocene I and II, 253–254
Proanthocyanidins, 195
Prostaglandin E2 (PGE$_2$)
inhibition of PGE$_2$ production by wogonin, quercetin, and oroxylin A, 115, 116f
(+)-Pseudoephedrine, 23–24
Pu-Erh tea
DPPH scavenging, 91, 94
effects on LDL oxidation in humans, 98, 99t, 101
effects on plasma lipids in humans, 98, 99t
effects on α-tocopherol content in LDL in humans, 98, 99t, 101
human study, 92
lovastatin, 89f, 91, 94, 95f, 100
manufacture and fermentation, 87, 88, 99
properties and biological activities, 88, 90, 99
See also Green tea; PET; Tea
Puerarin, 16
Punicalin, 16, 17f

Q

Quercetin
  biological activities, 114
  change in quercetin glycosides in onion powder by different heat treatments, 218
  change in quercetin glycosides in onions by different heat treatments, 218
  in berries, 195
  in Huang Qui, 114
  inhibition of iNOS and COX-2 gene expression, 115–116
  inhibition of iNOS and COX-2 gene expression, but not iNOS and COX-2 activities, 117f
  inhibition of nitric oxide production, 114, 116f
  inhibition of PGE$_2$ production, 115, 116f
  quercetin loss in onions during cooking, 216
  structure, 115f
Quinghao or sweet wormwood (*Artemisia annua* L., Asteraceae), 24

R

Radix Stephania tetrandra (RST)
  active components in RST, 43f
  anti-inflammatory effects, 38f–42f, 40f
  cardiovascular protective effects, 46
  cytotoxicity, 43, 43f
  effects on Phorbol-12-myristate-13-acetate (PMA)-induced adhesion by neutrophils, 37, 42f
  effects on Phorbol-12-myristate-13-acetate (PMA)-induced ROS production, 37, 41f
  extraction methods, effects on potency, 37, 40f
  tetrandrine, 37, 38f–39f, 43f
  See also Fang-Ji
Reactive nitrogen species (RNS), 67–68
Reactive oxygen species (ROS)
  biological effects, 203
  effects of rosemary polyphenols on intracellular ROS generation, 123, 127, 133–136
  effects of water extract of Hsian-tsao on intracellular ROS level, 210, 212f, 213
  measurement of ROS levels, 206
  production by neutrophils, 37, 41f
Resveratrol, 50f, 52, 68
Reverse-transcribed polymerase chain reaction (RT-PCR), 71, 77
ROS (reactive oxygen species). *See* Reactive oxygen species (ROS)
Rosemary (*Rosmarinus officinalis* Labiatae)
  antioxidative properties, 73, 75, 122, 192t, 198
  antitumor effects, 73, 122
  cleavage of poly (ADP-ribose) polymerase and DFF45/ICAD, 131, 133f
  cytotoxicity of phytopolyphenols, 127, 128f
  DPPH free radical scavenging by phytochemicals, 69, 73, 75
  effect on cytochrome c release, 136, 137f
  effect on intercellular ROS generation, 133–136, 138
  effect on mitochondrial transmembrane potential, 133–136
  effects of rosemary phytopolyphenols on caspace activities, 128, 130
  induction of apoptosis, 127–128, 136–139
  inhibition of covalent binding of benzo(a)pyrene to epidermal DNA, 122
  isolation of rosemary compounds, 68–69

phytochemicals, effect on cell
    viability, 73, 75f
phytochemicals, effects on LPS-
    induced NO production, 75–76
phytochemicals, free radical
    scavenging activity, 73, 82
phytochemicals, structure, 73, 74f
use and properties, 66, 67, 82
See also Carnosic acid; Carnosol;
    Rosmarinic acid; Ursolic acid
Rosmarinic acid
antioxidative properties, 73, 75
cytotoxicity, 127, 128f
effect on cell viability, 73, 75f
effect on LPS-induced NO
    production, 75–76
effect on nitrite formation, 76
induction of apoptosis, 127–128,
    129f, 130t
structure, 74f, 124f
RST. See Radix Stephania tetrandra

S

S-allyl-L-cysteine (SAC), 261, 262t,
    264t
Sage or Tan Shen (*Salvia miltiorrhiza*,
    Labiatae), 9–10
Schisandra chinensis (turcz) Baill
antihepatotoxic use, 235
capillary electrochromatography of
    Schisandra lignans, 241–242
chemical composition, 236
DDB (dimethyl-4,4'-dimethoxy-
    5,6,5,'6'-dimethylene-
    dioxybiphenyl-2,2'-dicarboxylate),
    235
gas chromatography (GC) and
    GC/MS of *Schisandra* lignans, 239
high performance liquid
    chromatography (HPLC) of
    Schisandra lignans, 236, 238–239,
    240f
high performance liquid
    chromatography-mass

spectrometry (HPLC-MS) of
    Schisandra lignans, 242–243
lignans detected in *Schisandra*, 236,
    237f, 239
medicinal use, 235–236
micellar electrokinetic capillary
    chromatography (MEKC) of
    Schisandra lignans, 241–242
schisandrin A and schisandrin B,
    structures, 237f
schisandrol A and schisandrol B,
    structures, 237f
total ion chromatography (TIC) of
    Schisandra lignans, 243–244
use in Traditional Chinese Medicine,
    235
Schizandrin, 11
Schizanhenol, 11
Scutellariae baicalensis George
    (Huang Qui), 113, 114
Scutellariae Radix, 105–106
See also Baicalein; Wogonin
Servant herbs, 4
Sesquiterpene lactones, 271–272, 273,
    277
See also Britannilide;
    Eremobritanilin;
    Oxobritannilactone
Shang Han Tsa Ping Lun (Treatise on
    Febrile and Miscellaneous
    Diseases), 4
Shen Ku mushrooms (*Agaricus blazei*
    Murill, Agaricaceae), 6
Shen Nung, 4
Shen Nung Pen Tsao Ching (The
    Book of Herbs by Shen Nung), 3–4,
    25
Shi Liu Pi or pomegranate (*Punica
    granatum*, Punicaceae), 16
Shiitake mushrooms [Hsian Ku
    mushrooms, *Lentinus edodes*
    (Berk.) Sing], 6
β-Sitosterol, 89f, 96, 97, 98t,
    100
Smilax china L var. *kuru* Sakaguchi ex
    Yamamoto, 168, 169

Smilax nervo-marginate Hayata, 168, 169
Sodium tanshinone II-A sulfonate, 10
Spices. *See* Tropical spices
Spinosin, 14*f*
Stephania tetrandra S. Moore. *See Radix Stephania tetrandra* (RST)
Streptococcus mutans, 142, 143, 145–148, 151*f*
Streptococcus sobrinus, 143, 148, 151*f*
Superoxide anion radical (O_2-), 37, 40*f,* 203, 207–208
Swertisin, 14*f*
Syringic acid, 228, 229–230

T

Ta Suan or garlic (*Allium sativum,* Liliaceae), 18–19
Tan Shen or sage (*Salvia miltiorrhiza,* Labiatae), 9–10
Tang Kuei or Dong Quai (*Angelica sirensis,* Umbelliferae), 14–15
TCM. *See* Traditional Chinese medicine
Tea *(Camellia sinensis)*
 biological and pharmacological functions, 88, 99–100
 DPPH free radical scavenging, 94
 See also Black tea; (-)-Epicatechin-3-gallate (ECG); (-)-Epicatechin (EC); (-)-Epigallocatechin-3-gallate (EGCG); (-)-Epigallocatechin (EGC); Green tea; Green tea polyphenols (GTPs); Oolong tea; Pu-Erh tea; Tea fungus fermented black tea; Tea polyphenols
Tea fungus fermented black tea
 acetic acid production, 308, 309–310
 antioxidant activities, 308, 312
 biological activities, 308
 cell mass development, 309, 310*f*
 DPPH free radical scavenging activity, 311–312, 313*f*
 gluconic acid production, 308, 310–311
 glucuronic acid pathway of *Acetobacter xylinum,* 314–315
 glucuronic acid production, 308, 310–311
 kombucha tea, preparation, 307–308
 organic acid production, 308, 309–311
 pH changes, 310–311
 reduced type ascorbic acid, 314–315
 reducing power, 312, 313*f,* 314
 tea fungus, background information, 307
 yeasts used, 307–308, 311
 See also Acetobacter xylinum
Tea polyphenols, 88, 94, 99–100
 See also (-)-Epicatechin-3-gallate (ECG); (-)-Epicatechin (EC); (-)-Epigallocatechin-3-gallate (EGCG); (-)-Epigallocatechin (EGC); Green tea polyphenols (GTPs)
1,2,3,4-tetrahydro-β-carboline-3-carboxylic acids, 265–266
Tetramethylpyrazine, 15
Tetrandrine, 37, 38*f*–39*f,* 43*f*
Theasinensin-D, 18
Total ion chromatography (TIC), 243–244
Traditional Chinese medicine (TCM)
 basic theory and principles, 3–4, 25, 33
 historical use, 3
 modes of eliciting pharmacologic effects, 33
 quality control of TCM, 35
 variability in plant extracts, 32, 34–35, 36*f*
Traditional drugs (Chung Yao), 3
Transient transfection and luciferase assay, 72–73
Trilobine, 37, 38*f*–39*f*
Triterpenoids, 5, 6
Trolox, structure, 89*f*
Tropical spices

antioxidative spices and herbs in Southeast Asia, 177, 178*t*
See also *Amomum sublatum; Murraya koenigii; Myristica argentea* Warb; *Zingiber officinale*
Tu Chung (*Eucommia ulmoides*, Eucommiaceae), 10
Tung Chung Hsia Tsao [*Cordyceps sinensis* (Berk.) Sacc. (Hypocreaceae)], 12–13

U

UDP-glucuronosyltransferase (UGT), 105, 110–111
Upper Class herbs
astragalus or Huang-qi (*Astragalus membranaceus*, Leguminosae), 11–12
coix seeds (*Coix lachryma-jobi* var. ma-yuen Stapf), 13
definition, 4
ginsengs, 7–9
Kou Chi Tzu (*Lycium barbarum*, Solanaceae), 12
medicinal mushrooms, 5–6
Tan Shen or sage (*Salvia miltiorrhiza*, Labiatae), 9–10
Tu Chung (*Eucommia ulmoides*, Eucommiaceae), 10
Tung Chung Hsia Tsao [*Cordyceps sinensis* (Berk.) Sacc. (Hypocreaceae)], 12–13
Wu Wei Tzu (*Schizsandra chinesis*, Schisandraceae), 11
ziziphus (*Ziziphus jujuba*, Rhamnaceae), 14
Ursolic acid
antioxidative properties, 75
cleavage of poly (ADP-ribose) polymerase and DFF45/ICAD, 131, 133*f*
cytotoxicity, 127, 128*f*
effect on cell viability, 73, 75*f*
effect on LPS-induced NO production, 75–76
from Hsia Ku Tsao, 20
induction caspace activities, 128, 130, 132*f*
induction of apoptosis, 127–128, 129*f*, 130*t*, 136–139
induction of DNA fragmentation, 131*f*
inhibition of tumor initiation and promotion, 122
structure, 124

V

Vaccaria segetalis (Neck.) Garcke (syn. V. pyramidata Medik)
alkaloids, 285*f*
bioactivities, 288
extraction and isolation of compounds, 280, 281*f*
flavonoids, 282, 287*f*
other compounds, 288
phenolic acids, 282, 286*f*
phytochemistry, 282, 283*f*–288*f*
steroids, 282, 287*f*
triterpene saponins, 282, 283*f*–285*f*
use in traditional Chinese medicine, 280
Vanillic acid
effect on lipid peroxidation, 208, 209*f*
effect on peroxide-induced DNA damage, 210, 212*f*
genotoxicity, 210, 211*f*
in cranberries, 193, 194*t*
Varapamil, 46

W

Wasabi components, anticaries effect. *See* Caries; Isothiocyanates
WEHT (water extracts from Hsian-tsao). *See* Hsian-tsao

Western blots, 70, 136, 137f
Wogonin
  beneficial effects against liver
    damage, 106
  dietary effects on benzo(a)pyrene
    DNA adduct formation, 108–109
  dietary effects on CYP, UGT, and
    GST in mouse liver, 110–111
  in Huang Qui, 114
  induction of apoptosis, 118
  inhibition of AHH and AFO
    activities, 107–110
  inhibition of iNOS and COX-2 gene
    expression, 115–116
  inhibition of iNOS and COX-2 gene
    expression, but not iNOS and
    COX-2 activities, 117f
  inhibition of nitric oxide production,
    114, 116f
  inhibition of PGE$_2$ production, 115,
    116f
  mutagenicity, 106
  presence in Scutellariae Radix as
    glucuronide, 105
  reduction of genotoxicities and
    oxidations of benzo(a)pyrene and
    aflatoxin B$_1$, 106–108
  structure, 115f

Wu Wei Tzu (*Schizsandra chinesis,*
  Schisandraceae), 11

X

Xiang fan, Xiang sha, Xiang shi,
  Xiang wei, Xiang wu, and Xiang
  xu, 33
Xuanfuhua, 271

Y

Yao Shan (functional foods), 3, 4
Yun-Chih mushrooms (*Coriolus
  versicolor* Quél, Polyporaceae),
  5

Z

(Z)-Ajoene, 19
Zingiber officinale (ginger), 16–17,
  182, 184f
Ziziphin, 14f
Ziziphus (*Ziziphus jujuba,*
  Rhamnaceae), 14